PHOTO ACKNOWLEDGMENTS: Frontispiece and Part I, USDA, *Soil Conservation Service*; Parts II, III, and IV, *Oregon State University Office of Agricultural Communications*; Part V, A. Gene Nelson; photo on p. 283, *Oregon State University Office of Agricultural Communications*.

Macmillan Publishing Company
866 Third Avenue, New York, New York 10022

Collier Macmillan Canada, Inc.

LIBRARY OF CONGRESS CATALOGING-IN-PUBLICATION DATA

Castle, Emery N.
   Farm business management.

   Includes index.
   1. Farm management. I. Becker, Manning H. (Manning
Henry),   II. Nelson, A. Gene. III. Title.
S561.C34 1987   630'.68   86-126
ISBN 0-02-320200-9

Printing: 1 2 3 4 5 6 7 8   Year: 7 8 9 0 1 2 3 4 5

ISBN 0-02-320200-9

Emery N. Castle

Manning H. Becker

A. Gene Nelson

OREGON STATE UNIVERSITY

Farm Business Management

The Decision-Making Process

THIRD EDITION

MACMILLAN PUBLISHING COMPANY

NEW YORK

COLLIER MACMILLAN PUBLISHERS

LONDON

Farm Business Management
The Decision-Making Process

Preface

This book is written for students of farm management and agricultural economics. The students may be college freshmen or sophomores, or beginning or established farmers seeking a better understanding of what they are already doing—managing a farm business. Prior knowledge of economic principles is not needed to comprehend the contents of this book. Based on our use of this material for short courses in farm management for farmers and extension agents, we have found that any person with an interest in the subject can master this material.

This third edition of *Farm Business Management* retains the organization and order of presentation of the earlier editions while incorporating new developments in farm management techniques and updated examples illustrating the techniques that have withstood the test of time. Our objective is to provide an easily understood framework for decision making based on economic and business principles appropriate to the management of the typical farm business. The family farm is the type of business unit that will make use of these management tools and techniques, but because they are quite general, they are applicable to numerous decision-making situations. Examples have been selected to demonstrate that the tools of farm business analysis have application to the wide variety of farming situations found throughout the United States.

Part I of this edition draws on the strengths of earlier editions to emphasize the decision-making function in farm management. The changing structure of farm businesses, as well as new developments in management techniques, required significant additions. In Chapter 1, for example, goal setting is recognized as the fundamental first step in identifying management problems.

Part II integrates the economic principles and management techniques that are basic to making the decisions endemic to the operation of a farm business. Chapter 2 introduces the economic principles relevant to farm management and shows how they guide the decision-making process. Chapter 3 explains the need for farm management information and accounting records.

Chapters 4 and 5 have been expanded to give additional consideration to the increasingly important tools of financial analysis. Planning, budgeting, and control are given greater emphasis in Chapter 6, and the basics of investment analysis are introduced in Chapter 7. Chapter 8 presents the latest techniques for risk management.

If the students have had a course in accounting, Chapters 3 and 4 may be omitted or assigned without devoting class time to discussion. Chapter 7 is optional and might be read by only the more advanced students.

In Part III the acquisition of farm resources is discussed in conjunction with the selection of the farm business arrangement and the factors affecting farm size. Two significant revisions were made in this part of the book. Guidelines for choosing from among the various farm business arrangements are outlined in Chapter 9. Chapter 13 explores the important aspects of recruiting, hiring, and managing labor.

Some instructors start early in a course with a laboratory problem that involves developing a complete organizational plan for a farm. In this case, Chapters 10, 11, and 12, dealing with size, growth, capital, and land acquisition, can be read earlier and out of sequence.

Part IV covers the specific problems associated with managing farm enterprises. Chapter 14 discusses the process of selecting and combining enterprises; then Chapters 15 through 17, which can be studied in any order, analyze particular enterprise and production input decisions using the principles and techniques introduced earlier. When students are well grounded in principles and techniques and understand acquisition and size problems, they can master enterprise management problems with confidence.

Part V examines the major technological, international, and public policy forces that will influence farm management and the opportunities for beginning farmers.

The use of microcomputers and computer software as farm decision aids is discussed throughout the text in conjunction with the techniques being presented. For example, computerized accounting systems are discussed in Chapter 3, the use of spreadsheet software for budgeting is suggested in Chapter 6, and a linear programming application is described in Chapter 14.

Two topics whose absence may be noticed are marketing and income tax management. Marketing is too commodity specific and complex to be covered adequately in one or two chapters, and so this important aspect of farm management is left to other books. Income tax management is also complex, but the reason for its omission is the pending major changes in tax law that would make any discussion of this topic obsolete. Again, good references are available in the *Farmers Tax Guide* and university extension publications.

In revising the book for this edition, we exercised some semantic prerogatives. *Manager* and *farm manager* are used to refer to the owner-operator or farm tenant—the decision maker for the farm business. In order to economize on words and space, ranches are included by implication in the term *farms*, and the term *farmers* includes ranchers.

We acknowledge the various contributions of a number of individuals to the writing of this book. We wish to express appreciation to the students of Oregon State University who offered suggestions for the improvement of the text, to our colleagues across the country who communicated to us their experiences with the first two editions, and to the faculty of the Department of Agricultural and Resource Economics at Oregon State University. We would like to express special thanks to the following reviewers: Marvin T. Batte, Ohio State University; Robin Childs, California State University; Donald C. Huffman, Louisiana State University; Robert A. Luening, University of Wisconsin; Albert G. Madsen, Colorado State University; Earl J. Parthenheimer, Pennsylvania State University; Joel C. Plath, Virginia State University; John A. Rogalla, California Polytechnic State University; M. Steve Stauber, Montana State University; and Odell L. Walker, Oklahoma State University. Special recognition is given to Mary Ann Gardner for the contribution of her editing and word processing skills. Of course, the contents of this book are solely the responsibility of the authors.

E. N. C.
M. H. B.
A. G. N.

Contents

Farm Business Management
The Decision-Making Process

PART I

Introduction

The management of the commercial farm calls for entrepreneur-ship—the ability to make decisions and the willingness to accept risks. The success of entrepreneurs is determined by their ability to use their resources to achieve both monetary and nonmonetary goals. In Part I, the decision-making process is described in the context of modern agriculture.

CHAPTER 1

The Decision-Making Process

What makes a successful farmer? What are the requirements that farming as an occupation imposes on an individual? Must the farmer be a good mechanic? No doubt this skill helps, but there are successful farmers who are not mechanically inclined. Must the farmer have substantial knowledge of crops and soils? Such knowledge is of value, but if the information is available from other sources, the farmer need not be a skilled agronomist. If success is measured in terms of profitability, successful farming is determined by the farmer's managerial ability. *Farm management* is concerned with the decisions that affect the profitability of the farm business. Obviously, this is a very broad definition of management, because there are many decisions that affect profitability. Such management concerns include the following questions:

1. What should be produced? What crops should be grown? What livestock should be raised?
2. How large should the farm be? How much should be produced? How many acres of land, dollars of capital, and hours of labor should be used?
3. How should the factors of production—land, labor, and capital—be acquired? Should the capital be borrowed? Should the land be owned or rented? Should new or used machinery be purchased? Should the labor be hired or supplied by the farm family?
4. What method of production should be used? What feeding, breeding, tillage, and other practices should be followed?
5. When and where should inputs be purchased and products sold?

The preceding list of decisions gives some idea of the scope of farm management. The technical fields of engineering, agronomy, and livestock husbandry are important, but farm management is specifically concerned with the aspects of these technical subjects that affect profitability.

No one definition of farm management suits all purposes equally well. However, this definition has certain characteristics that make it consistent with the approach used in this book. The definition does not rest on the assumption that every farmer, to be a successful manager, must consider profits. Farmers and their families have many goals, one of which is to make a profit on their operation. Successful management may also be concerned with providing a certain amount of leisure time, as well as with making money. A good analysis of the situation involves the development of alternative plans that provide for various amounts of leisure time. This approach allows managers to choose a plan knowing how it will affect both leisure time and profitability. In other words, they can choose the combination of leisure and income that is to their liking and consistent with their goals.

Another desirable characteristic of this definition of farm management is that it does not include the selection of an arbitrary time period over which profit is to be maximized. Some farmers and farm families are in a position to take a long-range point of view. Others, because of illness, poor capital position, or various other reasons, must think in terms of a shorter period. Unless they remain solvent in the short run, they will not be farming in the long run. They may, quite rationally, make a decision that involves a sacrifice of future income. Other farmers who are presently under less pressure may want to take a long-range view. All may be equally good managers, but their situations and goals are different; consequently, their actions are different.

Decision-Making Steps

Successful managers conduct their businesses affairs economically, using whatever resources are available and making them go as far as possible toward achieving those goals most desired. Management includes making decisions, executing, and subsequently evaluating them. The following steps comprise the decision-making process:

1. Setting goals.
2. Recognizing the problem.
3. Obtaining information.
4. Considering the alternatives.
5. Making the decision.
6. Taking action.
7. Accepting responsibility.
8. Evaluating the decision.

Successful farm management requires the ability not only to make decisions but also to make the right ones—decisions that will pay off over the long haul. Rarely is a person skilled in all aspects of the decision-making process. But it is the development of these skills in all phases of decision making that makes the farm manager successful.

Setting Goals

Successful farm managers are goal oriented. A *goal* is a target or desired condition that motivates the decision maker. Moreover, these successful farm managers use the resources at their disposal to work toward the attainment of not just one goal but a number of goals ranked according to their priority. Goals for farm managers and their families might include increasing net income or net worth, allowing for leisure time, avoiding low income or losses, and expanding the business. Defining goals clearly enough to write them down provides additional motivation, makes priorities more explicit, and encourages family cooperation.

Recognizing the Problem

Decision making always involves a problem or an opportunity. A *problem* is a discrepancy between the manager's goals and what is actually achieved. Managers are aware of a problem when their accomplishments and goals do not agree. For example, when a broiler producer finds that the desired level of feed conversion is greater than the level actually achieved, a problem has been recognized. Or a corn farmer may find that the yield this year is lower than expected. In both examples, a state of tension or uncertainty is created for the farmer because the results do not measure up to expectations.

Obtaining Information

Once a problem has been recognized, the next step is to deal with it. Information is needed regarding the alternatives available for solving the problem and their probable consequences. Also, more information may be needed to help define the characteristics of the problem and to determine what is preventing attainment of the goal. The search for this information involves a commitment of time and effort, and perhaps money. The amount of effort that goes into obtaining additional information depends on the perceived value of the information relative to its cost. The search for new information should continue as long as its perceived value is greater than the cost of obtaining it. An important factor affecting the information's perceived value is the importance of the decision to the farm manager.

Considering the Alternatives

The manager then uses the acquired information to compare the alternative courses of action. The information is organized to provide an appropriate basis for comparing the consequences of the alternatives. For example, the alternatives might be compared in terms of their predicted effect on net income, cash flow, risk, or debt. The choice is ultimately influenced by the manager's goals.

Making the Decision

A decision is a choice, or selection, from among various ways of getting a particular thing done or goal accomplished. At this stage in the decision-making process, the manager passes judgment in choosing the most desirable (or least undesirable) alternative.

Making a decision is difficult for certain people. Some believe that they never have enough information. Others are frightened when all available choices seem to involve some danger. Perhaps the word *courage* best describes the ability to take the plunge and make a decision. The ability to make decisions is related to appropriate timing. Managers of profitable businesses may make more mistakes than procrastinators, but by making more decisions and taking action they will be more successful. Postponing a decision may result in a lost opportunity. It is important that the manager determine at what point additional investigation is no longer warranted; then a decision should be made.

Taking Action

Unless conviction is matched by action, information and analysis are of little value to the manager. Implementing the decision requires energy and organizational ability. It is at this point that the executive ability to get things done pays off. The plant superintendent and foreman are experts in this area. Determining the magnitude of the job to be done, estimating the necessary resources, and then acting in time is the essence of this ability. There are many people who just never get around to getting things done. The farmer who remarked, that "I already know how to farm better than I do" belongs in this category.

Accepting Responsibility

Of course, managers must be willing to accept responsibility for their decisions. Farmers know that their livelihood and that of their families depend on the decisions they make. Many people are unwilling to accept this responsibility. Such people are undoubtedly happier and probably more productive in a salaried position, where either they are not expected to make many decisions or are not held strictly accountable for them.

Evaluating the Decision

The decision-making process is not complete until the outcome has been considered. This step involves reevaluating the decision on the basis of the outcome achieved. For instance, a farm manager is involved in post-decision evaluation when an income statement is prepared or a crop yield is measured. This phase of the decision-making process is often underestimated. It is through the evaluation of past decisions that managers can learn from experience and thereby improve the decision-making skills associated with setting goals, identifying problems, considering alternatives, making decisions, and taking action.

Problems Encountered by the Farm Manager

Even though it is true that those who are adept at all steps of the decision-making process are more likely to be successful farm managers, this is not very helpful. It is also necessary to understand how the decision-making steps are applied, and this depends on the kinds of problems farmers face.

Because farmers work closely with physical and biological phenomena, successful farmers must be able to obtain such technical information from reliable sources. As mentioned earlier, successful managers need not be engineers, agronomists, or nutrition experts, but they must have access to technical information. They must locate the source of such information and select relevant data. They may find county extension agents a fruitful source of scientific and practical information; in some cases, they may rely on farm publications. They may personally contact a research scientist at the state experiment station or a field representative from a feed or chemical company.

The technical information required for successful farming is not limited to the physical and biological areas. Knowledge of price trends, of labor and machinery availability, and of government programs and services is also necessary. Agriculture is an industry that involves economics, sociology, psychology, and political science, as well as physical and biological relationships. Again, it is not necessary for farm managers to be experts in these fields. It is, however, necessary for them to know the sources of information and to be able to judge the relevance of such information to a particular situation.

Lack of information is not the only obstacle to efficient management. Many other problems may be present. Some of these problems may be beyond the control of the individual. The farmer may be handicapped by a home environment that influences management performance. Family health, family attitudes regarding economic progress, education, and community life all influence the environment in which the farmer operates. This environment may be either favorable or unfavorable.

Other problems are more central to the realm of farm management. These problems involve the acquisition of the resources necessary for farming, the proper combination and management of these resources, and the selection and combination of enterprises. It is in these areas that farm management as a field of study has made its greatest contribution. However, farm managers cannot confine their attention solely to these internal problems; they must also consider the external factors that affect the profitability of the farming business. Although most of this book is concerned with internal management problems, off-farm forces cannot be ignored.

A Classification of Decisions

The following characteristics are helpful in classifying decisions commonly encountered in farm management:

1. Importance.
2. Frequency.
3. Imminence.

4. Revocability.
5. Available alternatives.

Each of these characteristics is discussed briefly below.

Importance

It is obvious that decisions vary in importance. The importance of a decision may be measured by the size of the potential gain or loss involved. For example, deciding whether to use chemically treated or untreated fence posts for grazing land is much less important than the decision to acquire grazing land in the first place. Both decisions occur infrequently, but they differ in the importance of their ultimate effect on profitability. The farmer is justified in spending much more time determining the carrying capacity of grazing land being considered for purchase than determining how long different types of fence posts will last.

Another example of the relative importance of decisions is the decision to engage in a swine enterprise, which is more important than choosing the breed of swine. It follows that sufficient attention should be given to important decisions even if this means neglecting less important areas.

Frequency

Decisions also vary with respect to the frequency with which they are made. For example, livestock must be fed every day. Choosing what to feed and how much to feed is repeated many times. Such decisions can be routinized by developing a feeding plan at the beginning of the feeding period and continuing to use it until conditions warrant a change. Many farm management decisions are not intrinsically important, but because they are made repeatedly, their cumulative effect is crucial.

Imminence

The penalty for delaying the decision is not the same for all decisions. When the cost of waiting is low, it may be desirable to wait until more information is available before committing resources to a particular course of action. In other instances, it is important to act promptly. Examples will help illustrate this characteristic. If a farmer is considering building a machine shed, the cost of waiting may not be great unless considerable damage is being done to the machinery. On the other hand, when the weather breaks in the spring, postponing a crop production decision—to plant oats or barley—involves a very high cost. By the same token, when the planting season arrives, it is not profitable to spend a great deal of time investigating and choosing the most suitable varieties. After the crop has been planted, however, the next decision of this kind need not be made until the next planting season. This interval allows ample time to conduct the proper kind of investigation.

Revocability

Once made, some decisions can be revoked only at a considerable cost. For example, the decision to plant an orchard may be revoked only at a very high cost once trees have been planted. On the other hand, changing a decision regarding the kind of animal feed to use is a simple matter. Therefore, farmers who cannot afford drastic, costly changes are wise to select the type of farming that permits flexibility as conditions warrant.

Available Alternatives

Some situations present a multitude of choices for decision making; others may provide perhaps only two alternatives. When a wide range of choices is available, techniques must be developed for eliminating less likely courses of action and concentrating on a few workable choices for careful study and analysis.

Tools of Farm Management

Factors and policies affecting profitability are emphasized in this book because they provide a workable index for evaluating a farmer's success as a manager. Even though profits may not be the only goal, the farmer must know what effect decisions will have on income if they are to be made rationally.

Decision making is a highly personal, complex process. There are tools, or aids, that help farmers improve the quality of their decisions. First, however, the skills needed to use these tools must be developed, which involve the reasoning process itself.

The formulation of ideas is largely deductive; that is, the reasoning process leads from certain known information to a tentative conclusion, an idea. Then information is collected to test the idea. This is empirical evaluation, the testing of an idea by observing specific evidence. Everyone unconsciously uses both processes to arrive at conclusions. However, by using rational rather than intuitive thought processes, most people can improve their effectiveness because they become more systematic. This systematic process, known as the *scientific method*, has been shown by experience to be the most effective means of investigating a problem.

Economic principles are an important tool guiding the manager's deductive reasoning process. They are helpful in pinpointing the type of information that should be collected and studied, and they provide the framework for using the information to analyze the alternative decisions properly.

Farm accounts and records are a useful source of information about the farm business. Although these records are essential for preparing income tax reports, their usefulness does not end there. They help the manager identify problems and opportunities for improving the profitability of the business. Records also provide the necessary raw data for testing ideas and alternatives.

Farm budgeting procedures are testing devices used to compare farming alternatives on paper or by computer before investing money in an idea. These procedures are no panacea for all farm management problems, but they have proved useful in making many farm management decisions.

Economic principles, farm records, and budgeting procedures are helpful in analyzing problem situations and estimating the consequences of alternative decisions. Farm budgets are used to evaluate the consequences of different decisions when prices, yields, and other factors outside the manager's control can be predicted with some degree of confidence. However, not all farming situations can be reduced to predictable outcomes. In such cases, managers should develop strategies for dealing with uncertain and risky situations. These strategies and contingency planning are tools that allow managers to anticipate and plan ahead for the adverse outcomes that can result from uncertain situations.

Setting Farm Business Goals

The way to get results and to be successful is first to set goals. If managers know where they are going and when they want to be there, their jobs are much easier. Airplane pilots and their passengers would be very frustrated if they did not know their destination and expected time of arrival before taking off. With these goals in mind, they periodically check to be sure the airplane is on course. Goal setting works in navigation, and it works in management.

What Is a Goal?

A goal is a target toward which the manager is willing to work. It is something that is desired; it provides direction for planning. The following points are important in defining goals:

1. Goals should be specific. For example, "to make more money" is not specific, but "to increase net farm income $2000 by increasing the size of the dairy cow herd" is.
2. Goals should be objective and realistic. Before setting a goal, ask yourself whether you can attain it. Each person has unique talents, and managers must know their abilities and resources and set their goals accordingly.
3. Goals should require effort; otherwise they are not really goals.
4. Goals should include deadlines. "Increasing net farm income $2000 by expanding the dairy herd *next year*" is a more useful goal.
5. Goals should be measurable. If they are not, how will you know that you've achieved them? (A system is required in order to measure your progress.)
6. Certain goals should be easy, and others hard. Accomplishing the easy goals will give you encouragement and incentive to tackle the more difficult ones.
7. Goals should be flexible so they can be altered as conditions warrant. Do not

box yourself in. As conditions change and you learn more about yourself
and your capabilities, be ready to revise your goals.

8. Finally, recognize that all goals will not be attained. Don't be too hard on
yourself. Learn from the experience and use it to establish new goals.

Although most people think about goals, they are much more useful if they're
written down. Putting them in writing requires the extra effort and discipline that
often mean the difference between success and failure.

There are short-term, intermediate, and long-term goals. When setting a goal,
ask, "Where do I want to be one year from now? Five years from now?" Consider
how the different goals are related to each other. Reaching short-term goals should
result in the achievement of some long-term goals later on.

Goals can be competitive, independent, or complementary. A farm family might
want to "invest more money in the farm business" and "increase family consumption expenditures." With a fixed amount of money available, these goals are competitive, at least in the short run. Examples of complementary goals are "more
time away from the farm" and "traveling to Canada." Putting goals down on paper
will show how they are related. Actions that help to achieve one goal can simultaneously help or hinder the achievement of other goals. For example, increasing
crop yields would result in higher income. This higher income could be invested
in new machinery or used for family vacations. However, it is clear that money
spent on a new tractor cannot be used to travel to Canada. Identifying competitive
goals helps in setting priorities and choosing which goals to reach first and which
ones to postpone.

It is important to remember that most farms are family operations, and different
family members have different goals and may rank the importance of the same
goals differently. The result can be conflict. It is important, therefore, that the
goals be developed by the family as a group; the goals of all family members should
be considered. Successful farm operations are those in which all parties involved
work together in setting the future course of the business.

Four main conclusions can be drawn from the many formal and informal studies
of farm family goals:[1]

1. Farm families often find it difficult to verbalize their goals and easier to rank
the goals on a predetermined list.

2. When families do express their goals, these tend to be short-term goals that
reveal diverse areas of interest.

3. Although a wide range of goals is indicated, there are some similarities
among many families.

4. Factors such as age, income, and net worth affect the relative importance of
certain goals.

Individual and family goals develop from personal needs, interests, past experi-

---

[1] G. F. Patrick and Brian Blake, *Setting Farm Family Goals*, Purdue University Cooperative Extension
Service EC-514, June 1980.

ences, and values. Examples of general goals that are shared by almost all farm families are as follows:

1. To increase net farm income.
2. To avoid years of low income or high losses.
3. To maintain or improve the family's life-style.
4. To increase the family's net worth.
5. To avoid losing the farm business.

Many farm families have expressed other general goals:

1. To reduce debt or to have no debt obligations.
2. To have leisure time for personal activities.
3. To participate in church and/or community activities.
4. To expand the size of the farm business.

Maximizing or increasing income is cited frequently as a goal of farmers; other major goals are "not being forced out of business" and "avoiding years of low income or losses." These two general goals—increased income and survival of the farm business—are often competitive, but farmers must consider the relative priority of each and decide how to manage the business in order to achieve both.

Steps in Setting Goals

There is no one best way to set goals, because different approaches work better for different individuals. The following list provides a starting point for developing your own system:

1. Reserve a definite time for goal setting and for reviewing progress. You might set aside some time each week to list ideas for possible goals. Then a few hours at the beginning of each month might be used to review your progress toward agreed-upon goals, to revise, and to set new goals based on the list of new ideas. Finally, once a year, a day might be set aside to involve the entire family and all others participating in the farm operation in the goal-setting process.
2. The next step in setting goals is to assess what has been done in the past. Think about the major decisions that have been made over the last three years and write down the circumstances surrounding them. What alternatives did you consider? This process forces you to examine precisely what your goals have been. Would you make the decision differently today? Have your goals changed?
3. Consider the alternatives that are available to you now. This is where you need to be creative; do not let past experiences limit future possibilities. For example, you might eliminate one enterprise from the farm and add another; you might buy 200 acres of land; or you might take a part-time, off-farm job. For each of these alternatives, list what would be required to implement it and the likely outcomes. These outcomes might reveal achievable goals. Next, look at the requirements for each alternative. If these requirements cannot be met now, set short-term goals to meet them eventually.

4. Now, by reviewing the requirements and outcomes of the alternatives you have listed, develop another list that sets forth your short-term, intermediate, and long-term goals. Then examine the relationships between them. Indicate whether each goal will help, hinder, or have no effect on the achievement of the others. That is, are the goals complementary, competitive, or independent?

5. Ask each family member to do this exercise independently. Afterward discuss the areas of agreement and disagreement. Based on these discussions, develop a list of family goals in a form similar to the one illustrated in Table 1-1. List the goals in order of priority. There are big ($10,000) goals and small ($10) goals. Your priorities should reveal these value differences.

TABLE 1-1. Worksheet for Setting and Attaining Goals

Short-Term Goals	Actions Needed to Attain Goals	Starting Date	Completion Date
1.			
2.			
3.			
4.			
Intermediate Goals			
1.			
2.			
3.			
4.			
Long-Term Goals			
1.			
2.			
3.			

6. The last and most important step is taking action, starting with those goals that have the highest priority. Because there are so many demands on your time and attention, this list of goals will help in setting priorities and in general planning. As a result, you will reduce frustration and achieve what you most want to accomplish.

Specialists in human behavior and motivation say that what you *think is possible* influences what you are *able to accomplish*. You are, or can be, who you think you are. Successful managers think ahead. By setting goals, they shape events. They don't let events shape them. Success in farming, as in any other enterprise, depends on setting and attaining goals.

Summary

Agriculture is a diverse and changing business. To be successful, young people entering agriculture must have realistic expectations. Enthusiasm is important, but management skills are crucial.

Farm management concerns the decisions that affect the profitability of a farm business. This broad definition is consistent with the assumption that profitability is the fundamental factor contributing to the successful operation of a farm business. But this definition also allows for the consideration of other goals.

The following steps guide the decision-making process: (1) setting goals, (2) recognizing the problem, (3) obtaining information, (4) considering alternatives, (5) making the decision, (6) taking action, (7) accepting responsibility, and (8) evaluating the decision. This process can be applied to a broad range of decisions faced by farm and ranch managers. Its application requires physical, biological, economic, and other technical information from reliable sources. In addition, considering decisions in terms of their importance, frequency, imminence, revocability, and available alternatives helps in establishing priorities for decision making.

The tools of farm management that are the focus of this book allow the decision-making steps to be executed rationally and efficiently. Although these tools cannot guarantee success, they do result in better decisions, which improve the chances for success. Success must be defined in relation to the manager's goals, and this first step in the decision-making process is often the most difficult. Putting goals in writing requires discipline and diligence but provides direction for more effective management of the agricultural business.

Recommended Readings

GESSAMAN, P. H., and K. PROCHASKA-CUE. *Goals for Family and Business Financial Management, Part I: Overview and Self Assessment.* University of Nebraska Cooperative Extension Service CC-312, Revised July 1985.

GESSAMAN, P. H. *Goals for Family and Business Financial Management, Part II: Identifying Your Goals.* University of Nebraska Cooperative Extension Service CC-313, Revised July 1985.

GESSAMAN, P. H. *Goals for Family and Business Financial Management, Part III: Your Priorities and Management Plan.* University of Nebraska Cooperative Extension Service CC-314, Revised July 1985.

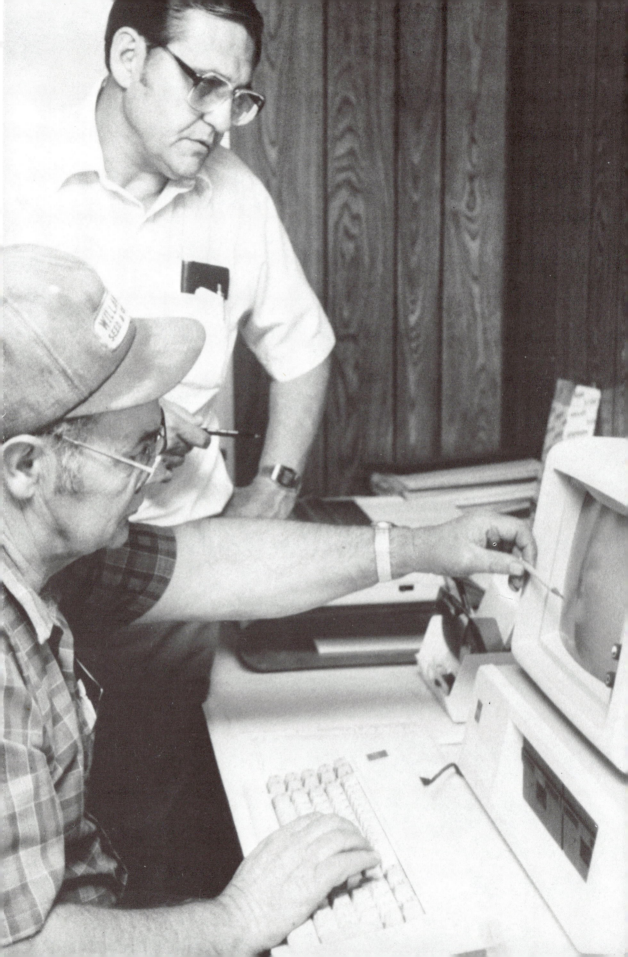

PART II

Tools of Decision Making

The commercial farmer often manages a large investment in a changing and highly technical environment. Part II of this book presents principles and procedures that farmers can use to help them make decisions in a dynamic world.

CHAPTER 2

Economic Principles

Economic principles serve as guidelines for making farm management decisions. Trial and error procedures may give the same results, but economic principles used properly are more efficient. They provide logical procedures for gathering relevant information and organizing it to analyze decisions. Ultimately, they save time and result in better decisions.

Economics has to do with making decisions regarding the use of time, money, and other resources. Because resources are limited or scarce, these choices must be made; that is, if you choose one thing, you must give up another. If you choose to use your time to care for hogs, you cannot use it to plant corn; if you plant the field to corn, you cannot plant it to soybeans; and if you use your money to buy a new truck, you cannot buy new tillage equipment. Economic principles provide the framework to guide the decision-making process.

A fundamental principle of economics is the concept of opportunity cost. The *opportunity cost* of a management decision is the income that could have been earned from the next best choice. The opportunity cost of producing a commodity is not the out-of-pocket expense; it is the income that could have been earned by using the same resources in the next best alternative. For example, when a farmer pays $100 per acre to rent land to grow corn, the opportunity cost of using this land to grow corn is not the $100 rent but the $120 net return that could be obtained by using the land to grow soybeans instead of corn. The concept of opportunity cost emphasizes the consideration of the alternative ways in which resources can be used when making decisions.

Farm managers must make a variety of decisions about the amounts and combinations of inputs, such as fertilizer, water, and feed, to use in the production process. They must also decide what products to produce and how much of each to produce. This chapter discusses the application of economic principles to the decision-making process.

Production Relationships

Physical production relationships form the basis for the application of economic principles by determining the range of possibilities available to the manager. A *production function* describes the physical relationship between the quantities of inputs (factors of production) and the quantities of output (product) that result.

For example, a production function shows how the daily milk production of a dairy cow depends on the amount of feed consumed. It reflects the ability of the cow to convert feed into milk. This production function assumes that the cow's condition and environment, the quality of the feed, and all other inputs remain constant; otherwise the production function would express not only the result of the quantity of feed but also the mixed effect of cow condition, environment, feed quality, and other inputs. The production function, therefore, indicates the maximum quantities of product resulting from different quantities of one input or from a combination of inputs with all other inputs held constant, or fixed. In this case, it shows the dairy farmer how milk production would be affected if the amount of feed input were varied.

The production function can be represented as a table, a graph, or an equation. A simple case with one variable input is used to illustrate a production function in Table 2-1. The data describe the relationship between the production of corn and the amount of seed used per acre. They show the annual output of corn resulting from various quantities of seed with all other inputs, such as irrigation water, land, fertilizer, and chemicals, held constant. These data indicate the obvious: If no seed is planted, corn production is zero. The data also indicate that the amount of corn produced increases with the amount of seed planted up to 30,000 seeds per acre. If more than 30,000 seeds are planted, corn production decreases because of the greater competition among the corn plants. Thus, knowing the production function helps the manager narrow the range of decisions.

The Total Physical Product Curve

Production functions can also be illustrated by graphs. For example, Figure 2-1 depicts the same relationship between seed input and corn output described by the data in Table 2-1.[1] This graph of the production function helps the manager visualize the relationship. The graph starts at zero—no seed, no output. Then, as more seed is applied per acre, corn production increases quite rapidly, then increases less rapidly, and finally decreases as the seeding rate becomes so high that the corn plants compete with each other.

This relationship is illustrated by the *total physical product* (TPP) curve in Figure 2-1. The curve begins at zero, (1) increases at an increasing rate, then (2) increases at a decreasing rate, and finally (3) declines. This particular example includes all three of the possible relationships. Although, in practice, production functions may not include all the relationships shown, they will exhibit diminishing returns.

Diminishing returns characterizes a production function for which the added output resulting from each successive addition of a variable input (with other inputs held constant) eventually decreases. In other words, over some range on the TPP curve, production increases at a decreasing rate. The importance of the variable input in the production process, the level at which other inputs are fixed, and the nature of the production function itself determine where the rate of increase in

---

[1] The equation for this production function is $Y = 5.5X + 0.36666X^2 - 0.010185X^3$.

TABLE 2-1. Data Showing a Hypothetical Relationship Between Corn Output and Seed Input Per Acre

Seeds Planted (no./acre)	Corn Produced (bu/acre)
0	0
2,000	12.4
4,000	27.2
6,000	44.0
8,000	62.3
10,000	81.5
12,000	101.2
14,000	120.9
16,000	140.2
18,000	158.4
20,000	175.2
22,000	190.0
24,000	202.4
26,000	211.9
28,000	217.9
30,000	220.0
32,000	217.7

output starts to slow. Diminishing returns result because one or more inputs, or factors of production, is fixed. They can be avoided only by increasing the level of all inputs, including management.

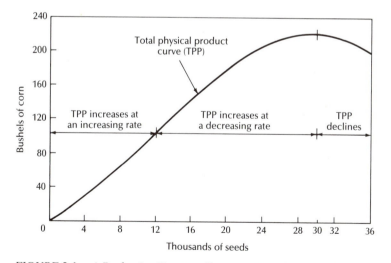

FIGURE 2-1.   A Production Function Showing a Hypothetical Relationship Between Corn Output and Seed Input per Acre.

Diminishing returns also place an upper limit on production where the rate of increase diminishes to zero. It is never profitable to use more inputs than necessary to get maximum total production. Only if the variable input costs nothing would it be profitable to produce the maximum total product where the return (added output) resulting from an additional unit of input has decreased to zero. Beyond this point the returns to additional inputs are negative; that is, total production is declining.

The concept of diminishing returns also determines the shape of the average physical product curve and the marginal physical product curve.

The APP and MPP Curves

The *average physical product* (APP) curve and the *marginal physical product* (MPP) curve are derived from the production function, or TPP curve. Both curves are useful in understanding and using the production function for decision making. The APP is the TPP divided by the units of input required to obtain that total product. If 12 units of variable input result in an output of 100, the APP at this point is 100/12, or 8.33. The APP first increases and then decreases, as illustrated in Figure 2-2, depending on the shape of the TPP.

An understanding of the concept of marginal change is necessary to comprehend fully the MPP curve. *Marginal change* refers to the increment or addition, or, more precisely, the increase or decrease at the margin. The MPP, then, is the incremental change in TPP (an increase or decrease) as the result of a one-unit change in input. The MPP curve increases to a maximum, then decreases, and eventually becomes negative. Its shape depends on that of the TPP curve. The MPP expresses the *slope* of the production function at every level of input. When the TPP is at its maximum, the quantity of product is not affected by a relatively small change in input; therefore, at that point, MPP is zero.

Where the production function shifts from increasing at an increasing rate to increasing at a decreasing rate, the increase in the quantity of total product resulting from a one-unit increase in input is greatest. The slope of the production function is greatest at this point. Therefore, MPP is at its maximum where production (TPP) changes from increasing at an increasing rate to increasing at a decreasing rate.

Two other relationships can be observed from Figure 2-2: When MPP is greater than APP, APP is increasing; and when MPP is less than APP, APP is declining. To explain the first relationship, a student who on each successive quiz receives a score above her average for all quizzes to date will increase her average score. In other words, when the marginal score is greater than the current average score for all quizzes, the all-quiz average increases. The student's marginal score, if it is greater than the average score, pulls up the new average score. The reverse is also true. If the marginal score is less than the average, it pulls down the current average.

MPP and APP are equal when APP is at its maximum. The amount added to TPP by an incremental increase in input (MPP) is the same as the average amount of product (APP) where APP is at its maximum. Understanding these relationships helps narrow the range of production alternatives that must be considered in decision making.

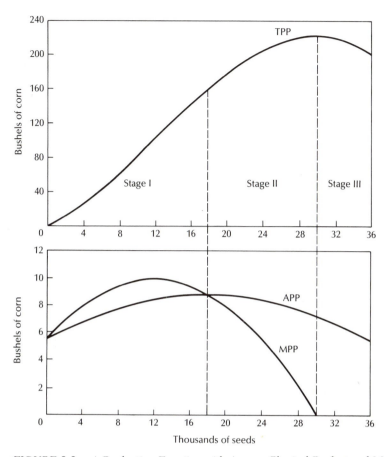

FIGURE 2-2.   A Production Function with Average Physical Product and Marginal Physical Product Curves.

The Three Stages of Production

Even without considering costs and returns, the production function and the APP and MPP curves provide some guidance in making decisions about input use. For example, it would be irrational to use so much of an input that the TPP decreases (unless the input has a negative cost). It is also irrational to limit the use of an input when a greater APP can be achieved from an increase in its use. A rational range for decision making is thus determined: the area between maximum APP and maximum TPP (or between the point where MPP and APP are equal and where MPP is zero). This area is referred to as stage II of the production function and is illustrated in Figure 2-2.

Stage III is clearly irrational because the use of more input decreases the TPP, or an increase in costs results in a decrease in revenue. Note also that the MPP is negative in stage III.

Stage I is irrational because the use of more input results in a greater APP, and if it pays to use the input at all, it pays to use it up to the point where the TPP per unit of input is at a maximum.

Therefore, the decision-making process is simplified because only stage II input levels need to be considered. Stage II will always be rational if it pays to produce at all, and if the prices of all inputs and products are constant and positive.

Choosing the Level of Input Use

The appropriate economic principle to use depends on the decision to be made and the situation. Principles are presented for use in deciding (1) how much of a single input to use, (2) how to allocate an input in limited supply, and (3) what combination of inputs to use.

A Single Variable Input

Marginal analysis is used to compare the cost of the variable input with the market price of the product in order to determine the most profitable level of an input to use. Profits are maximized when the increase in the total value of the product resulting from a one-unit increase in the variable input is equal to the added cost of using this additional unit of the input.

To illustrate this method of analysis, the production relationship between wheat and nitrogen is used to find the most profitable level of nitrogen to apply (Table 2-2). As the quantity of nitrogen applied increases (column 1), the TPP of wheat also increases (column 2). However, the TPP increases at a decreasing rate, as can

TABLE 2-2. A Marginal Analysis of the Most Profitable Application of Nitrogen to Dry-Land Wheat

(1) Input of Nitrogen (lb/acre)	(2) TPP of Wheat (bu/acre)	(3) Total Value of Product[a]	(4) Total Input Cost[b]	(5) Net Return above Input Cost	(6) Marginal Physical Product	(7) Marginal Value of Product	(8) Marginal Input Cost
0	32.0	$128.00	$ 0	$128.00			
					1.4	$5.60	$2.50
10	33.4	133.60	2.50	131.10			
					1.2	4.80	2.50
20	34.6	138.40	5.00	133.40			
					1.0	4.00	2.50
30	35.6	142.40	7.50	134.90			
					0.8	3.20	2.50
40	36.4	145.60	10.00	135.60			
					0.6	2.40	2.50
50	37.0	148.00	12.50	135.50			
					0.4	1.60	2.50
60	37.4	149.60	15.00	134.60			
					0.2	0.80	2.50
70	37.6	150.40	17.50	132.90			

SOURCE: Data based on actual field trials in central and eastern Montana, developed by Leroy
D. Luft, formerly at Montana State University, 1979.

[a] Wheat price at $4.00 per bushel.
[b] Nitrogen cost at $0.25 per pound.

be seen from the MPPs (column 6). In other words, the data in Table 2-2 correspond to stage II of the production function. In order to determine the maximum profit level of fertilizer to apply, the cost of the fertilizer and the price to be received for the product must be determined first.

Here the price of wheat is assumed to be $4.00 per bushel, and the price of nitrogen plus the cost of application is $0.25 per pound, or $2.50 per 10-pound increment. A commonsense method for choosing the optimum level of nitrogen is to subtract the total nitrogen cost (column 4) from the total value of wheat produced (column 3) at each level of fertilizer application and then to select that application level resulting in the greatest difference (column 5). It can be seen that the net return above the input cost is greatest with the application of 40 pounds of nitrogen. This is an acceptable and correct way to make this decision, but marginal analysis is more efficient.

Consider the MPPs resulting from the 10-pound increases in nitrogen applied (column 6). When no fertilizer is applied, the expected yield is 32 bushels of wheat per acre. When 10 pounds are applied, the yield is 33.4 bushels. Consequently the MPP, or the added yield resulting from this 10-pound application of fertilizer, is 1.4 bushels. Next, using 20 pounds of nitrogen, the yield is 34.6 bushels of wheat. By applying this second 10 pound increment, another 1.2 bushels of wheat (MPP) are produced.

In the next column (column 7), a value is placed on the MPP. To compute this figure, the MPP is multiplied by the $4.00-per- bushel price of wheat. The *marginal value of the product* (MVP) resulting from an input increase from 20 to 30 pounds of nitrogen is $4.00 (1 bushel × $4.00 = $4.00).

The last column in the table (column 8) gives the marginal or added cost associated with applying an additional 10 pounds of fertilizer per acre. This *marginal input cost* (MIC) of $2.50 is the same for each level because, in each case, the nitrogen application is increased by the same amount, 10 pounds. This figure is calculated by multiplying the cost of the fertilizer, $.25 per pound, by 10 pounds.

In order to determine the most profitable amount of input to use, the following principle is applied: *Continue to increase input as long as the MVP for the last increment is greater than the MIC.* In this example, the MVP exceeds the MIC when 40 pounds of nitrogen are applied. Increasing the nitrogen applied to 50 pounds would increase the value produced by $2.40, but at a cost of $2.50; therefore, no more than 40 pounds should be applied. The most profitable level may actually be somewhere between 40 and 50 pounds per acre, but given the variability in production due to weather and the difficulty of spreading an exact amount of fertilizer, 40 pounds is a reasonable approximation. (Variability and risk are considered in a later chapter.)

Note that marginal analysis produces the same result as the commonsense method referred to earlier. The advantage of marginal analysis, however, is the ease with which the manager can evaluate how this decision regarding the amount of fertilizer to apply would be affected by changes in the cost of the input and the price of the product. For example, what if the price of nitrogen increased from $0.25 to $0.39 per pound? The MIC would increase from $2.50 to $3.90. Without making any additional calculations, the manager can see that the MVP, $4.00, exceeds the MIC when 30 pounds of nitrogen are applied. But when the nitrogen application is 40 pounds, the MVP is only $3.20. Therefore, no more than 30 pounds should be applied. Likewise, with only a few calculations, the effect of a change in the price of wheat can be evaluated. The best fertilizer rate, if the wheat price were $4.20 rather than $4.00 per bushel, would be 50 pounds per acre.

The production function for corn in Table 2-1 can be used in a similar way to find the most profitable amount of seed to plant per acre. If the price per bushel of corn is assumed to be $2.00 and the cost of 1000 seeds is $4.00, what quantity of seed should the manager plant? To find the answer, the MPPs should be calculated first. The MPPs are then multiplied by the price of corn to get the MVPs. The MIC for seed is $8.00, which is the $4.00 cost per 1000 seeds multiplied by 2, the increment in the table. The most profitable quantity of seed to use is found, then, by increasing the amount of seed until the MVP no longer exceeds the MIC. That point is 28,000 seeds per acre.

It is somewhat unrealistic to assume that only one input is variable and all others are fixed, as in these examples. When more than one input is variable, the best combination of inputs must be determined. But first, what if the supply of input is limited and there are alternative uses for it?

Limited Input Supply

The total quantity of available input may be so limited that it is not possible to equate MVP with MIC. The farmer may be producing more than one crop and may have little capital for fertilizer and other variable inputs. Also, a supply shortage in a given year may restrict the availability of some inputs.

When supplies are limited, the principle of *equimarginal returns* (equating marginal returns) is used to allocate the supply of available input so that it is used when and where it makes the greatest contribution to profit. When there are alternative uses for the input, a choice must be made: Should fertilizer, for example, be divided equally between crops, or should one crop receive more than another? Should capital be used for fertilizer or for something else?

The principle of equimarginal returns is as follows: *An input in limited supply should be allocated among all of its alternative uses, so that the MVPs for the last units allocated to each use are equal.* To illustrate the application of the equimarginal principle, hypothetical fertilizer production functions are presented for three crops in Table 2-3. Assume that the farm consists of 120 acres and that there are 40 acres in each crop. Also assume that the quantity of fertilizer is limited to a total of 10 tons that must be allocated among the three crops. How much should be applied per acre to each crop? The crop prices are $0.06, 0.05, and $0.02 per pound for crops A, B, and C, respectively. The cost of fertilizer is $0.20 per pound.

The first step is to calculate the MPPs for each 50-pound increment of fertilizer per crop. The next step is to multiply these MPPs by the appropriate price to find the values of the marginal physical products (MVPs).

Now the fertilizer is allocated successively to each crop according to the highest return. Because there are 40 acres in each crop, each 50-pound increment per acre uses one of the 10 tons available. The first and second tons of fertilizer are allocated to crops B and C, where the MVPs are $20 per acre. The third and fourth tons are allocated to crops A and C, where the MVPs are $18 per acre. The

TABLE 2-3. Fertilizer Production Functions for Three Hypothetical Crops

	Crop A				Crop B				Crop C		
Fertilizer (lb/acre)	TPP (lb/acre)	MPP (lb/acre)	MVP ($)[a]	Fertilizer (lb/acre)	TPP (lb/acre)	MPP (lb/acre)	MVP ($)[b]	Fertilizer (lb/acre)	TPP (lb/acre)	MPP (lb/acre)	MVP ($)[c]
0	1200			0	1200			0	2112		
50	1500	300	18.00	50	1600	400	20.00	50	3112	1000	20.00
100	1750	250	15.00	100	1950	350	17.50	100	4012	900	18.00
150	1950	200	12.00	150	2250	300	15.00	150	4812	800	16.00
200	2100	150	9.00	200	2500	250	12.50	200	5512	700	14.00
250	2200	100	6.00	250	2700	200	10.00	250	6162	650	13.00
300	2250	50	3.00	300	2850	150	7.50	300	6762	600	12.00
350	2300	50	3.00	350	2975	125	6.25	350	7212	450	9.00
400	2325	25	1.50	400	3075	100	5.00	400	7512	300	6.00
450	2325	0	0	450	3100	25	1.25	450	7600	88	1.76
500	2300	-25	-1.50	500	3100	0	0	500	7650	50	1.00

[a] Crop A price is $0.06 per pound.
[b] Crop B is $0.05 per pound.
[c] Crop C is $0.02 per pound.

fifth ton returns a $17.50 MVP on crop B. This process is continued until all 10 tons are allocated.

Successively allocating the increments in this way satisfies the requirements of the equimarginal returns principle by approximately equating the MVPs. In this case, the most profitable use of the 10 tons of fertilizer is to apply 100 pounds per acre to crop A, 150 pounds per acre to crop B, and 250 pounds per acre to crop C. There is still, however, potential for additional profit: If another ton of fertilizer were available, an additional 50 pounds per acre could be applied to crop B, with an MVP of $12.50 per acre as compared to an additional cost (MIC) of $10. If there were no limit to the quantity of fertilizer available, what would be the most profitable quantity of fertilizer to apply per acre for each crop?

Combination of Inputs

When farm managers have more than one input to consider, they must select a combination that will yield a given level of production at the least cost. For example, what combination of nitrogen and phosphorus should be used to produce 36 bushels of wheat per acre? The concept of *input substitution* assumes that one input can be substituted for another, while maintaining a constant level of production, and that the physical relationship between the inputs is known. The input substitution principle provides the logic for determining the least-cost method of production: If the output is constant, it is economical to substitute one input for another when the cost of the substitute is less than the cost of the original input.

The data in Table 2-4 illustrate this principle. This table indicates 11 different combinations of alfalfa hay and grain concentrate; each combination produces a

TABLE 2-4. Concentrate and Alfalfa Hay Combinations Necessary to Produce 300 Pounds of Gain on a 400-Pound Beef Calf

(1) Concentrate (lb)	(2) Alfalfa Hay (lb)	(3) Marginal Rate of Substitution[a]
1316	1000	
1259	1100	57/100 = 0.57
1208	1200	51/100 = 0.51
1162	1300	46/100 = 0.46
1120	1400	42/100 = 0.42
1081	1500	39/100 = 0.39
1046	1600	35/100 = 0.35
1014	1700	32/100 = 0.32
984	1800	30/100 = 0.30
957	1900	27/100 = 0.27
932	2000	25/100 = 0.25

SOURCE: Oregon State University Agricultural Experiment Station Miscellaneous Paper 98, September 1960, p. 14.
[a] Change in concentrate consumed divided by change in hay consumed.

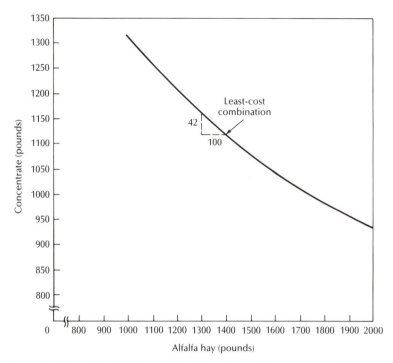

FIGURE 2-3.   Substitution Relationship Between Concentrate and Alfalfa to Produce 300
Pounds of Gain on a 400-Pound Beef Calf.

Source: Oregon State University Agricultural Experiment Station Miscellaneous Paper 98,
   September 1960, p. 14.

300-pound gain on a 400-pound beef calf. As the amount of hay is increased, the
amount of concentrate is decreased, and vice versa. This relationship is presented
graphically in Figure 2-3.

This example starts with a ration of 1000 pounds of hay and 1316 pounds of
concentrate to produce a gain of 300 pounds. If 100 pounds of hay were added to
the ration, the concentrate could be reduced by 57 pounds, and the ration would
still produce 300 pounds of gain. Note, however, that the substitution ratio is not
constant. Each additional 100 pounds of hay replaces less and less concentrate;
finally, increasing the alfalfa hay from 1900 to 2000 pounds replaces only 25
pounds of concentrate (957 to 932 pounds).

The least-cost combination of hay and grain can be found by a trial-and-error
procedure, that is, by calculating the cost of each combination and selecting the
one with the lowest cost. Suppose hay were $0.03 per pound and concentrate were
$0.075 per pound. The cost of 1000 pounds of hay combined with 1316 pounds
of concentrate would be $128.70 ($30.00 + $98.70). Performing similar calcu-
lations for all other combinations would determine the least-cost ration, but there
is a better way.

The least-cost combination can be determined more efficiently by using the
marginal rate of substitution (MRS), the marginal rate at which the added input

substitutes for the replaced input. More specifically, the MRS is the amount an input can be decreased as a result of increasing the other input by one unit without changing the level of production.

The procedure for applying the substitution principle is as follows:

1. Find the MRS by dividing the number of units of replaced input by the number of units of added input:

$$\text{MRS} = \frac{\text{units of replaced input}}{\text{units of added input}}$$

For example, consider feeding 1300 pounds of hay and 1162 pounds of concentrate versus 1400 pounds of hay and 1120 pounds of concentrate. That is, 42 pounds of concentrate are replaced by 100 pounds of hay:

$$\text{MRS} = \frac{42}{100} = 0.42$$

2. Compute the inverse price ratio as follows:[2]

$$\text{Inverse price ratio} = \frac{\text{price of added input}}{\text{price of replaced input}}$$

The cost of hay is $0.03 per pound and that of concentrate is $0.075 per pound, therefore:

$$\text{Inverse price ratio} = \frac{0.030}{0.075} = 0.40$$

3. Find the least-cost combination of inputs by continuing to substitute one input for the other if the inverse price ratio is less than the MRS:

$$\frac{\text{Price of added input}}{\text{Price of replaced input}} \leq \frac{\text{units of replaced input}}{\text{units of added input}}$$

or

$$\text{Inverse price ratio} \leq \text{MRS}$$

For this example:

$$0.40 \leq 0.42$$

For the next 100-pound increment of hay (1400 to 1500 pounds), the MRS is 0.39, which is less than the inverse price ratio. Therefore, the least-cost combination is 1400 pounds of hay and 1120 pounds of concentrate, as illustrated in Figure 2-3. (To prove that this is indeed the least-cost combination, calculate the total cost for this and other combinations in the table.)

The advantage of this procedure over trial and error becomes clearer when the price of one or both inputs changes. When the MRS principle is used, the cost of

---

[2] This is called the *inverse* price ratio because the price of the *added input* is the numerator, rather than the units of the *replaced input*, as is the case for the MRS.

each combination does not have to be recalculated. The new inverse price ratio is simply compared to the previously calculated MRSs to determine the least-cost combination of two inputs.

The input substitution principle, therefore, is an efficient tool for choosing the best combination of inputs to produce a given quantity of product. The quality of the product and the timing of production are *assumed* to be the same, or constant, for each combination. However, 300 pounds of gain on a beef calf may be obtained more slowly with a low-concentrate ration, and the grade and condition of the animal after gaining 300 pounds may change. It is possible to modify the procedure to take these changes into account. For example, the added cost of a lower rate of gain could be included with the cost of the alfalfa hay being substituted for concentrate.

In this example, the inputs are substituted at a decreasing rate; that is, smaller and smaller amounts of concentrate are replaced by the same amount of added alfalfa hay. Some inputs, however, can be substituted at a constant rate. This may be the case for barley and corn in livestock rations. It is also true of anhydrous ammonia and ammonium sulfate, sources of nitrogen. When inputs substitute at a constant rate, all of one input is normally used, usually the least costly one in the required quantity.

The least-cost method and level of production is not necessarily the most profitable. For example, it may be more profitable to feed the beef calf for 350 rather than 300 pounds of gain even though the cost per pound of gain is less for 300 pounds. In the foregoing analysis, the object was to determine the least-cost way to produce a given output. However, deciding how much to produce is equally important. These two aspects of the management decision are interrelated, but to simplify their explanation they are presented separately.

Choosing the Level of Production

Deciding whether to produce a certain product and choosing the level of production are decisions affected by the cost of production and the returns expected. First, the relationships between cost and the level of production are considered.

Cost and Production Relationships

A *cost* is defined as a charge that should be made for an item used in the production of goods and services. Costs may or may not involve a cash transaction.

Cash and Noncash Costs.   The total cost of production includes both cash and noncash costs. *Cash costs* require an out- of-pocket expenditure. For example, cash costs are incurred when a farmer purchases and pays for fertilizer, fuel, and feed. Because *noncash costs* do not involve a cash outlay, they are more difficult to identify. Noncash costs are incurred when inputs like unpaid family labor, the farmer's own land, and equity capital are used. These noncash costs are usually opportunity costs, which means that the cost is based on what the resource could have earned by using it for the next best alternative. Another example of a noncash cost is depreciation (defined in the next chapter).

Fixed and Variable Costs.   Costs are also classified as fixed or variable. Whether a cost is fixed or variable depends on the decision to be made and the period of time being considered. In general, *fixed costs* are those that are incurred whether anything is produced or not. These costs remain constant regardless of the decision. *Variable costs* are those that change with the level of production; therefore, if there is no production, variable costs can be avoided.

Consider the production of apples. The cost of the established orchard and interest on the equipment investment are fixed costs; the costs for fertilization, pruning, spraying, and harvesting are variable costs until these operations are performed. After the fertilizer has been applied and pruning has been done for the year, these become *sunk*, or fixed, costs. If the manager is considering expanding the apple orchard, the cost of establishing the new trees, the interest on added equipment investment, and other costs that are fixed for the established orchard are variable for the new orchard. Therefore, only variable costs are relevant to the decision, because fixed costs are not affected.

From a decision-making standpoint, the variable costs are the most relevant, and the particular situation and decision being considered determine what these costs are. The variable costs are those affected by the decision, and the fixed costs are those not affected. The inputs should always produce returns greater than their variable cost. If they do not, the inputs should not be used in this particular enterprise.

Suppose a farmer has a poor wheat crop ready to harvest. Because of bad weather, it is expected to yield only eight bushels per acre, and the price is $4.00 per bushel. If the total cost of production is $160 per acre, the farmer will lose money. Under these circumstances, the objective is to minimize losses rather than to maximize profits. One possibility is to abandon the wheat crop; the other is to harvest it. The wheat should be harvested if its value will at least cover the variable costs of harvesting. With eight bushels of wheat at $4.00 per bushel, or $32 per acre, and variable harvesting costs at $10 per acre, the farmer should harvest the wheat because the return is greater than the variable costs.

The total cost of production includes the variable costs, which depend on the level of production, and the fixed costs, which remain unchanged in total as production changes. Thus, for each specific level of production:

Total cost = total variable cost + total fixed cost

Dividing both sides of the preceding equation by the level of production expresses the relationship of the per unit or average costs.

Average total cost = average variable cost + average fixed cost

The relationships among these different measures of cost are illustrated in Figure 2-4. The total cost (TC) curve shows how the TC changes depending on the level of production. Both the TC and the total variable cost (TVC) increase as output increases; they are upward sloping curves. The total fixed cost (TFC) is constant regardless of the output and is shown as a straight line.

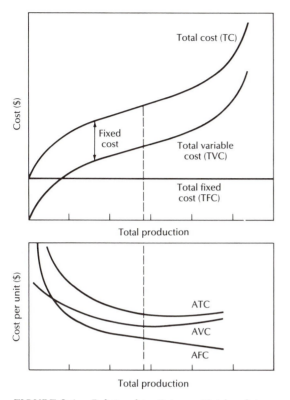

FIGURE 2-4.   Relationships Between Total and Average Cost Curves.

The average total cost (ATC) curve shows how the total cost per unit changes depending on the level of production. The ATC curve first declines and then increases with increasing production. The shape of the ATC curve is determined by the variable and fixed cost relationships and the production function from which they are derived. The average fixed cost (AFC) becomes smaller as output increases because the total fixed cost is spread over more units of output. Therefore, the AFC curve is downward sloping as long as total physical product is increasing. The shape of the average variable cost (AVC) curve corresponds to the average physical product (APP) derived from the production function. The AVC is declining where the APP is increasing, and increasing where the APP is declining.

These relationships between the cost curves and the production function determine the profit-maximizing level of production.

Profit-Maximizing Level of Production

Examples of the costs and revenues to be considered in making production-level decisions are illustrated in Table 2-5. This table is based on a production function similar to the one in Figure 2-1, except that there are three variable inputs rather than one. All other inputs are held constant. In Table 2-5, it is assumed that the least-cost combination of water, fertilizer, and seed is used to obtain each level of

TABLE 2-5. Costs and Revenues for Producing Various Quantities of Corn

(1) Total Product of Corn (bu/acre)	(2) Total Fixed Cost per Acre	(3) Total Variable Cost per Acre[a]	(4) Total Cost per Acre	(5) Average Total Cost per Bushel	(6) Marginal Cost per Bushel[b]	(7) Marginal Revenue ($3/bu)
100	250	108	358	3.58		
					.40	3.00
110	250	112	362	3.29		
					.80	3.00
120	250	120	370	3.08		
					1.60	3.00
130	250	136	386	2.97		
					2.80	3.00
140	250	164	414	2.96		
					4.00	3.00
150	250	204	454	3.03		
					5.20	3.00
160	250	256	506	3.16		
					6.40	3.00
170	250	320	570	3.35		

[a] Variable cost in this case is the cost of the least-cost combination of water, nitrogen, and seed for each level of production. In Figure 2-1, only seed is variable.
[b] Marginal cost is calculated by dividing the increase in total costs by the increase in yield.

production. (The method for finding this least-cost combination was described in the preceding section.) The total of the costs for these three inputs is the total variable cost per acre in Table 2-5.

The total fixed costs of $250 per acre include all of the costs associated with growing corn other than the cost of water, nitrogen, and seed. They include interest on the investment, depreciation, taxes, insurance, labor, and other costs incurred up to this point in the production process.

To maximize profits, increase production if the increase in total revenue resulting from a one-unit increase in production is greater than the increase in total cost required to produce it. The increase in total cost resulting in a one-unit increase in production is called the marginal cost (MC), and the corresponding increase in total revenue is called the marginal revenue (MR). The MR is the same as the price if the price remains constant regardless of the quantity sold. Profits are maximized when MC = MR.

For the example in Table 2-5, profit is maximized at the production level of 140 bushels per acre, where the MR, or the price of $3.00 per bushel, is greater than the MC of $2.80 per bushel. Increasing production by another 10 bushels increases the MC to $4.00, compared to the MR of $3.00. To verify that 140 bushels is indeed the most profitable level of production, calculate the total revenue at each production level and subtract the total cost.

Average total cost is important in making production decisions because, in the long run, the price received must exceed the ATC if a profit is to be earned. At 140 bushels per acre, the ATC is $2.96, and the profit therefore is $0.04 per bushel. The ATC and MC relationships for this example are illustrated in Figure 2-5.

Suppose the price drops to $1.80 per bushel. First, the farmer would adjust the level of production to 130 bushels per acre. At this point, the MR of $1.80 is

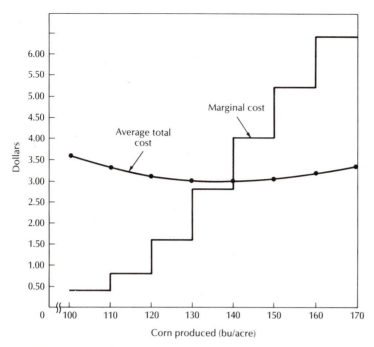

FIGURE 2-5.   Average and Marginal Cost Curves for Corn Production.

greater than the MC of $1.60. But at this lower price there are no profits; the ATC is $2.97 per bushel. So, in fact, the farmer would be trying to minimize losses in the short run. Whether the farmer would continue to produce corn depends on the long-run price expectations. If the lower price, $1.80, is viewed as only temporary, the farmer will probably continue to produce, but at the lower production level. If the price is not expected to improve, alternatives such as producing other crops, renting the land to someone else, or selling the farm might be considered.

Now suppose the price increases to $4.25 per bushel. The profit-maximizing level of production is 150 bushels per acre, where the MR of $4.25 exceeds the MC of $4.00. With this higher level of production, the ATC is $3.03 per bushel. Profits are maximized in this case where the ATC is increasing. This situation is contrary to the popular belief that a lower cost per unit means a higher profit. The profit increases because the additional production resulting from the use of more inputs is worth more than the cost of producing it. So, the objective is not to produce where the cost per unit is lowest, but where MC = MR.

A Combination of Products

The combination of products a manager chooses to produce is influenced by the technical relationship of these products. For example, the technical relationship of barley to alfalfa is different from that of barley to wheat. The relationship of timber production, which utilizes winter labor, to strawberries is different from that of dairy cows to strawberries. Product relationships can be classified as competitive, supplementary, and complementary.

Two products are *competitive* when an increase in one product results in a decrease in the other. They are competing for the same inputs at the same time. A farmer with 640 acres suitable for wheat and barley production can plant the land to all wheat or to all barley or to any combination of the two. Because barley and wheat use the same type of land, machinery, and other supplies at the same time of year, they are competitive. Any inputs used to produce one must be taken from the other.

Another relationship exists between wheat and winter feeding of cattle. Cattle feeding utilizes labor at one time of year, wheat production at another. If there is sufficient operating capital for both the cattle and the wheat, these products are *supplementary*, because an increase in one does not affect the other. Of course, cattle or wheat production might be expanded to the point where the products begin to compete for operating capital, management, or even land. At this point, they are no longer supplementary. They have become competitive.

A *complementary* relationship exists when an increase in one product causes an increase in the production of another. For example, increasing the winter grazing of sheep on grass seed fields may, if managed properly, cause an increase in the yield of grass seed harvested per acre. Likewise, a legume that fixes nitrogen in the soil might increase the yield of next year's corn crop. As with supplementary products, expanding the production of two complementary products causes them eventually to compete for one or more of the same resources. Table 2-6 illustrates these relationships; the data are hypothetical.

When determining the most profitable combination of products, these technical relationships between products should be considered. The first step in selecting the combination of products to be produced is to identify those products that are

TABLE 2-6. Complementary, Supplementary, and Competitive Products
and the Marginal Rate of Substitution

Product Relationship	Wheat Production	Alfalfa Production	Marginal Rate of Substitution
Complementary	7000	0	
			$-320/+320 = -1.00$
Supplementary	7320	320	
			$+\ \ \ 0/+320 = +\ \ \ 0$
	7320	640	
			$+320/+320 = +1.00$
	7000	960	
			$+500/+320 = +1.56$
Competitive	6500	1280	
			$+700/+320 = +2.19$
	5800	1600	
			$+1600/+320 = +5.00$
	4200	1920	

competitive and then to select the one or two that are most profitable. Next, the nature of the competitive relationship between them should be examined.

Some products substitute at a constant rate; that is, a one-unit increase in one results in the same amount of decrease in the other regardless of the level of production. In this case, the manager should devote all resources to the product that gives the highest profit from the use of those resources. However, when a constraint such as a marketing contract limits the sale of this product, the manager should then produce up to this limit and add a second product to utilize the remaining resources.

The second type of competitive relationship is illustrated by the competitive range in Table 2-6, where alfalfa replaces wheat at an increasing rate; that is, larger and larger amounts of wheat output are replaced by the same amount of added alfalfa output. In this case, the most profitable combination of products can be found by using the marginal principle as follows:

1. Compute the marginal rate of substitution (MRS) by dividing the number of units of replaced or substituted product by the number of units of added product:

$$\text{MRS} = \frac{\text{units of replaced product}}{\text{units of added product}}$$

2. Compute the inverse price ratio of the products:

$$\text{Inverse price ratio} = \frac{\text{price of added product}}{\text{price of replaced product}}$$

3. Continue to substitute one product for the other if the inverse price ratio is greater than the MRS:

$$\frac{\text{price of added product}}{\text{price of replaced product}} \geq \frac{\text{units of replaced product}}{\text{units of added product}}$$

or

$$\text{Inverse price ratio} \geq \text{MRS}$$

Once the products with competitive relationships have been considered, the next step is to identify those products that are supplementary and complementary. Remember that these products can be added to the combination of products being produced without decreasing the output of the other products. In fact, adding a complementary product actually increases output. These supplementary and complementary products should be added to the combination of products up to the point where they become competitive. Referring to Table 2-6, at least 640 units of alfalfa should be produced to take advantage of the complementary and supplementary relationship between alfalfa and wheat in this hypothetical example. It may be profitable to produce alfalfa beyond 640 units (in the competitive range), but this decision depends on the price ratio of alfalfa and wheat compared to the MRS, as explained earlier.

There are other factors to be considered in choosing the best level and combination of farm products. These factors are discussed in more detail in later chapters.

Inadequate Information and Economic Principles

A frequent criticism is that the use of economic principles requires more information than farmers have available. It is probably true that most farmers do not know the exact response of their crops to fertilizer. Crop yields vary with the weather. Prices are uncertain, and farmers may not know early in the season, when fertilizer is applied, what the price of the crop will be at harvest time.

Considerable attention is devoted to risk and uncertainty in later chapters. However, farmers—by one means or another—must decide whether to apply fertilizer and, if so, how much. What has been said about fertilizer–yield relationships applies to all input–output relationships. A lack of information does not invalidate the principle. Economic principles serve as guides in gathering and analyzing information for better decision making.

The principles outlined here—opportunity costs, marginal analysis, and substitution—are being used more frequently by researchers in agronomy, animal husbandry, and other fields. Research trials are being conducted at experiment stations and in cooperation with individual farmers to provide information for farm management decisions. This information, used in conjunction with individual farm records, will facilitate the application of economic principles in farm management.

Summary

Economic principles guide the process of making farm management decisions. Choosing among alternatives is complicated by limited resources—with unlimited resources, managing the farm business would present no challenge. In fact, it would be very dull. Because resources are limited, the concept of opportunity cost is relevant. The opportunity cost of producing a commodity is not the total expenses incurred but the income that could be earned by using the same resources in the next best alternative. Thus, economic principles stress the importance of considering the alternatives when making decisions.

Production relationships—the total physical product (TPP), average physical product (APP), and marginal physical product (MPP)—define the context within which farm managers make decisions about the amounts and combinations of inputs to use. By studying these relationships, the manager can narrow the possibilities to the rational range (stage II of the production function). By applying marginal analysis to the production relationships, the amounts and combinations of inputs to use can be determined:

1. For a single input, profits are maximized where the marginal value of product (MVP) equals the marginal input cost (MIC).
2. When the supply of input is limited, the input should be allocated so that the MVPs for each use are equal.
3. The least-cost combination of inputs for a given level of output is found

where the inverse price ratio equals the marginal rate of substitution (MRS) of the two inputs.

Deciding whether to produce and determining the combination of products and the level of production depend on the cost relationships and the technical relationships between the alternative products. The distinction between fixed and variable costs is important because only variable costs, those affected by the decision, are relevant. The technical product relationship—competitive, supplementary, or complementary—determines the optimal combination of products. Profit-maximizing levels of production and combinations of products are determined by applying the following marginal concepts of economics:

1. To maximize profits, the level of production is increased to the point where the marginal cost (MC) equals the marginal revenue (MR).
2. The maximum profit combination of competitive products for a given amount of resources is found where the inverse price ratio equals the MRS of the two products.

The application of economic principles is sometimes limited by a lack of adequate information. The next chapter introduces procedures for acquiring this information, but because of the vagaries of weather and markets, the information about future production relationships and prices will never be completely adequate. Nevertheless, economic principles provide the framework for logically considering and making farm management decisions.

Recommended Readings

Epp, D. J., and J. W. Malone, Jr. *Introduction to Agricultural Economics*. New York: Macmillan Publishing Company, Inc., 1981. Chapters 2 through 4.

Leftwich, R. H. *The Price System and Resource Allocation*, 8th ed. Hinsdale, Ill.: Dryden Press, 1982. Chapters 8 and 9.

Doll, J. P. and Frank Orazem. *Production Economics: Theory with Applications*, 2nd ed. New York: John Wiley & Sons, Inc., 1984. Chapters 2 through 5.

APPENDIX TO CHAPTER 2

Mathematical Presentation of Economic Principles

The knowledge and use of calculus make it possible to present the economic principles discussed in this chapter in a more concise form.

A simple production function can be expressed algebraically as follows:

$$y = f(x_1 | x_2, x_3, \ldots, x_n)$$

This equation says that y, the TPP, is dependent upon $x_1, x_2, x_3, \ldots, x_n$, but in this case only x_1 is variable and $x_2, x_3, \ldots, x_n$ are set at some constant level. The TPP can then be defined as the total physical product resulting from every quantity of x_1 with $x_2, x_3, \ldots, x_n$ set at some constant level, or TPP $= y = f(x_1)$.

The APP is the quantity of y per unit of x_1 at some given level of x_1:

$$\text{APP} = \frac{y}{x_1}$$

or because

$$y = f(x_1)$$

the APP can be expressed as follows:

$$\text{APP} = \frac{f(x_1)}{x_1}$$

The MPP is the change in TPP, or y, resulting from a unit change in x_1. In calculus this is referred to as the derivative of the function, $y = f(x_1)$. The MPP expressed in calculus notation is:

$$\text{MPP} = \frac{dy}{dx_1}$$

To illustrate how to find the most profitable amount of input to use, a specific example is used. The general form of the production function is:

$$y = a + bx - cx^2$$

where y = total product; x = input; a, b, and c = constants; in this case, $a = 4$, $b = 3$, and $c = 0.4$

The MPP is the derivative of the production function.

$$\text{MPP} = \frac{dy}{dx} = 3 - 0.8x$$

The point of most profitable production is found where the MVP is equal to the MIC. The MVP is determined by multiplying the MPP, dy/dx, by the price of Y, Py. The MIC is the price of x, Px. When profit is at a maximum:

$$(dy/dx)\, Py = Px$$

For the example:

$$(3 - 0.8x)\, Py = Px$$

Assume that $Py = 2$ and $Px = 2$, and solve for the most profitable level of x:

$$(3 - 0.8x)\, 2 = 2$$
$$6 - 1.6x = 2$$
$$1.6x = 4$$
$$x = 2.5$$

To find the output of y, substitute 2.5 in the production function wherever x appears:

$$y = 4 + 3(2.5) - 0.4(2.5^2)$$
$$y = 9$$

Now the general form of the production function is examined. This function can be characterized as follows:

$$g(y_1, y_2, y_3, \ldots, y_m) = f(x_1, x_2, x_3, \ldots, x_n)$$

where y_i is a product and x_j is an input. Px_j is the price of input x_j and Py_i is the price of product y_i. These prices are assumed to be constant regardless of the quantities purchased or sold.

The profit-maximizing quantity of input x_j is at the point where the marginal input cost of x_j ($MICx_j$) is equal to the marginal value product from x_j ($MVPx_j$) or where

$$Px_j = \frac{dy_i}{dx_j} \cdot Py_i$$

which is the same as

$$MICx_j = MPPx_j \cdot Py_i \equiv MVPx_j$$

The profit-maximizing quantity of product is at the point where MCy_i is equal to MRy_i or

$$\frac{d(Px_1 \cdot x_1 + Px_2 \cdot x_2 + \cdots + Px_n \cdot x_n)}{dy_i} = Py_i$$

assuming that $x_1, x_2, x_3, \cdots, x_n$ are combined in the least-cost manner to produce any given quantity of y_i. The first term in the preceding equality is MCy_i, or the first derivative of TCy_i, and the second term is MRy_i.

The least-cost combination of two inputs given a fixed level of production is where the $MRSx_j,x_{j+1}$ is equal to the inverse ratio of the input costs:

$$\frac{dy_i/dx_j}{dy_i/dx_{j+1}} = \frac{Px_j}{Px_{j+1}}$$

which can be rewritten as

$$-\frac{dx_{j+1}}{dx_j} = \frac{Px_j}{Px_{j+1}}$$

and is equivalent to

$$MRSx_j, x_{j+1} = \frac{Px_j}{Px_{j+1}}$$

The profit-maximizing combination of two products, given a fixed bundle of inputs, is where the $MRSy_i, y_{i+1}$ is equal to the inverse ratio of the product prices:

$$\frac{dy_i/dx_j}{dy_{i+1}/dx_j} = \frac{Py_i}{Py_{i+1}}$$

which can be rewritten as

$$-\frac{dy_{i+1}}{dy_i} = \frac{Py_i}{Py_{i+1}}$$

and is equivalent to

$$\mathrm{MRS}y_i,\ y_{i+1} = \frac{Py_i}{Py_{i+1}}$$

Although knowledge of calculus is an advantage in applying the marginal principles of economics, it is not essential. These principles can also be applied using graphs and tables as illustrated in this chapter.

CHAPTER 3

Farm Management Information

Effective farm management requires an extensive base of financial and economic information. Decisions can be no better than the information used to make them. Thus, managing a farm business requires a system that will collect, store, retrieve, and analyze the relevant information. Information is available from outside and inside the farm business. External sources include firms with which the manager does business, as well as public and private agencies such as university extension services and experiment stations.

The emphasis here is on internally generated accounting information. Although the problem for many managers is the lack of adequate accounting information, too much irrelevant, useless information can also be a problem. Therefore, the first step in designing the information system is to decide which information is relevant. Then the manager must decide how to organize it and have it available at the right time and place.

In designing the information system, the manager must keep in mind the economic principles discussed earlier. The time and money invested in the system should be considered in the same way as fertilizer, feed, or any other production input. That is, additional time and money should be invested in gathering information up to the point where the added value of the information is equal to the added cost. In other words, it is worthwhile for the manager to continue to improve the information as long as the marginal value of additional information exceeds its marginal cost. The cost will depend on the opportunity costs of the manager's time and money. The value is more difficult to estimate. It must be based on the anticipated returns from making better management decisions as a result of having the additional information. For information to have value, it must have the potential to affect the management decisions.

This chapter presents the procedures and components involved in designing and using a farm accounting system. The next chapter describes how these components are put together to prepare financial statements that summarize the information critical to the successful management of a farming enterprise.

Accounting and Its Purpose

Accounting is a set of principles and procedures for systematically recording the transactions of the business, preparing financial statements to summarize the records, and interpreting the financial statements. Economic principles provide the framework for managing the farm business. Accounting principles provide the basis for generating the information needed to understand the activities of the business and to apply the economic principles. Accounting information also provides the basis for the preparation of financial statements used in analyzing and monitoring the progress of the business.

Purposes of Farm Accounting

Farm accounts provide financial information about the farm business. There are three general uses for this accounting information: reporting, diagnosis, and planning.

Reporting.   Many farmers keep accounts solely for the purpose of filing income tax returns. The records, therefore, must provide at least the data to verify tax returns. "A farmer, like other taxpayers, must keep records to prepare an accurate income tax return, and be sure only the tax owed is paid."[1]

A good accounting system is also important for making loan applications and reporting to lenders. Farm accounts help document the credit worthiness of a loan application by providing the information the lender needs to assess the profitability, repayment capacity, and risk-bearing ability of the farm business. The lender may also require periodic reports to monitor the loan. For this reason, the accounts must be complete and up-to-date.

Farm accounts are needed to meet a number of other business reporting requirements. For example, accurate accounts are required to substantiate the division of income for share lease arrangements, partnerships, corporations, and profit-sharing agreements. They are also necessary for settling estates and for participating in certain types of government farm programs.

Diagnosis.   Farm accounts can help identify the strengths and weaknesses of the business. One useful approach in identifying problems—or opportunities for improvement—is the process of *management by exception*, which involves using the accounting system to identify exceptions or deviations of the actual performances from the expected. These exceptions and deviations are the impetus for planning and action. The key element in this process is an accounting system capable of locating unacceptable performance—the exception. This process involves more than determining whether the business made or lost money. It includes examining each enterprise and determining what caused the exceptions. For example, suppose a

---

[1] U. S. Department of the Treasury, *Farmer's Tax Guide*, Internal Revenue Service Publication 225, (revised annually).

cattle feeder's accounting system indicates that the cost per hundredweight of gain is consistent with expectations. However, the system might have done better. If it allowed for closer inspection in order to evaluate the price paid per ton of feed, the manager might have discovered that a higher feeding efficiency was offsetting a weakness in feed purchasing; that is, the feed could have been bought at a lower price.

Planning. The accounting system can aid planning by providing data that would help to estimate the effects of potential decisions on the profits and financial position of the business. For budgeting purposes, the system can provide cost information and production coefficients unique to the individual farm.

Although these records are a good source of information for planning purposes, this information must be used with caution. After all, records are by definition a matter of history; they cannot predict the future. Information about past performance, however, is valuable because it describes the unique situation of the individual farm operation. However, it must be used in conjunction with other data in projecting future consequences. Records of physical inputs and outputs such as those entered in crop and livestock records are especially valuable. A few years of records provide a sound basis for estimating the yields expected from a particular field and the rate of gain from steers fed in a particular way. These physical coefficients will remain fairly constant even though the price of the crop and the cost of the feed may change.

The Activities of a Farm Business

To understand the design and role of the accounting system, it is useful to consider the three activities involved in the management of a business: investment, financing, and production activities (Table 3-1). Each type of activity involves different types of transactions that are handled differently in the accounting system.

TABLE 3-1. Farm Business Activities and Accounting Transactions

Activities	Sample Accounting Transactions
Investment activities	Purchase capital assets
	Depreciate capital assets
	Sell capital assets
Financing activities	Borrow money
	Repay loans
	Pay interest
Production activities	Sell products and services
	Incur expenses
	Earn net income

Investment Activities

Investment activities are related to the purchase, depreciation, and sale of assets. These assets are the items owned by the business, such as supplies, feed, machinery, breeding animals, and land that are used in production activities to generate net income. In the production process, supplies such as fertilizer may be entirely consumed. Other assets, such as machinery and buildings, are not used up in production but do depreciate. *Depreciation* refers to an asset's decrease in value as a result of its age, use, and obsolescence. This depreciation becomes a production expense.

The total of all assets determines the size of the business. The mix of assets—the various amounts invested in land, machinery, and breeding animals, for example—determines the most profitable production activities.

Financing Activities

The financing activities of the business determine the total liabilities, that is, the sum of obligations or debts owed to others in the form of notes, mortgages, and accounts payable. The difference between assets and liabilities represents the investment of the manager's equity or net worth in the business. The ratio between debt and equity financing of the business has important implications for its potential profitability and riskiness.

Financing activities involve borrowing money, repaying loans, and paying interest on outstanding principal. Interest payments are charged as expenses against production activities. On the other hand, borrowed money allows for the acquisition of additional assets to contribute to production activities.

Production Activities

The production activities of the farm business are intended to generate revenue through the sale of products and services. In the process of producing these products and services, expenses are incurred, such as the cost of feed, veterinary services, herbicide applications, and so forth. Depreciation and interest on borrowed money are also expenses. Subtracting all these expenses from total revenue yields the net income for the business. This net income can be reinvested back into the business or withdrawn to meet family living expenses. Reinvesting the net income increases the net worth, or equity, of the farmer.

Designing a Farm Accounting System

Designing accounting systems for farm businesses presents some special problems because of the interdependence of the home, farm business, and nonfarm business interests. This interdependence makes it more difficult to identify the financial transactions that properly belong to only the farm business. Home and family transactions include food, clothing, home improvements, family auto expenses, family life insurance, and so forth. Nonfarm business interests might be investments in stocks and bonds, an off-farm job, or part ownership in a nonfarm business such as a vegetable processing cooperative. Cash often flows back and forth between farm, home, and nonfarm interests. These transactions must be separated to allow for proper tax reporting and management analysis.

Initial Design Considerations

An accounting system should (1) be easy to keep, (2) provide necessary information, and (3) provide the information when it is needed. At times, these requirements may be in conflict; that is, records that are the easiest to keep may not satisfy the other two essential requirements. In accordance with the economic principle outlined earlier, the manager should continue to spend time on accounting records until it becomes more profitable to spend time elsewhere. It is possible to record information that has less value than would be earned if the time had been spent in another way.

Farmers have a number of alternative record-keeping systems from which to choose, or they may design their own. A number of university extension services offer record-keeping services; banks, production credit associations, and farm supply companies also provide such services for their customers. A wide variety of printed forms for farm accounting are available from commercial office supply firms and other sources. This chapter presents a generalized design for an accounting system that can be adapted to the forms best suited to a particular farm business. The examples used here are merely illustrative.

Managers must be able to interpret and analyze their records properly in order to make the best management decisions, but this does not mean that they must do the actual posting of entries; other members of the family or an employee can do it. Some farmers use a professional accountant to keep and summarize their financial records, thereby freeing their time for managing the business.

Accounting Period

The accounting period determines how frequently inventories are taken and the time interval over which other accounts, such as cash receipts and expenses, are summarized. The most common accounting period for farm businesses is the calendar year. Some farm managers, however, choose a fiscal year that corresponds to their production year. For example, cow-calf ranches may find it useful to close their accounts in November or after calves are sold. Farmers who are leasing land may find it advantageous for their fiscal year to coincide with their annual production and lease plans. Because the accounting year should be the same as the tax year for reporting income taxes, this decision also has important tax planning implications.

Accounting Method

Farmers can choose the *cash* or *accrual* method of accounting. Most farmers use the cash method because it is easier to keep and is consistent with the way they report income taxes.

A complete accrual system is a double-entry system in which offsetting entries, a debit and a credit, are made for each transaction. For example, if cattle are sold for cash, the cattle inventory account will be debited and the cash account will be credited. One advantage of the double-entry system is its built-in checks and balances; another is its accuracy in reflecting the revenue and expenses associated with a year's production. With double-entry accounting, revenue is recognized when it accrues (when the sale is finalized or the product is delivered), regardless of when the cash is received. Expenses are recognized when they are incurred, rather than when they are paid. The objective is to match the expenses with the production that occurred during the year. Thus, the accrual method more truly reflects the net income earned by the business during the year.

With a cash, or single-entry, system of accounting, entries are made when the cash transactions occur. For example, an expense is recognized when payment is made, and revenue is recognized when money is received. The disadvantage of the cash method is the erroneous impression it can give regarding the profitability of the business, unless appropriate adjustments are made. Even though the cash method with a single-entry system is used, it is possible to calculate the business's net income as it would have been calculated by the accrual system. This process involves making adjustments for the changes in inventory values and accounts payable and receivable between the beginning and end of the year.

Farmers are allowed to use either the cash or the accrual method for income tax reporting. The cash method offers more flexibility for tax planning purposes. The choice of reporting method, cash or accrual, is made when the first tax return is filed. It is possible, of course, to use a double-entry accounting system and report income taxes using the cash method.

The general system of accounting presented here is a single-entry system that assumes the cash method for income tax reporting and allows for appropriate adjustments to estimate the accrual net farm income, the same as would be calculated using the double-entry system.

Parts of a Farm Accounting System

Because of the diversity of farm businesses, not all records are of equal importance on all farms. Although individual forms and schedules are different, certain basic information is needed to provide an accurate financial accounting of the business:

1. *Capital and depreciation schedules.* These schedules are used to record purchases and sales of capital assets, estimated market values, and depreciation. Examples of these capital assets include machinery, equipment, trucks, purchased breeding livestock, land, buildings, and real estate improvements.
2. *Liability schedules.* This information includes records of money loaned and borrowed, principal and interest payments, and unpaid principal balances. Amounts payable and receivable on open accounts are also recorded.
3. *Farm inventory schedules.* The quantities and values of crops, feed, animals

for sale, raised breeding stock, and supplies on hand at the end of the year are recorded here. Depreciable assets and land are normally not included.

4. *Cash receipt and expense schedules.*   These forms are necessary to record the cash transactions of the farm business and to classify them systematically and consistently.

Each accounting system uses slightly different procedures for handling the various types of transactions. Some general examples are used here to illustrate the types of information that must be kept for all accounting systems.

Capital and Depreciation Schedules

The capital and depreciation schedules are related to the investment activities of the business. Here the manager records information about the depreciable and real estate assets that contribute to production for more than one year. Examples include machinery, equipment, purchased breeding stock, land, and buildings. Two types of records need to be kept: (1) capital purchases and sales and (2) depreciation.

For capital purchases, the following information should be recorded:

1. Date of purchase.
2. Description of the item purchased.
3. Status (new or used).
4. Purchase cost if no trade was involved.
5. Amount paid "to boot" if a trade was involved.
6. Description of the item traded.
7. *Book value* (the undepreciated balance) of the item traded.
8. Cost or basis of the item purchased (4, or 5 plus 7).

Information recorded for capital sales should include the following:

1. Date of sale.
2. Description of the item sold.
3. Amount received through the sale.
4. Book value of the item sold.
5. Gain or loss through the sale (3 minus 4).

Depreciation is an important accounting concept related to the investment activities of the business. The net income from production activities is also affected by depreciation because it is treated as an expense.

To compute depreciation, the original cost of an asset must be prorated over its useful life. The amount of depreciation charged should correspond to the asset's loss in value over time. The original cost is a prepaid expense. However, if the asset will be used in more than one accounting period, this cost should be allocated to those accounting periods that correspond to the productive life of the asset.

The loss in value of an asset is determined by its age, use, and obsolescence. The importance of each of these factors depends on the asset and the use it receives. For example, a loss in value of buildings is usually due to age and obsolescence rather than to use. A building will decline in value almost as rapidly when not used as when used. Also, two trucks of the same make and model purchased at the same time would have the same value if value were based on age and obsolescence alone. However, if one truck has been driven 20,000 miles during the year and the other one only 2000 miles under comparable conditions, it is obvious that they would not have the same value. Use is more important than age or obsolescence in determining the remaining value of the two trucks.

Annual cost and performance should also be considered in determining the amount of depreciation. In the case of machinery, repair costs tend to be low and use tends to be high when the machine is new. But as the machine gets older, repair costs rise and performance is reduced. Untimely breakdowns consequently mean lost time as well.

Because depreciation affects taxable income, the income statement, and the balance sheet, various objectives may be in conflict when depreciation is calculated. There may be income tax advantages to rapid depreciation. At the same time, a substantially undervalued inventory may seriously weaken the apparent financial position of the business—a disadvantage when applying for credit.

Depreciation charges can be spread evenly over the life of an asset, or they can be higher during the early part of its life. An even spread is appropriate for durable assets, such as fences or farm buildings, which have uniform repair costs throughout their lives and are not subject to significant design improvements. On the other hand, some assets depreciate rapidly when still relatively new; for instance, automobiles are redesigned every two or three years, and machinery manufacturers frequently develop improved technology.

Repair costs rise as an asset ages. Assuming each renders the same service, a new machine is consequently more valuable to the business than an older one. In general, therefore, a method concentrating depreciation near the beginning of an asset's life is preferred. Three common methods are illustrated and compared.

The Straight-Line Method.   Annual depreciation is computed by dividing the original cost of the asset, less any salvage value, by the expected years of life. A $12,000 asset with a $4000 salvage value and an expected life of eight years would depreciate $1000 each year:

$$\frac{\$12,000 - \$4000}{8} = \$1000$$

The Double-Declining Balance Method.   A constant rate of depreciation is used every year and applied to the value of the asset at the beginning of the year. The salvage value is not subtracted from the cost. The rate is applied to the unrecovered balance (book value) until the salvage value is reached, and then no further depreciation is taken. The rate used for figuring the annual depreciation is twice the rate of the straight-line method. An asset expected to last for 10 years depreciates at the rate of 10 percent using the straight-line method; twice this amount, 20 percent, is the rate for the double-declining balance method. For an eight-year life, the double-declining balance rate would be 25 percent (see Table 3-2).

The Accelerated Cost Recovery System.   This system is used by the Internal Revenue Service for computing taxable income. It replaces the old depreciation methods for property purchased after December 31, 1980. With this system, the annual

TABLE 3-2. Double-Declining Balance Depreciation of a $12,000 Asset with an Eight-Year Life and a $4000 Salvage Value

Year	Value at Beginning of Year	Annual Depreciation	Remaining Book Value
1	$12,000	25% × $12,000 = $3,000	$12,000 − 3,000 = $9,000
2	9,000	25% × 9,000 = 2,250	9,000 − 2,250 = 6,750
3	6,750	25% × 6,750 = 1,688	6,750 − 1,688 = 5,062
4	5,062	5,062 − 4,000 = 1,062[a]	4,000
5	4,000	0	4,000
6	4,000	0	4,000
7	4,000	0	4,000
8	4,000	0	4,000

[a] Here 25 percent of the beginning book value would be $1265; but because the salvage value is $4000, only $1062 is charged as depreciation in the fourth year.

recovery allowance (depreciation) depends on the class of the property. The majority of farm assets are classified as five-year property. Examples include farm machinery and equipment, breeding and dairy cattle, grain bins, and silos. Following are the recovery rates for each year for five-year property:

Year	Recovery Rate (%)
1	15
2	22
3	21
4	21
5	21

Which Depreciation Method to Use. Table 3-3 compares these three methods for computing depreciation and illustrates how the choice of method can affect the value of the asset on the balance sheet.

In deciding which of the three depreciation methods to use, there are two considerations: (1) the method that best approximates the loss in value or service of the asset and (2) the income tax implications. Obviously, the straight-line method is the easiest to use and understand, and it provides an average cost for budgeting purposes. However, most farm assets depreciate more rapidly than the straight-line method indicates. The double-declining balance and the accelerated cost recovery methods are therefore more suitable.

The choice of depreciation method may result in potentially wide discrepancies in asset values. The double-declining balance and accelerated cost recovery methods may have the effect of undervaluing assets and overstating expenses. Conversely, applying the straight-line method to an asset with a long life may overvalue

TABLE 3-3. Comparison of Three Depreciation Methods for an Asset with a
$12,000 Cost, Eight-Year Life, and $4000 Salvage Value

	Straight Line		Double-Declining Balance		Accelerated Cost Recovery	
Year	Depreciation	Book Value	Depreciation	Book Value	Depreciation	Book Value
1	$1,000	$11,000	$3,000	$9,000	$1,800	$10,200
2	1,000	10,000	2,250	6,750	2,640	7,560
3	1,000	9,000	1,688	5,062	2,520	5,040
4	1,000	8,000	1,062	4,000	2,520	2,520
5	1,000	7,000	0	4,000	2,520	0
6	1,000	6,000	0	4,000	0	0
7	1,000	5,000	0	4,000	0	0
8	1,000	4,000	0	4,000	0	0

NOTE: To simplify this example, investment credit and other cost recovery options (important income tax considerations) have been omitted.

it and understate the depreciation expense. For the purpose of analyzing the farm business, the depreciation should approximate the decline in the asset's value due to age, use, and obsolescence. However, this method of depreciation may not be allowed or be advantageous for income tax purposes; in this case, the manager should use one method for income tax purposes and another for analysis and management.

Machinery, equipment, purchased breeding stock, buildings, and other depreciable assets should be entered on a depreciation schedule (see Table 3-4). This schedule allows the recording of the total depreciation for several years based on the cost, salvage value, depreciation method, and the life established for each asset.[2]

Liability Schedules

Liability schedules provide a record of the business's financial activities, which include accounts payable, notes, sales contracts, mortgages, and land contracts. The purpose of these schedules is to keep track of the total unpaid principal on loans and amounts payable on accounts, required payments, and due dates.

An accounts payable schedule is not necessary with a cash system, because purchases are recorded only after payment has been made. However, a file should be maintained for each unpaid account. By adding the statements from all the files, the manager can determine the total amount owed on credit accounts at any time. Because this information is needed in preparing financial statements, the manager should always do this on the last day of the accounting year. As payments are made on the accounts, the statements are transferred from the "payable" files to a "paid" file.

[2] To simplify this example, information relative to investment credits and other cost recovery options—important income tax considerations—has been omitted.

TABLE 3-4. A Simplified Depreciation Schedule

Item and Description	Date Acq'd.	Cost or Basis	Salvage Value	Deprec. Method	Life or Rate	End 1980		End 1981		End 1982	
						Deprec.	Book Value	Deprec.	Book Value	Deprec.	Book Value
Tractor #3	9-1-77	4000	0	SL	10	400	2667	400	2267		
Truck	1-10-79	2418	150	DDB	6	537	1075	358	717		
Tractor #4	10-1-80	4567	0	SL	8	143	4424	571	3853		
Beef cow	10-4-79	500	150	SL	5	70	262	70	192		
Silo	1-2-75	4560	0	SL	15	304	2736	304	2432		

Schedules are needed for both multiple- and single-payment loans. Multiple-payment loans can be recorded on a schedule like that of Table 3-5. The schedule provides columns for the name of the lender, the purpose of and security for the loan, the interest rate, and the repayment schedule. Other columns of this schedule allow for recording the date(s) payments are made, the unpaid principal balance at the beginning of the year, and the annual amounts applied to interest and principal.

Other information that might be included in the schedule is the accrued but unpaid interest as of the end of the accounting year. Additional columns also might be included for the portion of the principal balance due within 12 months and the portion due after 12 months. This latter information on accrued interest and principal balances is useful when preparing the balance sheet. Single payment loans require the same information, but additional columns are needed to record the money borrowed and the dates.

Farm Inventory Schedules

Farm inventory schedules are needed to record and keep track of both investment and production activities. These inventory items are a portion of the business's assets. Also required are the changes in inventory values from the beginning to the end of the year in order to record the net income from production activities. The list of the assets in inventory, including quantities and values, should be completed at the end of each accounting year; for many farmers, this is December 31. The inventory taken at the end of the accounting year is the beginning inventory for the next year. Past inventory records can serve as a checklist. It is important that inventories be conducted as close as possible to that date to ensure the greatest degree of accuracy. It is easier to get accurate physical counts and measurements if the manager does not have to rely on memory.[3] It is also important to use market price quotations from newspapers and other reliable sources for placing values on the inventory. Certain inventory items—the quantity, the price or value per unit, and the total value—should be indicated as of the last day of the accounting year.

Livestock and Poultry to Be Sold.   Livestock and poultry to be sold are the animals that will be sold within one year of the inventory date. Those animals to be kept for breeding purposes are listed in another category. Also included here are any livestock and poultry products that are intended for sale, such as milk, eggs, and wool. The animals should be grouped by kind and weight, and the value estimated as if the animals were to be sold the day of the inventory.

Raised Breeding Livestock.   Included in the raised breeding livestock category are farm-raised breeding animals and replacements that will be kept during the year.[4] A description of the animals, including their age and kind, should be recorded, as

---

[3] Some references giving detailed instructions on physical measurement are Robert A. Luening and W. P. Mortenson, *The Farm Management Handbook*, 6th ed. Danville, Ill.: Interstate Printers and Publishers, Inc., 1979, Appendix B; and Ralph R. Botts, *Farmers' Handbook of Financial Calculations and Physical Measurements*, U.S. Department of Agriculture Handbook No. 230, 1966.

[4] Purchased breeding stock is depreciated and therefore is included in the capital and depreciation schedules for tax purposes.

TABLE 3-5. A Simplified Liabilities Schedule

To Whom Owed	Purpose or Security	Interest Rate	Payment Date(s)	1984			1985		
				Principal Balance	Interest Payment	Principal Payment	Principal Balance	Interest Payment	Principal Payment
Federal Land Bank	Land mortgage	12%	12/1	$132,000	$15,840	$2,782	$129,218	$15,506	$3,116
Local State Bank	Tractor	13%	3/1	7,938	1,032	2,330	5,608	729	2,633

well as the number in each category. Including their estimated weights provides additional information when valuing the animals.

The beginning-of-year inventory can be used to check the new end-of-year inventory. The totals for each kind of livestock should balance as follows: Beginning inventory plus purchases plus births equals deaths plus sales plus consumed livestock plus ending inventory. A large discrepancy in the totals indicates that livestock have been lost, stolen, or unreported.

Stored Crops. A list should be prepared of all crops stored on the farm or in commercial storage as of the end of the accounting year. Examples of stored crops include corn, beans, wheat, cotton, fruit, nuts, and forages such as hay and silage. Again, quantities should be estimated using the most common units of measure—bushel, ton, hundredweight, and so forth. It is important to consider any grading or other quality factors that may affect the value of these commodities.

Purchased Feeds. The purchased feeds inventory includes any purchased grains, forages, prepared commercial feeds, vitamins, minerals, and protein supplements that are being stored for feeding to livestock. The price paid can be used as the estimated value. Quantities of each feed should balance as follows: Beginning inventory plus purchased feed plus produced feed equals consumed feed plus sales plus ending inventory. A large discrepancy in the totals indicates that feed has been lost, stolen, or unreported.

Growing Crops. The growing crops inventory provides a record of the value of the crops that are growing as of the last day of the accounting year. The cash costs for bringing each crop up to its current state of growth should be totaled separately. These costs include fertilizer, seed, herbicides, fuel, hired labor, and so forth. This cash cost approach provides a conservative estimate of the value of the growing crops.

Supplies. The supplies category includes gas, fuel, oil, veterinary medicines, herbicides, pesticides, fertilizer in storage, seed, and other supplies.

The major purpose of the inventory is to obtain a reasonable estimate of the year-end value of the foregoing assets. The use of this inventory information for preparing financial statements is discussed in the next chapter. It is important to be realistic in setting these values; if anything, they should be conservative. The various methods and guidelines for valuation are also discussed in the next chapter.

Cash Receipt and Expense Schedules

The checking account of the business provides the primary source of information for cash receipt and expense schedules that are used to record the production activities of the business. To use the checking account in this way, however, requires that all income received be deposited in the account and that checks be used to pay all expenses. When this is not possible, managers should keep track of purchases made with cash and then write themselves a check to provide a record of the transactions. Deposits should clearly show the source of deposited income, which may include borrowed money, savings certificates that were cashed, interest earned, and gifts, as well as income from farm product sales. Two checking accounts can be set up to keep business and nonfarm transactions separate. However, when this procedure is not practical, the system described here provides for their separation.

Tables 3-6 and 3-7 are examples of multicolumn schedules that are used for recording cash receipts and expenses, respectively. As each deposit is made or each check written, it is recorded on the appropriate schedule. Several columns are provided for dividing receipts and expenses into various categories for tabulation. In each schedule there is a column for nonfarm transactions. If there is only one checking account, all the checks for family living expenses and deposits of off-farm income are entered there. If two checking accounts are maintained, these columns record the transfers of money between the two accounts.

In Table 3-6, all receipts are classified by source: (1) operating receipts, (2) breeding stock and other capital sales, (3) money borrowed, and (4) nonfarm receipts. A complete entry with the date, explanation, quantity, and total value should be made for each transaction. Gross receipts should be recorded, and the deductions should be entered as expenses. For example, statements from dairy processors might show the quantity and price, along with deductions for hauling, handling, and other charges. The gross return should be shown under receipts, and the deductions should be listed in the appropriate expense section.

Expenses in Table 3-7 fall into four main categories: (1) operating expenses, (2) debt payments, (3) capital purchases, and (4) nonfarm expenses. The expense in dollars, date of the transaction, and quantity should always be recorded. The details of employee expenses are usually better kept in a separate labor and payroll ledger and the summaries transferred to the general farm record. Operating expenses such as those for feed, seed, fertilizer, and chemicals are usually recorded under the different headings. Expenditures for items that have a useful life of more than one year are classified as capital purchases. Some examples are machinery, breeding stock, buildings, and land terracing. (The details related to these capital purchase transactions are recorded in the investment schedules.)

Using entries commonly encountered on farms, the following transactions are entered on the sample receipt and expense schedules of Tables 3-6 and 3-7:

Receipts and Expenses for April

April 1	Hired labor (paid by check)	$ 924
April 2	Paid fertilizer bill	7,200
April 3	Sold wheat (for cash)	4,000
April 5	Sold feeder calves (for cash)	11,000
April 6	Loan payment ($3,600 principal, $3,400 interest)	7,000
April 7	Borrowed money from bank to purchase tractor	12,000
April 8	Purchased new tractor	13,600
April 15	Purchased refrigerator for home	400
April 20	Sold old tractor (for cash)	4,000
April 30	Received interest on savings certificate	120

TABLE 3-6. A Simplified Cash Receipts Schedule

| Date | Total Cash Received | Explanation | Quantity | Operating Receipts | | | | Sales of Breeding Stock | Capital Sales | Money Borrowed | Nonfarm Receipts |
				Crop Sales	Sales of Livestock	Livestock Products	Other				
4/3	$ 4,000	Wheat sales	950 bu	$4,000							
4/5	11,000	Feeder calves	30 fat		$11,000						
4/7	12,000	New tractor loan								$12,000	
4/20	4,000	For old tractor							$4,000		
4/30	120	Interest on certificate									$120

TABLE 3-7. A Simplified Cash Expenses Schedule

| Date | Total Cash Expended | Explanation | Quantity | Operating Expenses | | | | Debt Payments | | Capital Purchases | Nonfarm Expenses |
				Hired Labor	Seed	Fert. and lime	Other	Interest	Principal		
4/1	$ 924	To: A. Worker, wages		$924							
4/2	7,200	To: Growmore Co				$7,200					
4/6	7,000	To: Farmland Bank						$3,400	$3,600		
4/8	13,600	To: Big wheel for tractor								$13,600	
4/15	400	To: Arctic for refrigerator									$400

(The reader is encouraged to study the entries in the tables and to reread this section in order to develop a clear understanding of the reasoning behind these entries.)

To balance or check the records, the expense and receipt schedules can be compared to the checking account. The cash on hand at the beginning of the month plus cash received less cash expended should equal the cash balance in the account at the end of the month. Assume that there was a cash balance of $1000 in the account at the beginning of the month. Then:

$ 1,000	Beginning balance
+ 31,120	Total cash received
= 32,120	Total cash to account for
− 29,124	Total cash expended
= $ 2,996	Ending balance

There is another way to check the records: The total cash received should equal the sum of the totals of all other receipt columns, and the total cash expended should equal the sum of the totals of all other expense columns.

Computerized Accounting Systems

Many computerized accounting systems are now available. Some are relatively simple, emphasizing income tax reporting, and others are very comprehensive. Some operate on the farmer's own microcomputer; others involve mailing the monthly records to a central processing center. The principal advantage of a computerized system is that it reduces the tedious and time-consuming work involved in entering, organizing, and totaling transactions. This time could be spent studying the reports and using them as a decision-making tool. Other advantages include the following:

1. More accurate information, because computers do not make arithmetic mistakes and are capable of handling more detail.
2. More timely data, because the computer can process data faster and more often.
3. Better analyses, because complex calculations are performed quickly and easily.

Available Systems

Computerized systems available today offer managers a wide range of sophistication and usefulness. Some computerized services offered by universities, community colleges, and other organizations process individually recorded transactions on a monthly basis, whereas others process record-book summary totals annually. Some offer additional services such as income tax return preparation and farm management consultation. The information provided by the systems varies from simple systems listing cash expenses and cash receipts to comprehensive systems providing financial statements, analysis reports, and enterprise accounts. Some services include the following:

1. *Year-end farm summary programs.* The farm manager keeps the records at the farm, and at the end of the year the totals are taken from the record book and entered into the computer for year-end summary analysis.
2. *Mail-in accounting programs.* Farmers record and code their transactions, mailing them monthly to the central processing center.
3. *On-farm microcomputer programs.* A large number of accounting programs are designed for personal microcomputers. Both single-entry and double-entry systems are available. With their own computers, managers can have immediate access to accounting reports. On the other hand, they must assume responsibility for entering the records in the computer. This procedure takes time and requires a knowledge of basic accounting principles and computer operation.

How a Mail-In System Works

Most mail-in systems have two or more reporting forms or record sheets for on-farm use. These reports are usually designed to make preparation of the basic information as easy as possible, leaving the organizing and sorting to the computer. On a typical monthly reporting form, the farm manager enters the date, the transaction code (receipt, deductible expense, mortgage payment, capital purchase), a description of the transaction (payee, purpose, and so forth), the enterprise code (crops, livestock), the quantity involved in units (tons, head, bushels), the dollar amount, and the check number. The forms are normally sent once a month to the computer center, where they are processed with data from other farms.

The typical computerized system includes reports such as an income statement and balance sheet, a depreciation schedule, and a listing of checks written and deposits made. It may also provide a detailed listing of the monthly transactions for individual enterprises as well as income and expense summaries by enterprise. Physical quantities can be accumulated as well.

Some systems use specially printed checks and deposit slips with a code number to identify the purchase or receipt. This practice greatly simplifies the preparation of records for processing, but these systems can only provide cash flow summaries unless inventories and other noncash transactions—on separate forms—are also sent.

How an On-Farm System Works

A wide variety of microcomputer accounting systems is available for on-farm use. They have different capabilities for generating accounting reports and require different procedures for entering and retrieving data. The systems also differ in the degree of accounting expertise required of the user. Most accounting systems come with user manuals, but these manuals vary in content and quality. The manual should show the user step-by- step how to operate the system. It should also include an index to allow quick reference to the proper page when questions arise.

The following description of the process for entering transactions illustrates how a microcomputer system might work. When ready to pay a bill, the farm manager turns on the computer and initiates the accounting program. A "menu" on the computer screen asks for the transaction code—for example, pay expense. The manager responds, and the computer asks for the name of the payee. The manager then enters the code for a name and address already on file in the computer or enters the name and address so that they can be added to the file.

Next, the manager enters the date and check number. The screen then displays a list of bank accounts (farm, personal, and so forth) that can provide funds for the payment, and the farmer enters the account number and the amount. Then the manager sees a display of the possible expense categories (fertilizer, loan payment) to which the payment is to be charged. After the expense category is entered, the computer prints the check, complete with address, ready to put into a window envelope for mailing.

When the entry is completed, the computer saves the transaction on a file. Then it adjusts the bank account balance for the check that has been written and adds the payment to the appropriate expense category total. Microcomputer accounting systems vary in their capabilities, but most allow the manager to select the reports for cash monitoring, income tax return preparation, and financial statements.

Discovering Data Entry Errors

The information generated by a computerized accounting system, whether a mail-in or an on-farm system, is no more accurate than the data entered. Errors in handling numbers are inevitable; numbers are transposed, the wrong numbers are written, and the wrong computer keys are struck. The accounting program should be designed to discover these errors whenever possible. However, it cannot catch all of the errors. The first step in verifying the data is to double-check the entries and the reports generated. The entries in the computer should be checked against the bank statement as well. The computer program should provide an audit trail so that if a particular total appears to be unreasonable, the manager can check to be sure that the entries that determine this total are correct.

Selecting a Computerized System

In selecting the most appropriate computerized system, it is advisable to do the following:

1. Determine one's information needs, which depend on such factors as the size, type, and complexity of the farming operation; the education of the user and his or her training and experience in accounting; the time available to use it; and how the information will be used, that is, the decisions that must be made.
2. Consider the input requirements of the system. For a mail-in system, the number of input forms and the complexity of the coding system determine the time needed to provide the required data. For a microcomputer system, its design and ease of use determine the time needed to enter data.

3. Review the formats of the reports. The accounting system may provide many different types of reports. Are they readable and easily interpreted? How frequently will the various reports be provided?
4. Finally, consider the cost of the system.

Computerized accounting systems help relieve managers of some of the chores of record keeping and accounting. But the systems do not eliminate the need to understand the data input requirements and the interpretation of the accounting reports generated by the system.

Summary

This chapter has broadly considered farm accounts and their importance to the farm manager's information system for reporting, problem diagnosis, and planning purposes. The investment, financing, and production activities of the farm business must be considered when designing a farm accounting system. The elements of the accounting system's design include the accounting period chosen and whether the cash or accrual method of accounting is used. A complete farm accounting system includes schedules for recording (1) capital purchases, sales, and depreciation; (2) liabilities; (3) farm inventory; and (4) cash receipts and expenses.

Depreciation refers to the decrease in an asset's value as a result of age, use, and obsolescence. The method of depreciation to use depends on the purpose; one method might be used for income tax reporting and another for analysis and management.

Liability and farm inventory information is needed to adjust cash receipts and cash expenses so that the net farm income can be calculated on an accrual basis. This way, the cash method with single-entry accounting can be used to give the same result as a more complicated double-entry system.

Computers can be used to facilitate farm accounting. Farmers have several alternatives, including programs that provide year-end summaries, mail-in services that generate monthly reports, and accounting packages to be used on personal microcomputers. In selecting an accounting system, the first step is to understand the farmer's management information needs.

Recommended Readings

JAMES, SYDNEY C., and EVERETT STONEBERG. *Farm Accounting and Business Analysis*, 3rd ed. Ames: Iowa State University Press, 1986. Chapters 1 through 7 and 10.

SONKA, STEVEN T. *Computers in Farming: Selection and Use*. New York: McGraw-Hill Book Co., 1983. Chapter 7.

CHAPTER 4

Farm Financial Statements

Financial statements integrate and summarize the information from the farm accounting system discussed in the preceding chapter: The balance sheet summarizes the asset values, the liabilities, and the net worth position of the business; the income statement summarizes cash receipts, expenses, and inventory schedules; and the cash flow statement provides an overview of the business's cash flow, including debt payments, family living expenses, and income taxes.

This chapter is devoted to the preparation and interpretation of these financial statements and their interrelationships.

The Need for Financial Statements

As farms have grown in size, the requirements for capital and financing have increased, and financial management has become a more critical component of the farmer's management responsibilities. Three of the most important tools for financial management are the balance sheet, the income statement, and the cash flow statement. Before looking at each financial statement individually, it is useful to have an overview of all three statements, their purposes, and their relationships to one another.

These financial statements provide the link between the accounting system and the financial analysis of the farm business. They allow the farmer to assess the business's financial position and performance and to chart its progress. These statements are also used as a basis for obtaining credit, because they provide lenders with the information needed to evaluate loan applications.

Financial statements are used to answer three critical questions:

1. Is the farm business solvent?
2. Is the business earning an adequate profit?
3. Does the business have sufficient liquidity?

Solvency refers to the farm's total capital structure and its ability to meet its liabilities. The business is solvent if the sale of all assets would be sufficient to pay off all liabilities. The business's solvency position depends on its net worth and the relationship between its net worth and its total liabilities. Not only must the net worth be positive, but it must be large enough, relative to the total liabilities, to assure that it will remain positive. If the net worth is negative (liabilities exceed assets), the business is said to be insolvent or bankrupt. The balance sheet is used to determine solvency at a given time.

Profitability, on the other hand, is based on earnings returned to resources invested. Are these returns adequate to reimburse the farmer for the opportunity cost of the resources? This performance measure reflects the number of units sold, the price received per unit, and the total expense involved in producing these units. The income statement is used to assemble and organize this revenue and expense information for a given time period.

Liquidity is the farm business's capacity to generate sufficient cash to meet its financial obligations as they come due. An additional financial statement, a cash flow statement, is needed to determine liquidity. Because the income statement may include some noncash sources of income and does not include all cash requirements, a cash flow statement is also needed to summarize all sources and uses of cash for a given time period.

The construction of these three financial statements is not dependent on a particular accounting system. Any system that provides the necessary information described in the preceding chapter is adequate for preparation of these three financial statements. Much of the required information is the same as that used for completing income tax returns. A more comprehensive accounting system, however, allows for more accurate financial statements.

Although each financial statement is discussed in turn here, all three must be used together for a thorough financial analysis. The financial statements are interrelated. These interrelationships provide crosschecks to ensure the accuracy of the analysis. The statements must pertain to the same accounting year, as defined in the preceding chapter. The balance sheet should be completed as of the last day of the accounting year, and the time period for the income and cash flow statements should coincide with the same accounting year.

The Balance Sheet

The balance sheet summarizes the investments that have been made in the business (assets) and how these investments have been financed (liabilities and net worth) as of a given point in time. Other names used for the balance sheet are net worth statement and financial statement.

The balance sheet balances the assets against the liabilities and net worth. An *asset* is defined as anything of value owned by a business or person, including obligations by others to the individual or the business. Cash, farm inventory, machinery, accounts receivable, and land are examples of assets. Assets that are not owned, but leased, by the business are not included in the balance sheet.[1] A *liability* is an obligation or debt owed to others by the individual or the business.

---

[1] To reflect the size of the operation accurately, these leased resources should be described in a footnote to the balance sheet.

Examples of such claims are accounts payable, notes, and mortgages. *Net worth* is the difference between assets and liabilities:

Total assets − total liabilities = net worth

The net worth is the value that would remain if the business were liquidated, or sold, and all liabilities were paid. The net worth represents the total equity owned by the farm business or its operator; it is that portion of the asset value not claimed by parties outside the business.

The balance sheet provides a picture of the solvency of the business at a given point in time. This "snapshot" of the business is taken the last day of the accounting year, which is December 31 for farmers who use the calendar year. The same date should be used every year. In addition, interim balance sheets may be needed at other times during the year, such as when applying for a loan.

Before starting to prepare a balance sheet, the farmer must decide what the balance sheet is to represent. Is it a *business balance sheet*, including only the business's assets and liabilities? Is it a *personal balance sheet*, including only those assets and liabilities relating to the family? Or is it a *consolidated balance sheet*, including both personal and business assets and liabilities? If the farm is a partnership or corporation, a business balance sheet should be prepared. Because many farm operations involve complex business arrangements, including corporations and partnerships, the balance sheet should clearly identify the entity it represents.

Lenders prefer a consolidated balance sheet because it presents the total financial picture. Sometimes it is difficult to separate personal items from business items; this is true particularly for typical family farm operations. It is important, however, to list personal and business items separately for analysis later. Therefore, preparation of both a business balance sheet for analysis purposes and a consolidated balance sheet for a summary of the total financial picture is recommended.

Balance sheets can be constructed on either a cost or a market-value basis. If the cost method is used, assets are valued at the price originally paid for them less any accumulated depreciation up to the balance sheet date; liabilities are valued at their actual dollar amounts. With the market-value method, values represent the current market price for which the asset could be sold on the date of the balance sheet.

The choice of methods depends on the reason for preparing the balance sheet. For borrowing money, it is important to know that lenders are more interested in the market-value balance sheet because it shows the borrower's current equity position. However, the cost-based approach also plays an important role, which is discussed later.

Table 4-1 illustrates a simplified balance sheet for a hypothetical farm business operated by Frank Pharmer. For this example, assets and liabilities have been valued according to their market values as of the end of the preceding accounting year. The balance sheet in Table 4-2 is a consolidation of Mr. Pharmer's personal and business interests.

Assets

Assets are classified according to how easily they can be converted to cash—their liquidity. The categories are current, intermediate, and fixed assets (see Table 4-1).

TABLE 4-1. Business Balance Sheet for Frank Pharmer, December 31

Assets (market value)		Liabilities and Net Worth (market value)	
Current Assets		Current Liabilities	
Cash	$ 2,000	Accounts payable	$ 1,000
Savings	19,000	Note payable	2,000
Livestock to be sold	26,000	Principal due next year	12,000
Crops and feed	68,000	Accrued interest	7,200
Growing crops	6,000	Accrued tax liabilities	9,000
Supplies	3,000	Contingent income taxes	
Prepaid expenses	2,000	on current assets	31,800
Subtotal	$126,000	Subtotal	$ 63,000
Intermediate Assets		Intermediate Liabilities	
Machinery and equipment	$112,000	Notes payable	$ 25,000
Breeding stock	16,000	Sales contract	10,000
Life insurance value	4,000	Contingent income taxes	
		on intermediate assets	8,400
Subtotal	$132,000	Subtotal	$ 43,400
Fixed Assets		Long-Term Liabilities	
Farmland	$247,200	Real estate mortgage	$132,000
Buildings	30,000	Contingent income taxes	
		on fixed assets	14,400
Subtotal	$277,200	Subtotal	$146,400
		Total liabilities	$252,800
		Net worth	282,400
Total assets	$535,200	Total liabilities and net worth	$535,200

Current assets are those that will be used up within the next accounting year or that can be easily converted to cash without affecting the operation of the business. Mr. Pharmer's current assets include cash in checking and savings accounts. Interest on savings that has accrued but not been paid should be included when valuing these assets. Livestock being held for sale and inventories of crops and feed are valued according to the market quotes for the balance sheet date. The value placed on crops that were planted and growing as of December 31 should be based on the cash costs incurred. Supplies are those items on hand at the end of the year that will be used next year, for example, fuel, medicine, chemicals, and seed.

Prepaid expenses are for services already paid for but not fully rendered by the end of the year; examples might include an advance payment for insurance, a deposit paid to a custom operator for an herbicide application to be done next year, or a retainer paid to an accountant. The amount shown here reflects the value of the service not provided as of the date of the balance sheet.

TABLE 4-2. Consolidated Balance Sheet for Frank Pharmer, December 31

Assets (market value)			Liabilities and Net Worth (market value)		
Farm business assets		$535,200	Farm business liabilities		$252,800
Personal Assets			*Personal Liabilities*		
Household goods	$ 6,000		Personal loans	$ 0	
Personal vehicle	2,000		Contingent tax		
Residence	46,800		on residence	2,400	
Total personal assets		54,800	Total personal liabilities		2,400
			Consolidated liabilities		$255,200
			Consolidated net worth		334,800
Total consolidated assets		$590,000	Total consolidated liabilities and net worth		$590,000

Other current assets that might exist include hedging accounts, short-term notes receivable, and income tax refunds due. Marketable securities, such as stock in corporations, government bonds, and mutual funds, that could be easily sold for cash to finance the farm business are also possiblities.

Intermediate assets are those that support production and that are not normally sold in the ordinary course of doing business. These are the assets, such as machinery and equipment and breeding stock, that are used to produce assets for sale. Intermediate assets generally have a 1- to 10-year life and are less liquid, which means that they are not easily converted to cash. Although breeding stock and machinery might be easily converted to cash, their sale would seriously affect the productivity of the farm. It is this feature that most distinguishes intermediate assets.

Frank Pharmer's intermediate assets also include the cash value of his life insurance because this is a potential source of financing for the farm business. Another asset that might be included in this category is nonmarketable securities, such as shares of stock in a cooperative and other securities that cannot be readily sold, but might be used as security to borrow funds for the farm business. Loans and accounts (due after one year) that are owed to the business in exchange for farm assets sold are other examples of intermediate assets. Retirement accounts should also be included here on a business balance sheet if they are a potential source of funds for the farm business.

Fixed assets, by definition, are permanent and benefit the farm on an ongoing basis. In the normal operation of the business, these assets have a useful life of over 10 years, and their sale would completely or seriously disrupt the business. Farm real estate is usually the major fixed asset on the balance sheet. The value of any land leased by Frank is not included on the balance sheet but the acreage should be listed in a footnote.

Other possible fixed assets are business loans and contracts owed to the farm business on which final payment will be made more than ten years beyond the balance sheet date. An example would be an installment land contract for farmland sold.

The total value of Frank's assets on a current market-value basis, found by adding the current, intermediate, and fixed assets, is $535,200. (Note that the value of his real estate at $277,200 represents over 50 percent of his assets.)

Liabilities

Liabilities are classified by due date. As with assets, the categories are current, intermediate, and long-term liabilities.

Current liabilities are those that are owed as of December 31 and that must be paid within the next 12 months. Accounts payable refers to the amounts Frank owes (but has not yet paid) on 30- to 60-day credit accounts provided by suppliers for the purchase of inputs, such as feed, fertilizer, repairs, and so forth. As is typical of most farm operations, Frank's note payable is the annual operating loan that was obtained from a commercial lender for the purpose of financing the year's production expenses. As of December 31, there is an unpaid balance on this loan that is being carried to the next year. This carryover is not of serious concern because the amount is relatively small and there may be a good reason for not repaying the loan completely. For example, with good prospects for a higher price in the future, Frank may have postponed marketing a portion of this year's crop.

Also included as current liabilities are the principal payments that will be due within the next twelve months on intermediate and long-term loans. The total accrued interest on all of Frank's loans is $7200. This accrued interest is the unpaid interest that has accumulated since his last payments. It is the amount of interest he would owe if he were to pay off all his loans on December 31. For example, the real estate loan payment due this next November 1 will include $12,600 in interest. By the date of the balance sheet, interest has accrued for 2 months, so the accrued interest on this loan is $2100 ($12,600 \times 2/12$).

A current liability that Frank does not have is accrued lease expense. An accrued lease expense is the amount that would be due if the lease were terminated on the balance sheet date. For example, suppose a farmer began leasing a grain storage building on October 1, at a cost of $3600 per year with an initial payment of $600 and the remainder due after January 1. By December 31, the date of the balance sheet, the building would have been used for 3 months, and the accrued lease expense would be $900 ($3600 \times 3/12$). Subtracting the $600 already paid, leaves an accrued lease expense of $300. Including these accrued expenses such as interest and lease payments is necessary to give an accurate picture of the farm business's solvency.

There are two types of current liabilities related to taxes. The first is *accrued tax liabilities*, which consist of property taxes, employer payroll withholdings, and income and social security taxes that have accumulated but have not been paid. The balance sheet date usually coincides with the last day of the tax year, so the accrued tax liability is simply the taxes due for that year minus any payments that have already been made toward these taxes. The accrued property taxes can be calculated from this year's tax statement or estimated from last year's statement. Employer payroll withholdings that have accrued but have not been paid come from the farmer's records. The liability for accrued income and social security taxes as of December 31 consists of the income and social security taxes on farm earnings that will be due with the tax return to be filed in March. As this is a business balance sheet, only those accrued taxes owed by the business are included; taxes on personal assets and nonfarm income are not included.

The other tax-related current liability is *contingent income taxes*. Farmers who report income using the cash method owe income taxes on assets produced for sale only after they have been sold. Therefore, it should be recognized that income taxes would be due if the assets were in fact sold as of the date of the balance sheet. That is why this liability is called *contingent*—it is contingent on the sale of the assets. Figuring the contingent tax liability involves (1) finding the taxable amount, (2) determining the tax rate that would apply, and then (3) multiplying the taxable amount times the tax rate.[2] Because this calculation requires a good understanding of income tax regulations, an accountant can be very helpful. As shown in Table 4-1, contingent liabilities are an important factor in assessing the financial position of the business; Frank would owe $31,800 in income taxes if the current assets were sold on the balance sheet date. Unless this contingent income tax liability is included, Frank's net worth will be overstated.

Other contingent liabilities might include pending lawsuits, co-signed loans, guarantees, and penalties for early termination of lease agreements. There may

----

[2] The following steps are used to estimate the contingent income tax liability for all the current assets (except marketable securities):

1. To find the amount of taxable income that would result from the sale of these assets, add the following values from the current assets section: livestock to be sold (less their purchase cost if any), crops and feed, growing crops, supplies, and prepaid expenses. From this total, subtract the total of the following from the current liabilities section: accounts payable, accrued interest, and accrued tax liabilities (less accrued income tax).

2. To estimate the tax rate, find the amount of total taxable income on the most recent federal income tax return. If this amount was higher or lower than normal, adjust it to an expected normal level. Now use the appropriate tax schedule—single, married filing jointly, married filing separately, head of household, or corporation—to locate the tax bracket for the normal taxable income. The percentage figure that is used to calculate the taxes on income in this bracket is the federal ordinary income tax rate. To this tax rate, add the Social Security self-employment tax rate (if the taxable income is below the Social Security maximum) and the state income tax rate (if applicable).

3. To calculate the contingent tax liability, multiply the taxable amount found in step 1 by the combined tax rate estimated in step 2.

To estimate the contingent income tax liability for marketable securities, the same steps are used. The taxable amount is the capital gains income, which is the difference between their current market value and their original cost. The tax federal rate to use is the capital gains tax rate which is equal to the ordinary federal tax rate used for the other current assets times the difference found by subtracting the percent of capital gains income excluded from 1. The percent excluded and the requirements to qualify for this lower tax rate are described in the tax return instructions. Capital gains income is not subject to Social Security self-employment tax, but state income taxes may apply. If so, the appropriate state rate should be added to the federal capital gains rate to give the combined rate.

also be contingent liabilities associated with the dissolution of a partnership (division of assets) or with a divorce (alimony payments). These liabilities should be disclosed on the balance sheet by listing them in the appropriate category—current, intermediate, or long-term—according to when the liability would be due. If there is doubt about whether and where to include the contingent liability, describe it in a footnote. This is particulary important when preparing the balance sheet for a loan application.

Intermediate liabilities are those due from 1 to 10 years beyond the date of the balance sheet. For Frank, these liabilities consist of a note payable, a sales contract, and contingent income taxes on intermediate assets. The note and contract were used to finance machinery and equipment purchases. (To avoid double counting, it is important to remember that this is the principal due *after* the next accounting year, because the principal due next year was included in current liabilities.)

Contingent income taxes are $8400, which would be due if the machinery, equipment, and breeding stock were sold on the balance sheet date. This estimate of the taxes that would be due is based on the difference between the market value of these intermediate assets and their book value (original cost less accumulated depreciation).[3]

Other intermediate liabilities that might exist include loans from parents, relatives, or friends; loans on life insurance policies; and contingent income taxes and early withdrawal penalties on retirement accounts.

Generally, balance sheets include only owned assets and the liabilities incurred in their purchase. However, financial lease arrangements (Chapter 16) are being used more frequently in agriculture as an alternative for financing the ownership of machinery and equipment. Instead of borrowing money and purchasing the asset, the farmer enters into a 3 to 10 year lease agreement. The result is a financial obligation very similar to debt financing. This obligation for lease payments should be listed as an intermediate liability (the amount due within the next year would be a current liability). There should also be a corresponding entry in the intermediate asset section. If the leased asset will belong to the farmer at the end of the lease period, the value should be the asset's current market value on the balance sheet date. However, if the asset will revert to the leasor, the value should be equal to the outstanding obligation for the lease payments.

Long-term liabilities are those that, when originally acquired, had a maturity of more than 10 years. The long-term liability associated with Frank's real estate mortgage consists of the unpaid principal balance minus the total principal due next year. Land contracts, another arrangement that can be used to finance real estate purchases, are also long-term liabilities. In addition, there may be long-term debts to parents, relatives, or friends that were used to initiate the farm business, or to buy into an existing operation.

---

[3] To estimate the contingent income tax liability for machinery and equipment, the taxable amount is the difference between the current market value and the book value (original cost less accumulated depreciation). The tax rate to use is the same one used to calculate contingent taxes on current assets (except marketable securities).

For raised breeding stock, the taxable amount is the current market value from the intermediate assets section. The tax rate to use is the capital gains rate used for marketable securities. For purchased breeding stock, the contingent income tax liability is calculated the same as for machinery and equipment.

Income tax would be due on any capital gain resulting from the sale of the land. This estimated contingent tax liability is based on the difference between the real estate's market value and its cost basis (usually the original cost less accumulated depreciation on the buildings and improvements).[4]

Adding Frank's current, intermediate, and long-term liabilities gives total liabilities of $252,800, using the market-value approach.

Net Worth

To calculate his net worth, Frank subtracts his total liabilities from his total assets, giving $282,400. In other words, if Frank were to sell his farm business at its current market value on December 31 and pay all his liabilities, including the income taxes that would be due, he would have $282,400 left. Knowing his net worth helps him both to assess his capacity to take risks and to estimate his total wealth upon retirement. Net worth is also useful to lenders in indicating additional available collateral.

It is important to understand how various transactions affect the balance sheet. For example, if cash is used to purchase new farm machinery, total assets and net worth remain unchanged. However, current assets and current liabilities are changed. If money were borrowed to purchase the machinery, total assets and liabilities would both be affected, but net worth would remain unchanged. Using cash for a family vacation, however, would decrease net worth, whereas an inheritance used to purchase farm assets would increase it. It is helpful to think through various transactions that might occur and to determine how assets, liabilities, and net worth would be affected by them.

Frank Pharmer's consolidated balance sheet, a combination of his business and personal assets and liabilities, is presented in Table 4-2. Frank's personal assets consist of household goods (furniture and appliances), a personal vehicle, and the farm residence. The only personal liability is the contingent income tax that would be due if the residence were sold. The consolidated assets, the total of farm business and personal assets, have a total market value of $590,000. Frank's consolidated net worth is $334,800.

Net Worth on a Cost Basis

The balance sheet can be prepared using either the market- value or cost approach. The market-value approach is presented in Tables 4-1 and 4-2. The double-column balance sheet uses both market and cost valuations (see Table 4-3). As a result, there are two net worth figures: The cost approach gives a net worth that is consistent with traditional accounting practice; and the market value approach gives a net worth that is realistic for financial analysis. The market value approach provides the data used for analyzing solvency, profitability, and liquidity. The cost net worth is needed, however, to explain why market net worth has changed. Did it change because of the profitability of the business, or did it change because of the effects

---

[4] To estimate the contingent income taxes on real estate, the taxable amount (current market value minus total book value) is multiplied times the same capital gains rate used for marketable securities and raised breeding stock.

TABLE 4-3. Double-Column Balance Sheet of the Farm Business for
Frank Pharmer, December 31

Assets	Cost	Market	Liabilities and NW	Cost	Market
Current Assets			Current Liabilities		
Cash	2,000	2,000	Accounts payable	1,000	1,000
Savings	19,000	19,000	Note payable	2,000	2,000
Livestock to be sold	26,000	26,000	Principal due next year	12,000	12,000
Crops and feed	68,000	68,000	Accrued interest	7,200	7,200
Growing crops	6,000	6,000	Accrued tax liabilities	9,000	9,000
Supplies	3,000	3,000	Contingent income taxes		
Prepaid expenses	2,000	2,000	on current assets	0	31,800
Subtotal	126,000	126,000	Subtotal	31,200	63,000
Intermediate Assets			Intermediate Liabilities		
Machinery and			Notes payable	25,000	25,000
equipment	93,700	112,000	Sales contract	10,000	10,000
Breeding stock	16,000	16,000	Contingent income taxes		
Life insurance value	4,000	4,000	on intermediate assets	0	8,400
Subtotal	113,700	132,000	Subtotal	35,000	43,400
Fixed Assets			Long-Term Liabilities		
Farmland	155,300	247,200	Real estate mortgage	132,000	132,000
			Contingent income taxes		
Buildings	16,000	30,000	on fixed assets	0	14,400
Subtotal	171,300	277,200	Subtotal	132,000	146,400
			Total liabilities	198,200	252,800
			Net worth	212,800	282,400
			Total liabilities and		
Total assets	411,000	535,200	net worth	411,000	535,200

of inflation or deflation of the value of the assets? The double-column format also
helps in estimating contingent income taxes.

Because of the difficulty of estimating cost figures for some assets, a modified
cost approach is suggested here. For example, the cost values for livestock are
particularly difficult to estimate. When the cash method is used for reporting taxes,
raised livestock has a zero cost—for tax purposes—because all the expenses for
raising the livestock have been deducted. The cost value for purchased livestock to
be sold (not used for breeding) is the original purchase price; for breeding livestock
the cost value is the original purchase price less accumulated depreciation. To
provide a more reasonable value for raised livestock and to reduce the confusion
involved when there are both purchased and raised livestock, the modified cost
approach values livestock at current market values.

For most current assets, the cost and market values are the same. An exception
would be marketable securities such as stock in corporations, government bonds,

and mutual funds. The cost value would be the price originally paid for the security when it was purchased. The market value would be the price for which it could be sold on the date of the balance sheet.

Under intermediate assets, the cost figure for machinery and equipment is the original cost less the accumulated depreciation. For example, the original cost for machinery and equipment amounts to $144,200, and subtracting accumulated depreciation of $50,500 leaves the remaining book value of $93,700 that appears in the cost column of Table 4-3. Because of inflation, this cost value is likely to be lower than the current market value of the machinery and equipment. For breeding stock and life insurance, the market value is the same as the cost.

The biggest difference between the cost and market values is in the fixed asset category. For farmland owned for several years it is not unusual for its current market value to be substantially higher than its original cost. For buildings, as it is for machinery and equipment, the cost value is found by subtracting the accumulated depreciation ($15,000) from the original cost ($31,000) to give the book value ($16,000) as of the balance sheet date.

The total in the cost column for current liabilities is lower because contingent income taxes are not included. The same is true for intermediate and long-term liabilities. Therefore, the net worth, calculated using the modified cost approach, is less than the current market-value net worth found earlier.

Inflation and deflation in the values of assets complicate the measurement of financial progress. Knowing the net worth based on cost is necessary to differentiate the effects of income earned from the effects of inflation or deflation on the change in net worth from year to year. To illustrate, suppose Frank compares his market and cost net worths for this year with last year's figures:

	Cost	Market
Net worth (this year)	$212,800	$282,400
Net worth (last year)	−181,800	−231,400
Change in net worth	$ 31,000	$ 51,000

The change in market-value net worth ($51,000) is due to both inflation and income earned. The change in the cost net worth ($31,000) is due only to the impact of earned income. The change in net worth resulting from inflation, then, is $20,000 ($51,000 minus $31,000).

Asset Valuation Concepts

Placing accurate market and cost values on farm assets is essential for the preparation of the balance sheet. This section provides some general guidelines for setting values.

Estimating Market Values.   Some of the more common guides to market values are as follows:

1. Net market price.
2. Purchase cost.
3. Replacement cost less depreciation.
4. Income capitalization.

The most appropriate method to use for a particular asset depends on its nature and the purpose for which it is held. Assets held primarily for sale are readily valued in terms of their contribution to the business. However, assets such as machinery, breeding stock, and land are not sold or purchased as often, so a less direct valuation method must be used.

The *net market price* is applied to those assets held primarily for sale. This net price, which is the market price less transportation and marketing costs, provides a realistic measure of their current value to the farm business.

Purchase cost is commonly used for valuing farm supplies such as feed, fertilizer, and fuel. For items that have been recently purchased, the original purchase price is an appropriate indicator of market value. When prices are subject to significant change, the cost of replacing the asset might be more appropriate. A variation of this purchase cost method is the *cost or market, whichever is lower,* approach, which compares the two values and uses the lower and more conservative one. This approach is appropriate when the item is, or could be, both bought and sold by the business.

Buildings and related real estate improvements are difficult to value because comparable sales information is not usually available. For these assets, the *replacement cost less depreciation* method may best determine the value. Long-lived assets, such as buildings, may be valued more realistically with this method, particularly when there have been major changes in cost and technology. The replacement cost should be for the same asset or for one that will accomplish the same function. This approach takes into consideration the changing use of some farm assets. For example, a building constructed to house 100 dairy cows 40 years ago at a cost of $20,000 may now be used as a machinery storage shed. The cost of a shed to replace the old barn may be only $5000. This method emphasizes the asset's present value in terms of its use and its contribution to earning potential.

The *income capitalization* method was developed to estimate the value of those assets whose contribution to the income of the business is made over a longer period of time. To illustrate this method, assume that a manager wants to value an asset that will yield $140 per year forever into the future and that the current rate of interest is 10 percent. The value of the asset (V) can be calculated using the following formula:

$$V = \frac{R}{r}$$

Farmland is an example of an asset with an infinite series of annual incomes. In this case, R is the expected net return to the land after all other costs have been subtracted. The interest rate, r, is usually the farm mortgage rate. According to this formula, at an interest rate of 10 percent, land returning $140 per year would be valued at $1400 per acre.

This formula or some variation of it is often used by professional appraisers. In practice, however, neither the precise annual income nor the actual interest rate is known. As a result, the formula is often used in combination with other appraisal methods, such as the comparable market sales method.

Following are some specific suggestions with respect to the valuation of farm assets:

1. All assets that will be sold within the year can be valued at the net market price.
2. Supplies generally should be valued at their purchase cost or market price, whichever is lower.
3. Market values of assets such as machinery and breeding livestock should be close to their cost less depreciation during periods of stable prices. In periods of persistent inflation or deflation, market prices should be used to ensure realistic values.
4. The market value of farm buildings that were constructed only a short time ago should be approximated; those that were constructed a long time ago should be valued by their replacement cost less depreciation.
5. Farmland may be valued by means of the capitalization formula in combination with comparable sales values.

Estimating Values Based on Costs.   Cost less depreciation is commonly used for intermediate assets such as machinery and equipment. By using a systematic method of charging depreciation to the current accounting period, financial position and changes in net worth are not affected by market price fluctuations.

These cost values for most assets can be taken directly from the income tax records. For supplies, the cost value would be the same as the market value—the price paid at the time of purchase. For machinery and equipment, the cost value would be the book value—the difference between its original cost and its accumulated depreciation. For crops and livestock in inventory, a modified cost approach is used that bases the value of these assets on their current market values. For farm real estate assets, the cost is the amount paid for them less accumulated depreciation on buildings and improvements; there is no depreciation on land.

The Income Statement

The income statement, also called a profit and loss statement or an operating statement, shows the earnings of the business for a given period of time, usually one year. In addition, it shows the sources of revenue, describes the nature of expenses, and calculates the net farm income. *Net farm income* is the return to the operator for unpaid labor, management, and net worth invested; this is because most farm businesses do not pay salaries to the operator. Farms organized as partnerships or corporations are exceptions. Regardless of how the farm is organized, the net income indicates what is available from the year's operations to pay income taxes, to pay family living expenses, and to reinvest in the business.

An income statement can be prepared in different ways. The best approach depends on the nature of the farm business. There are, of course, certain items that must be included in any income statement if it is to reflect the profitability of the business accurately. Items that are not associated with the farming operation should be excluded when calculating the net farm income.

To determine the farm's profit accurately, revenue and expenses should be accounted for on an accrual basis. This means that the revenue should reflect the value of the year's production and that the expenses should be those incurred in generating that revenue. Thus, changes in inventories and accounts receivable and payable must be used to adjust cash receipts and cash expenses to determine the actual revenue and expenses. With these adjustments, the income statement will reflect net farm income as if it had been computed from a double- entry accounting system.

As indicated in the preceding chapter, most farmers use a single-entry cash method of accounting. As a result, their income tax returns are not a good indication of the profitability of the business. For example, a farmer's tax return might show a cash revenue of $80,000 and expenses and depreciation of $60,000. The net income for tax purposes, then, would be the difference, $20,000. These figures do not reflect the fact that $15,000 of this $80,000 gross revenue was from the sale of crops produced in the preceding year. Therefore, the farmer would have to adjust the cash revenue to reflect the value of production for the current year by subtracting $15,000 from $80,000, yielding $65,000. Now when the expenses and depreciation are subtracted from the value actually produced, the true net farm income is $5000 ($65,000 minus $60,000).

The general format of the income statement is based on the following relationship:

Total cash revenue
+ revenue adjustments
= value of production
− cash operating and interest expenses
− expense adjustments
− depreciation
= net farm income

Table 4-4 is the annual income statement for Frank Pharmer's operation.

Revenue

The revenue section of the income statement indicates the value of production for the year. Included are the cash receipts from the sale of crops, livestock, and livestock products, minus livestock purchases. Also included are income from custom work and government payments. Only sales resulting from the normal operation of the farm business should be included under revenue. Sales of capital assets should not be included here. To determine the value of production, cash receipts must be adjusted for changes in the inventories of those items produced for sale.

TABLE 4-4. Income Statement for Frank Pharmer, January 1 to December 31

Revenue		
Crop sales	$155,000	
Livestock	24,000	
Breeding stock	3,000	
Revenue adjustment (Table 4-5)	−2,000	
Value of Production		$180,000
Expenses		
Cash operating	$ 97,000	
Interest paid	18,000	
Paid expense adjustment (Table 4-6)	+500	
Payable expense adjustment (Table 4-7)	−1,500	
Depreciation	20,000	
Total Expenses		$134,000
Net Farm Income		$ 46,000
Plus: Net nonfarm income		3,000
Minus: Income taxes[a]		9,000
Net Family Income After Taxes		$ 40,000

[a] Includes income and self-employment social security taxes, calculated as follows: cash payments made during the year plus accrued tax liability at the end of the year, minus accrued tax liability at the beginning of year.

These inventory figures can be taken from the balance sheets for the end of the current year and the end of the preceding year.

The inventory figures should include livestock to be sold, crops and feed on hand, and breeding livestock (see Table 4-5). It is also necessary to consider any accounts receivable for items sold but for which payables have not been received.

The year's production that was not sold will show up as an increase in inventory at the end of the year. Conversely, a previous year's production that is sold this year will be represented as a decrease in inventory. These revenue adjustments, based on the change in inventory values from the beginning to the end of the year, are illustrated in Table 4-5. Frank had an ending inventory of livestock to be sold

TABLE 4-5. Revenue Adjustments to Frank Pharmer's Income Statement

Item	Ending Value	Beginning Value	Change
Accounts receivable	$ 0	$ 0	$ 0
Livestock to be sold	26,000	27,000	−1000
Crops and feed	68,000	69,500	−1500
Breeding stock	16,000	15,500	+ 500
Revenue Adjustment			−2000

of $26,000, according to his balance sheet. The inventory of livestock to be sold from last year's balance sheet was $27,000. Subtracting $27,000 from $26,000 gives −$1000, meaning that cash receipts need to be reduced by $1000 to reflect the year's production accurately. Overall, Frank sold more than he produced this year, so the revenue adjustment is a negative figure (−$2000). After it is subtracted from the cash receipts, the value of production is $180,000.

Expenses

The farm business has both cash and noncash expenses. Cash operating expenses include hired labor, repairs, leases, seed, fertilizer, chemicals, feed, veterinary services, and so forth. Interest payments are listed separately and represent those amounts paid by cash or renewal. It is not appropriate to include principal payments on loans as a cash expense; they do not affect profit and are taken into account in the cash flow statement, which is discussed later in this chapter.

In order to reflect accurately the expenses associated with the value of production calculated previously, adjustments must be made for inventories and accounts payable. These adjustments are illustrated in Tables 4-6 and 4-7. The calculations must be made carefully, using the proper addition and subtraction signs. In Table 4-6, to determine the adjustments for growing crops, supplies, and prepaid expenses, the ending value is subtracted from the beginning value. In Table 4-7, to determine the adjustments for accounts payable and accrued property taxes, rent, and interest, the beginning value is subtracted from the ending value. The beginning and ending values in these two tables come from the balance sheets for the end of the preceding year and the end of the current year, respectively.

Expenditures for capital assets such as machinery, breeding stock, and real estate should not be included with expenses, because these assets will be kept for more

TABLE 4-6. Paid Expense Adjustments to Frank Pharmer's Income Statement

Item	Beginning Value	Ending Value	Change
Growing crops	$5500	$6000	$ − 500
Supplies	4100	3000	+ 1100
Prepaid expenses	1900	2000	− 100
Paid Expense Adjustment			$ + 500

TABLE 4-7. Payables Expense Adjustments to Frank Pharmer's Income Statement

Item	Ending Value	Beginning Value	Change
Farm accounts payable	$1000	$ 700	$ + 300
Accrued property taxes	800	600	+ 200
Accrued rent	0	0	0
Accrued interest	7200	9200	− 2000
Payable Expense Adjustment			$ − 1500

than one year. The annual cost associated with these capital assets is represented by depreciation, a noncash expense. (The calculation of depreciation is explained in the preceding chapter.)

After adding the cash operating and interest expenses, making the appropriate expense adjustment, and adding depreciation, Frank Pharmer's total expenses for the year amounted to $134,000 (Table 4-4).

Net Farm Income

Frank Pharmer's net farm income is the difference between his revenue and expenses, or $46,000. This figure represents what could be withdrawn from the business without affecting its net worth, that is, if Frank did not have to pay incomes taxes or withdraw money from the business for such things as family living expenses.

One source of farm income that has not been considered is the gain or loss on the sale of farm assets. This gain or loss is found by subtracting the original cost less accumulated depreciation (book value) from the net sale price. Because income from the sale of capital assets varies considerably from year to year, these amounts should be kept separate and not included in the calculation of net farm income.

Frank's net farm income represents the earnings on his labor, management, and net worth. It is the return on the operator's contribution of unpaid resources. Interest paid on borrowed capital has been included in the expenses, but no charge has been made for the net worth invested by the operator. Hired labor expenses have been included, but no allowance has been made for the contribution of the operator's unpaid labor. Likewise, no expense has been charged for management. Thus, the net farm income represents what is left over to reward the farm operator for the use of labor, capital, and management.

It is possible to use this net farm income figure to calculate a return to the operator's labor and management. This calculation involves subtracting a charge for the net worth invested, leaving a residual return to labor and management. The charge for the net worth investment should be based on what this money could earn if invested in another way—its opportunity cost.

Frank Pharmer believes that a 10 percent charge for his net worth is appropriate. The market value of Frank's net worth in the business was $282,400 at the end of the year. At the beginning of the year, it was $231,400. The average net worth invested during the year was $256,900 [($231,400 + $282,400)/2]. This can be used to make the following calculations:

Net farm income	$46,000
Ten percent charge for farm net worth	− 25,690
Return to labor and management	$20,310

This return to labor and management can be compared with off-farm employment opportunities to indicate how profitably labor and management are employed.

Using a similar approach, Mr. Pharmer can also calculate the return to his average net worth invested in the farm business. Mr. Pharmer feels that he could get a job at a nearby factory and receive an annual salary of $18,000. Using this figure as a charge for his labor and management:

Net farm income	$46,000
Labor and management	−18,000
Return to farm net worth	$28,000

Expressing the difference as a percentage of Mr. Pharmer's average net worth in the farm gives the following:

$$\frac{\$28,000}{\$256,900} = 10.9\%$$

To calculate the return to the total capital investment, Mr. Pharmer adds back the interest on borrowed capital (except that on operating loans repaid before the end of the year):

Return to farm net worth	$28,000
Total interest expense	+16,000
Operating capital interest[5]	− 1,400
Return to total farm assets	$42,600

The total interest expense of $16,000 is calculated by adding the interest paid, $18,000 (Table 4-4), to the change in interest accrued, −$2000 (Table 4-6). Then the $1400 interest on operating capital loans repaid before the end of the year is subtracted. The result of these adjustments is a return to total farm capital of $42,600. This return to total farm capital is then calculated as a percentage of the average value of the farm business assets for the end of each of the two years:

$$\frac{\$42,600}{\$522,950} = 8.1\%$$

This calculation indicates the return to total assets regardless of whether they are financed through debt or through equity and allows comparison of the farm's profitability with different debt levels.

It must be remembered that these are residual returns—what is left over as returns to management, to farm net worth, and to total farm assets. These residual returns

---

[5] The interest on operating loans repaid before the end of the year is subtracted to provide consistency between the return to total farm capital and the average farm capital investment. Averaging the beginning-of-year and the end-of-year farm asset values understates the investment over the year, because the beginning and ending assets do not include the increase in asset value that occurred during the year as a result of investing the funds from operating loans in feed, supplies, growing crops, livestock, and prepaid expenses. An average of the asset values from monthly balance sheets would correct the problem, but monthly balance sheets are not usually available for farm businesses. Thus, it is easiest to make the return to total capital consistent with the average of the total beginning and ending capital investments by subtracting the interest paid on operating loans that were received and repaid during the year.

are only as accurate as the opportunity costs arbitrarily placed on the manager's net worth, labor, and management. Therefore, managers must be realistic when determining potential salaries and must be certain that the figures reflect what could actually be earned elsewhere, rather than what they think should be earned.

Nonfarm Income and Taxes

The family may have sources of income other than the farm business. To reflect more completely the total income of the family, this nonfarm income can be added to the net farm income; then income taxes can be subtracted to arrive at the *net family income after taxes.*

The net nonfarm income is the total amount of income earned from sources other than the farm business minus the expenses incurred in earning it. The income taxes are the total paid and due for the year just ended, including federal and state income taxes and self-employment social security taxes.

With this information on nonfarm income and taxes, the relationship between the balance sheet and the income statement can be shown. Because most farm businesses do not pay salaries to the farm operator, family living expenses and other withdrawals must be considered in comparing the income statement with the balance sheet.

Using the cost net worth, it is possible to reconcile the income statement with the two balance sheets—one prepared at the end of the previous year and the other at the end of the current year—as follows:

> Beginning cost net worth
> + net family income after taxes (see Table 4-4)
> + gifts, additions, and so forth
> − net withdrawals for family living and nonbusiness purposes
> = ending cost net worth

If the figures do not check out, the discrepancy may occur in accounting for family living and nonbusiness withdrawals or in making adjustments needed to calculate net family income after taxes.

The Cash Flow Statement

Cash flow refers to the movement of dollars into and out of the farm business. The cash flow statement summarizes the farm business's cash sources and indicates how these funds were used during the accounting period. This statement is also known as a flow of funds statement or a source and application of funds statement.

The major purpose of the cash flow statement is to provide information about the liquidity and loan repayment capacity of the farm. Even though the income statement may show that the farm is earning a reasonable profit, the business may be headed toward financing problems. This can happen when the operation is not generating sufficient cash to meet all business, debt repayment, tax, and family obligations. Lenders are especially interested in this financial statement.

TABLE 4-8. Cash Flow Statement for Frank Pharmer, January 1 to December 31

Sources of Cash		Uses of Cash	
Beginning balance	$ 1,500	Ending balance	$ 2,000
Operating receipts	182,000	Operating expenses	115,000
Capital sales	7,000	Capital purchases	33,000
Money borrowed	23,500	Principal payments	50,000
Nonfarm income	3,000	Family living	9,000
		Income taxes	8,000
Total Sources	$217,000	Total Uses	$217,000

It is important to distinguish the cash flow statement from the cash flow budget (Chapter 6). A cash flow budget is more detailed; it includes a monthly or bi-monthly breakdown of cash income and expenses for the next year. By projecting the cash flows for these shorter time periods, it is possible to estimate the seasonal variations in cash inflows and outflows and to plan for borrowing operating funds as needed. The cash flow statement presented here, which summarizes the cash transactions for the past year, is useful in preparing the cash flow budget for next year's operation.

The basic relationship underlying the construction of the cash flow statement is that cash inflows must equal cash outflows; in other words, sources of cash must equal uses of cash (see Table 4-8).

Sources of Cash

The sources of cash in the cash flow statement include cash on hand at the beginning of the period; operating receipts from the sale of farm products, custom work, and government payments; income from the sale of capital assets; money obtained from new loans; and income from nonfarm sources.

The cash from operating receipts is the same as the revenue section of the income statement without the adjustment for inventory change. Capital sales represent the cash received from the sale of capital items, in this case, an old tractor. New loans are also a source of cash—for example, loans used to purchase capital assets or money representing operating funds to be repaid at the end of the year. Nonfarm sources of cash are also included in the cash flow statement, since they represent a part of the total pool of cash available to the farm family.

Uses of Cash

The cash balance at the end of the year is considered one of the uses of cash. In addition, cash expenses for operating the business, cash used to purchase capital assets, principal payments, family living expenses, and income taxes are included.

The operating expenses equal total expenses from the income statement except for depreciation, adjustment for inventory change, and accounts payable, which are not cash transactions. Principal payments represent the amounts applied to outstanding loan balances during the year. The expenditures for family living expenses are the funds expended for such nonbusiness purposes as food, clothing, entertainment, transportation, and personal investments. The cash payment for income taxes would be the amount paid this year against the previous year's income tax liability.

Cash Flow Relationships

The relationships of entries on the cash flow statement can be examined to identify possible problem areas. For example, the difference between operating receipts and operating expenses is the cash contribution made by the operations of the farm business. This figure needs to be evaluated, however, in relation to the change in inventory and accounts payable. The difference could be unusually large due to a reduction in the inventory of products for sale; likewise, it could be understated by a large reduction in accounts payable or an increase of supplies in inventory. In Frank's case, the adjustments to revenue and expenses shown in the income statement (Table 4-4) are not large enough to distort significantly this cash contribution from the operation of the business (Table 4-8).

The amount of cash used for capital purchases is larger than the cash received from capital sales, which indicates that Mr. Pharmer is investing capital in the business. A comparison of the difference between capital purchases and capital sales to the amount of depreciation in the income statement indicates that he is maintaining his investment in capital assets.

It can be seen, too, that Mr. Pharmer repaid more than he borrowed during the year. He was able to accomplish this while, at the same time, investing in new capital assets. Also, the relationship between nonfarm income and family living expenses indicates that the farm business must provide most of the cash to meet the family's requirements; income taxes must also be paid out of the cash generated by the business.

Examination of the cash flow statement can answer certain questions, such as whether family living requirements are too high, whether sufficient cash is available to maintain and increase the value of the capital assets, and whether this investment must be made with borrowed funds or with cash generated by the business. The debt repayment capacity of the business should also be considered. In Frank Pharmer's case, the ending balance of $2000 indicates that there was some margin of safety after making debt payments and paying expenses that year. Another important consideration is the disposition of the $50,000 in principal payments: How much represents scheduled payments on intermediate and long-term loans? How much represents money repaid on operating loans? How much represents advance payments on intermediate and long-term loans? The lower the portion that is scheduled, the greater the flexibility for managing the liquidity of the farm business.

Summary

This chapter has dealt with the preparation and interpretation of the balance sheet, the income statement, and the cash flow statement in analyzing the financial condition and performance of the farm business. Important tools for farmers, these financial statements are are also helpful in documenting their credit worthiness to potential lenders.

The balance sheet demonstrates the solvency of the farm business by comparing the value of assets to the liabilities and calculating net worth. The balance sheet typically assumes current market values for the assets and liabilities. However, the cost-basis balance sheet is also useful in determining the separate effects of inflation (or deflation) and earned net income on the yearly change in net worth.

The income statement of the business is used to analyze the profitability of the farm operation. The ending balance sheet for the current year and the one for the preceding year provide the information for the noncash adjustments to make this an accrual net income figure. The income statement allows farmers to compare the returns to their labor, capital, and management to the opportunity costs of these resources.

The cash flow statement summarizes all sources and uses of cash and provides the information needed to assess the liquidity of the business. The relationships within this summary can identify trends and possible problems.

These three financial statements must be used together as parts of a whole. The checks and balances between statements improve accuracy, and the relationship among entries provides valuable information to the farm manager. When consistently prepared over a number of years, these statements allow the analysis of long-term trends in the financial position and performance of the farm business.

Recommended Readings

FREY, THOMAS L., and D. KLINEFELTER. *Coordinated Financial Statements for Agriculture*, 2nd ed. Skokie, Ill.: Agri Finance, 1980.

OTTE, J. A. *Farm Profit Planning: A Step-by-Step Guide to Good Financial Management.* Lombard, Ill.: Wallace-Homestead Book Co., 1984.

CHAPTER 5

Business and Enterprise Analysis

Some of the most productive work farm managers can do is to analyze their farm accounts and financial statements. This information is of little value unless it is interpreted and used in decision making. Managers must take care when interpreting the results of the analysis and must remember that the analysis is not an end in itself but merely a means to an end—the more profitable operation of the farm business.

The analysis process should reveal both strong and weak aspects of the organization and operation of the business. The problems or weak points represent opportunities to increase profits; these opportunities can be identified by comparing the actual performance of the business, measured in different ways, with its objectives or expected standards. This process is called *management by exception.* The exceptions or deviations of the actual performance from the expected result are symptoms of a problem. These symptoms are then analyzed to pinpoint the cause of the problem, and alternative actions are identified for correcting it and increasing profits.

In the analysis process, symptoms must be distinguished from causes. The tendency to view a particular discrepancy as the problem can prevent the discovery of the real problem. For example, on a dairy farm, low milk production per cow may appear to be the cause of reduced profits when the actual cause is an inadequate feed ration. The low milk production is a symptom. Determining the most likely cause involves the following:

1. Obtaining more information about the symptom (the exception). Is it new? Has it been observed before? Is it related to other symptoms?
2. Examining possible causes that could explain the symptoms. Those explanations not consistent with all of the symptoms are eliminated.
3. Once the likely cause has been identified, it should be checked again to see that it explains all of the symptoms involved.

Analysis of the farm business can be conducted on two levels: (1) a financial analysis of the total business and (2) a more detailed analysis of the factors for each major enterprise of the business. An *enterprise* is any segment of the farm business that can be isolated by accounting procedures so that revenue and expenses can be allocated to it. Some typical farm enterprises are corn production, wheat production, hay production, crop machinery, hog finishing, and raising dairy-heifer replacements.

This chapter concentrates on the measures and factors useful in conducting an analysis of the farm business and later explains enterprise accounting and the procedures for the analysis of individual enterprise accounts. Guidelines are offered for applying an analytical procedure to the diagnosis of management problems. The next chapter deals with prescribing solutions to these problems.

The Essentials of Business Analysis

Three types of information are essential for conducting an analysis of the farm business: the information from the farm's accounts, performance factors and measures, and standards and objectives for comparison.

Accounting Information

A prerequisite for business analysis is a farm accounting system that provides the information for calculating the various factors and measures used to analyze the performance of the business. (The components of such a farm accounting system were described in Chapter 3.) The farm financial statements discussed in Chapter 4—the balance sheet, income statement, and cash flow statement—are also needed in the analysis process. A comprehensive analysis of the business is impossible without a complete and consistent set of records that have been maintained over time.

Performance Factors and Measures

Certain measures have been developed to help isolate the strong and weak points of a farm business. The number of factors that might be calculated using accounting and production information is almost limitless. The factors presented in this chapter are those that studies and experience have proven useful for identifying high-priority problems. These measures are related to (1) profitability, (2) capital position, (3) cash flow adequacy, (4) size, and (5) productivity and efficiency.

The various factors are closely interrelated. No one factor adequately identifies the cause of the problem. The manager should look at several factors, examining their interrelationship. The manager should also focus on those factors that have the greatest potential impact on the business. For example, it makes little sense to collect the information and compute factors for costs that account for only a very small percentage of the total cost of doing business.

Standards and Objectives for Comparison

Once an analysis factor has been calculated, it must be compared with some standard or objective. The comparison will indicate whether the performance measured by this factor is good, bad, high, or low. The appropriate objectives will vary with the time, economic conditions, type of farm, and location. Farm managers should develop their own objectives related to their own unique situation. These objectives can be derived from three different sources:

1. *Past experience.* The records from past years can be used to set objectives for comparison purposes. Comparisons between years for various performance factors indicate trends. Significant deviations from one year to the next are useful in pinpointing where performance has improved or deteriorated. Objectives based on history are somewhat conservative and may not adequately reflect the potential that can be achieved.

2. *Records for other farms.* Objectives can also be based on the accounting and record information of other farm businesses. To be most useful, these records should be for the same year from farms that have similar characteristics and an above-average level of management. However, it is not always possible for an individual farmer to obtain this information from other farms. This approach is also complicated by the different accounting procedures used by farmers. In some states, record-keeping associations allow members to obtain such comparisons. This type of comparative analysis requires a coordinated effort to obtain the information from the cooperating farmers. In some cases, there may not be enough similar farms to make valid comparisons.

3. *Budgeted projections.* Objectives can also be obtained from the farm plan that was developed at the beginning of the year. These budgeted projections are based on a combination of information from the farmer's personal experience and from that of others. (Farm budgeting procedures are discussed in the next chapter.) This approach allows farmers to set objectives consistent with their particular farm resource situations, the current economic outlook, and family goals.

Performance Factors for Business Analysis

The factors useful for analyzing the farm business's position and performance are organized into five categories. The first three categories—profitability, capital position, and cash flow adequacy—correspond to the three financial statements. These three aspects of the farm business are interrelated. A business with a high level of liquidity (more than adequate cash) is holding excess cash, thereby reducing profitability. Another business desiring to improve its capital position (solvency) by repaying debt ahead of schedule may experience liquidity problems. These analysis factors plus those relating to size and productivity and efficiency provide a complete overview of the business for analysis purposes.

Measures of Profitability

Several different measures of financial success can be computed for a farming operation. The calculations for these measures are indicated in Table 5-1. They are based on the example financial statements from the preceding chapter. Most

TABLE 5-1. Sample Calculations for Measures of Profitability

Factor	Definition	Calculation[a]	Result
Net farm income	Revenue minus expenses	$180,000 - 134,000$	$46,000
Return to labor and management	Net farm income minus the opportunity cost of the average of beginning and ending farm net worth	$46,000 - 0.10(231,400 + 282,400)/2$	20,310
Rate of return to net worth	Net farm income minus value of unpaid labor and management, divided by the average of beginning and ending farm net worth	$(46,000 - 18,000)/(231,400 + 282,400)/2)$	0.109
Rate of change in net worth	Ending net worth minus beginning net worth divided by beginning net worth	$(282,400 - 231,400)/231,400$	0.220

[a] The data are from Tables 4-1 and 4-4.

of these measures start with net farm income, which, of course, must be positive for continued business success.

Net Farm Income. As calculated in the income statement (Table 4-4), net farm income represents the return to the operator for unpaid labor, management, and net worth. In order to compare the farm's results for the year with the results of other farms for the same year or with the results for the same farm for different years, this measure should be broken down further. For example, two farms might have the same net farm income, but one might have a much larger net worth. With everything else the same, the farm with the smaller net worth is more profitable, because it is earning the same net income with a smaller investment by the farmer.

Return to Labor and Management. To calculate the return to labor and management, a charge is made for the average net worth invested, at a rate representing what the capital could earn if invested elsewhere, and the residual is attributed to labor and management. Comparing the return to labor and management with other opportunities for employment indicates the profitability of the farm business. To facilitate the comparison, this total annual return might be divided by the hours the operator worked on the farm in order to calculate the return per hour. If the return to labor and management exceeds the opportunity cost, the business is profitable.

Caution is needed in interpreting these residual profit figures because of the arbitrary procedures that must be used to allocate returns to the specific resources of labor, management, and capital. To consider the effects of these assumptions, the farm manager might calculate the return to labor and management by first assuming a 10 percent charge on net worth and then assuming a 15 percent charge.

Rate of Return to Net Worth. The rate of return to net worth is a measure of profitability computed by subtracting an allowance for labor and management from the net farm income. This return is then related to the net worth, usually the average for the year. This ratio is of interest to managers who want to maximize the return to their own capital investment in the business, their net worth.

The minimum standard or objective for the rate of return to net worth will depend on its opportunity cost—what it would return if invested elsewhere. It is important to remember, however, that the calculation of this rate of return to net worth does not take into account the effect of inflation (or deflation) on the value of farm assets, particularly land. This effect is considered in the change in farm net worth.

Rate of Change in Net Worth. The rate of change in net worth is calculated by subtracting the net worth at the end of the preceding year from that at the end of the following year and dividing the difference by the net worth at the end of the preceding year. This indicator takes into account not only the net farm income but also the withdrawals from the business for family living and income taxes. If the balance sheets are based on market values, this rate of change also includes the impact of inflation (or deflation) on asset values. If the net worth figures are taken from a balance sheet with assets valued at cost, then the impact of inflation (or deflation) is not

TABLE 5-2. Sample Calculations for Measures of Capital Position

Factor	Definition	Calculation[a]	Result
Net worth	Total assets minus total liabilities	535,200 − 252,800	282,400
Debt to asset ratio	Total liabilities divided by total assets	252,800/535,200	0.47
Debt to net worth ratio	Total liabilities divided by net worth	252,800/282,400	0.90

[a] The data are from Table 4-1.

included. A positive rate of change indicates that the farmer is making financial progress; a negative rate indicates that the farmer's net worth is being used up.

Measures of Capital Position

A set of capital position measures refers to the farm's solvency, its capital structure, and the relationship between its liabilities and assets. Solvency is a measure of the long-term financial stability of the business. These measures allow for an examination of the relationships between the various classes of assets and liabilities in the balance sheet (Table 4-1). Their definitions and calculations are presented in Table 5-2.

Net Worth.   The net worth is the difference between assets and liabilities; it is the value that would remain if the business were liquidated and all liabilities were paid. This net worth figure indicates the availability of collateral to borrow additional funds in order to expand the business or meet an emergency. However, it is not just the size of net worth, but also its relationship to other factors, that determines the capital position of the business. These relationships are often best examined by means of ratios.

Debt to Asset Ratio.   The debt to asset ratio shows the relationship of total liabilities to total assets. Calculated by dividing total liabilities by total assets, it measures the solvency, or degree of safety, of the entire business. The debt to asset ratio indicates the proportion of total assets that is financed by borrowed capital. A ratio of 0.5, or 50 percent, indicates that for every $1 in assets there is $0.50 in liabilities.

To illustrate the importance of this relationship between assets and liabilities, consider two farms, each with a net worth of $100,000. If farmer A has $200,000 of assets and $100,000 of liabilities, the total debt to asset ratio is $100,000 divided by $200,000, or 50 percent. If farmer B has assets of $400,000 and liabilities of $300,000, the ratio is 75 percent. A 25 percent decline in the value of assets would wipe out farmer B's net worth, whereas farmer A's operation would remain solvent.

If changes in the farm business would require additional liabilities, the influence of this increased debt on the financial solvency of the farm business should be considered. A safe debt to asset ratio depends on the type of farm and the prevailing degree of risk. If crop failures and wide price variations are common, a lower ratio is needed for safety than is required when conditions are more stable. Also, the wider the range over which the farm asset values might vary, considering general economic conditions and agricultural price fluctuations, the lower the ratio must be for safety.

Debt to Net Worth Ratio.   The debt to net worth ratio is also used to evaluate the capital position of the farm business. This ratio (also called the leverage ratio) is calculated by dividing the total debt by the net worth. It contrasts the lender's contribution to the business to that of the farm manager. A ratio of less than 1.0 indicates that the owner's net worth exceeds the total amount of borrowed capital. A lower ratio indicates a greater degree of solvency. The higher the ratio, the larger the share that lenders have in the farm's assets and the greater the risk exposure to both the farmer and the lenders.

Measures of Cash Flow Adequacy

Cash flow adequacy, or liquidity, refers to the business's ability to meet its financial obligations as they come due. Even though the business may exhibit good profitability, it may have cash flow problems. The reason is that cash flow is influenced by several other factors, such as capital sales and purchases, money borrowed and debt repayment, family living expenses, income taxes, and nonfarm income. Thus, it is important to consider measures of liquidity in the analysis process, along with measures relating to profitability and capital position. See Table 5-3 for the calculation of these measures, based on the example from Chapter 4.

Current Ratio.   The current ratio relates the liquid assets (cash and assets easily converted to cash) to the liabilities due within the next year. The measure is expressed as a ratio—current assets divided by current liabilities. Generally, a ratio of 2 to 1 is desired. Even though the overall financial position of the business may be secure, a low ratio is cause for concern, particularly if large bills are due within the next few months. A high current ratio indicates that some of the capital invested in current assets might be reallocated to more profitable uses. A chronically low current ratio may indicate the need to refinance or restructure debt arrangements in order to improve the situation.

Ratio of Cash Operating Receipts to Expenses.   The potential for generating cash through the operations of the business is indicated by the ratio of cash operating receipts to expenses. It is computed by dividing total cash operating receipts by total cash operating expenses less interest. Interest is deducted to allow for consistent com-

TABLE 5-3. Sample Calculations for Measure of Cash Flow Adequacy

·Factor	Definition	Calculation[a]	Result
Current ratio	Current assets divided by current liabilities	126,000/63,000	2.0
Ratio of cash operating expenses to receipts	Cash operating receipts, divided by cash operating expenses less interest payments	182,000/(115,000 − 18,000)	1.88
Debt service ratio	Principal and interest payments divided by value of production	(50,000 + 18,000)/180,000	0.38

[a] The data are from Tables 4-1, 4-4, and 4-8.

parisons among years and between farms with different amounts of debt. The change in inventory and this ratio should be examined at the same time. The ratio may be higher than normal due to the sale of inventory, or it may be lower as a result of an increase in the inventory held.

Debt Service Ratio.   Debt service is the total annual principal and interest payments that must be met from the revenue generated by the value of farm production. The ratio is computed by dividing debt service by the value of production. An operation with a high proportion of its total revenue required for debt service will experience financial pressure restricting the availability of funds for financing production, asset replacement, expansion, and family living. Also, the greater the burden of debt service on the value of production, the more susceptible the business is to income fluctuations and risk. This ratio is useful in determining the upper limit of the business's debt-carrying capacity.

Measures of Business Size

A number of measures have been developed to compare the sizes of different farms. When comparing farms, measures of size may be useful for two reasons. First, they are needed to indicate the minimum farm size required to obtain a satisfactory income. Second, if the performance factors of different farms (for example, machinery costs per acre) are to be compared, it may be desirable to hold size nearly constant in order to eliminate the effect of farm size on machinery costs.

Physical quantities of output produced can be used to measure size if the farms are producing the same commodity. But when the type of product varies among farms, difficulties are encountered. Size measured by the value of production is expressed in terms of the value of the farm's output. The other factors presented here are based on inputs to measure size.

There is no "best" measure of size that will apply under all circumstances. Comparison of farms on the basis of only one output does not always work. The measure chosen should be based on the purpose for which the comparison is being made and on the type of farm being considered.

Value of Production.   The value of production is a measure of size. It is total production multiplied by price. To the extent that the value of production really measures output, it is a better measure of size than input indicators are. However, as a measuring device, dollars may not be reliable from one year to the next or from one commodity to another. For example, if a year of poor prices follows one of good prices, the value of production might be down even though the farmer produced more, in physical terms, the second year. Yields may also fluctuate, causing variation in the value of production even though the basic farm organization remains the same. The value of production may be used effectively to compare the volume of business of similar farms for the same year or for a period of years. Because such farms are selling similar products, price changes will affect all farms in much the same way.

Total Invested Capital.   The total amount of money invested in the land, land improvements, buildings, machinery, livestock, crops, and supplies is often used to measure farm size. This measure has its limitations because no consideration is given to the amount of labor or operating capital used, nor is any attention given to the distribution of the capital.

Acres of Land.   The number of acres of land may be an accurate measure of size if the comparison involves land that is homogeneous and if the type of farming is similar. For example, in areas of specialized wheat production, acres of land may be a fairly good indication of size. Acreage loses its value as a measure of size when farms of different types are compared. Obviously, the acreage of a strawberry farm should not be compared with that of a wheat farm. An irrigated farm can hardly be compared with one having no irrigation. The amount of other inputs, such as fertilizer and labor, used in combination with land should also be considered.

Total Labor.   Measured in hours, days, or months, total labor includes the total input of operator, family, and hired labor. This measure is satisfactory when comparing farms of a given type, although it has serious weaknesses when used to compare types of farms that require different proportions of capital and labor. This situation is likely to exist when the farms are of different types, such as cattle ranches and tree fruit farms. It may also be true of farm operations of the same type if one is mechanized more than the other.

Number of Animals.   It may be possible to use animal numbers as a measure of size in comparing farms that produce the same species of animals. Basically, the same limitations apply here that apply to acres of land. Neither measure of size accounts for the efficiency or intensity of production. Thus, the amount of milk produced per dairy cow and whether the feed is purchased or raised are not considered when the number of animals is used to measure farm size.

Total Annual Inputs.   The total annual inputs of all resources—land, labor, capital—is a good measure of size when comparing farms of different types. Cash operating expenses, depreciation, the value of the operator's labor, and interest on the total investment are combined into a total dollar figure. This amount gives a relatively good single measure of total inputs, to be used when comparing farms that have widely different resource combinations or that sell different kinds of farm products.

Measures of Productivity and Efficiency

Most measures of efficiency specify the relationship of output to one input without any consideration of the quality of the output and the quality and quantity of other inputs. For this reason, they must be carefully interpreted. Efficiency measures can be compared on the individual farm from year to year or compared with published information for similar farms.

Computing the efficiency factors enables managers to identify management changes that will enhance their chance of success, but they must be cautious and not rely on a single efficiency factor. The relationships among these factors must be considered for the proper diagnosis of problems. Some of the more common measures of efficiency are now examined.

Labor Efficiency.   Labor efficiency is often measured by dividing total labor into output—for example, bushels of corn per hour and pounds of milk produced per annual full-time equivalent of labor. The limitations of such measures is that they ignore the relationship with other factors such as land quality, livestock productivity, building layout, and degree of mechanization. Time cards provide information on the hours of labor used by job performed and by enterprise. The information from these time cards can be accumulated by job, and on multiple-enterprise farms it can be accumulated by enterprise. Decisions regarding the substitution of machines for labor will be more accurate if there are reliable estimates of the amount of labor that could be saved.

Machinery and equipment investment per person employed by the farm business measures the use of capital (machinery and equipment) relative to the use of labor. This is calculated by dividing the market value of farm machinery and equipment by the annual full-time equivalents of labor employed in the farm business. This measure of efficiency varies by farm type and by the extent of labor-saving technology used. A high investment per person might indicate that too much capital has been invested relative to the personnel available. On the other hand, poor labor efficiency can be a result of too little capital invested in labor-saving machinery and equipment.

The most appropriate measure of efficiency varies with the type of farm. On farms with large machinery investments and relatively small amounts of labor, the efficiency with which the machinery is used may be more relevant than labor usage. Conversely, on farms where large labor inputs are combined with relatively small amounts of machinery, labor efficiency becomes the most important measure.

Crop Productivity and Efficiency.   The analysis of crop productivity and efficiency requires additional records, including such information as land use, yields, and agronomic practices.

These records are usually organized by field, with each field identified on a farm map. Yearly production and yields, varieties and rates of seeding, fertilizer applications, and chemicals used should be recorded. Dates of seeding, tillage, and harvesting are also useful for planning a calendar of operations. Crop records can be expanded to include data such as the amount of time spent on each of the various field operations, the amounts of rainfall during specific periods of the growing season, irrigation schedules, and any other information useful to the farmer.

Crop productivity is analyzed by comparing crop yields over time or with those of other farmers. Besides comparing the yields for each crop, a crop yield index can be used to assess overall crop management. The crop yield index expresses the yield of all crops as a percentage of an average for the farm for a number of years or an average for a group of comparable farms for one year. This measure gives proper weight to the amount of acreage devoted to each crop. The crop yield index is computed for Farmer Jones in Table 5-4. If average yields had been obtained, it would have taken 1041 acres to produce the quantities shown. The crop yield index is 104 which is found by dividing 1041 by 1000.

TABLE 5-4. Calculation of a Crop Yield Index

Crop	Jones's Acres	Jones's Current Yield	Jones's Total Production	Group Average Yield	Acres at Average Yield
Wheat	400	100 bu.	40,000 bu	95 bu	421
Potatoes	300	25 tons	7,500 tons	24 tons	313
Alfalfa hay	200	7 tons	1,400 tons	7.5 tons	187
Irrigated pasture	100	18 aum[a]	1,800 aum	15 aum	120
	1000				1041

Crop yield index = (1041 acres/1000 acres) 100 = 104%

[a] aum = animal unit month.

Machinery cost per acre is used to measure the efficiency of machinery. Total machine costs, including depreciation, interest, taxes, housing, repairs, insurance, fuel, and oil, are divided by the number of crop acres. Computing the fixed and variable costs per acre separately will save time later in budgeting. Machinery cost per acre is useful when comparing farms of similar types but has limitations if farms are of different types.

On many farms, costly and more specialized equipment requires special records that might include hours of use, repair expenses, and maintenance needs. This information can then be used to analyze repair costs, maintenance schedules, and trading decisions.

Animal Productivity and Efficiency.   Generally, feed is a major expense item for animal production enterprises. In order to determine feed efficiency, it is necessary to know how much is being used to produce what quantity of animal product. On farms where more than one animal enterprise use the same kind of feed, records by enterprise are needed to allocate the feed among the enterprises. To know how much feed is used by each animal enterprise may require keeping daily records.

It is also important to know the actual physical production of the animal enterprises. At the time of sale, it is easy to record the amount of beef, milk, or eggs sold in physical terms. The farmer may also want to weigh cattle or hogs during the feeding period in order to check on their efficiency of gain and feed conversion.

The physical records for enterprises such as dairy production may require more detailed record keeping. Production data and breeding dates, for example, should be kept on each cow for management to be most effective. The Dairy Herd Improvement Association (DHIA) system is one of the most comprehensive examples of animal production record systems available to monitor animal productivity and efficiency.

Animal efficiency can also be expressed in physical terms, such as pounds of milk per dairy cow, pigs per litter, pounds of daily gain and lamb or calf crop percentage. These measures may be useful for analysis purposes on certain farms. Returns per $100 of feed fed is a commonly used measure of animal efficiency, but it can lead to erroneous conclusions. For example, although the return per $100 of feed fed may be higher one year than another, efficiency in terms of the meat produced per pound of feed fed may actually be lower. Moreover, this measure is incomplete because it considers only one of the inputs. For example, assume that the same relationship exists between the price of the product and the price of feed for two years. Suppose a farmer increases the feed efficiency by building a shelter for the animals during the second year. The records show that the returns per $100 of feed fed have increased by $5. This increase may appear to be desirable, but if it cost $6 for buildings for each $5 gained, the farmer's profits have actually decreased.

A Procedure for Farm Business Analysis

Now, with the various factors for measuring the farm business's performance available, it is time to turn to the analysis process itself. Over the years, the procedure for farm business analysis described here has proved useful.

The various measures of profitability should be studied first. Farmers must decide whether net farm income is adequate, considering their capital investment, their labor used in the business, and their contribution of management ability. Assumptions should be made about the return the capital would receive if invested in another way, and then conclusions can be drawn about the returns to labor and management. Armed with these figures, a farmer can choose one of three courses of action: (1) liquidate the business, invest the capital elsewhere, and seek off-farm work; (2) continue the farming business largely as it is; or (3) make adjustments or changes in the farming business that will result in a higher net farm income in the future. The last alternative requires a more thorough analysis in order to pinpoint those parts of the business where improvements can be made.

The next issue to be resolved is size. Is the farming operation large enough (1) to take advantage of modern technology and (2) to produce enough units for sale to provide the operator with an adequate income? If the farm is too small, opportunities for expanding the operation should be considered first. Crop improvement may also help generate an adequate volume of business.

If the physical volume of business appears adequate, the prices received and paid are next examined. Is the manager producing and selling the products to best advantage—getting them to the market at the time and in the quantity and quality that will maximize net farm income? Opportunities for improvement may be limited with some commodities. After examining the prices received for products sold, the prices paid for the inputs are evaluated. Are supplies purchased at the most advantageous time and in quantities that enable the manager to get them at a favorable price? Good "buymanship" can be quite important on those farms where large quantities of inputs are purchased.

The next step is to examine the input–output relationships. Here the efficiency measures, properly interpreted, are of use. The efficiency of labor, machinery, buildings, and animals is examined. Amounts of production per acre or per animal are considered. In locating trouble spots, it is helpful to consider the efficiency measures simultaneously because their interrelationships are important. For example, machine efficiency may be high but labor efficiency low because machine capacity is not adequate for the size of the farm. The reverse could also be true—

labor efficiency could be high because of many labor-saving machines. But the output per machine might be so low that the cost of owning and operating the machinery offsets the higher labor efficiency. Keeping these factors in proper balance is the real challenge to the farm manager.

After the efficiency factors are considered in relation to one another for the whole farm, it may be desirable to examine individual enterprises. Some of these efficiency measures, discussed previously for the entire farm, may also be used for individual enterprises. Such efficiency measures may suggest that certain enterprises be expanded, curtailed, or eliminated. If records of variable costs have been kept by enterprise, it is possible to determine whether the returns from the enterprises have been covering the variable costs. An enterprise should be discontinued if there is no prospect of the returns exceeding the variable costs. A comparison of variable costs and returns by enterprise may suggest possible changes in the business, which can be tested by the budget methods discussed in the next chapter. As emphasized before, the primary value of analysis is to identify those parts of the farm business where further study is needed.

Note that consideration of business size, marketing, production efficiency, and enterprise mix should also include some analysis of managerial ability and skills. Does the manager have the desire and ability to operate a larger business? Does the manager have adequate knowledge to manage a new enterprise? Is the manager informed about new technology and practices? Continuing education, short courses offered by university extension services and community colleges, and consulting services can help provide this knowledge and ability.

Enterprise Accounting and Analysis

Enterprise accounts are an important source of information for analyzing individual enterprises and determining those that might be expanded and those that might be eliminated. But this information must be supplemented with other information such as the outlook for prices and budget projections to make it useful for planning.

Defining Enterprises

In order to single out a segment of the farm business for analysis as a separate enterprise, it is necessary to isolate the revenue and expenses for this segment by accounting procedures. This is the purpose of enterprise accounting. However, the procedures for isolating the enterprise's revenue and expenses also depends on the characteristics of the enterprise. Enterprises can be classified as one of three types:

1. Production enterprises—hay, corn, beef, turkeys—that actually produce a marketable product.
2. Service enterprises—tractors, combines, buildings—that provide services to each other and to the production enterprises but do not normally produce a marketable product.
3. Holding enterprises—storage, capital, feed mill—that hold inputs and products until they are used by a service or production enterprise or until they are marketed.

For accounting purposes, each enterprise is considered a separate business within the farm. Each enterprise purchases inputs from, and sells products to, other enterprises as well as sources off the farm. For example, the beef cow enterprise may purchase truck services from the truck enterprise, labor from the labor enterprise, and hay from the hay or storage enterprise. It may sell calves to the yearling enterprise, and it may sell cull cows to the off-farm auction. The inputs purchased by the beef cow enterprise are normally charged to the cows at the current market prices. For example, hay would be charged to beef cows at $75 per ton, the market price, even though the cost of production may be only $70 per ton. Such purchases and sales from one enterprise within the farm to another are referred to as *noncash transfers*.

In addition to recording noncash transfers, physical quantities such as the hours of labor, hours of machine time, tons of hay, and pounds and head of calves are essential in analyzing the enterprises and the farm. The first step in setting up an enterprise accounting system is to identify the enterprises.

Production enterprises should be separated to the extent that it is useful to do so. For example, if two wheat fields are similar in production practices, soil, and yield, there is probably little reason to treat each as an individual enterprise; but if one is on poorer soil and leased land, it might be useful to know if the leased field is making more or less profit than the one that is owned.

Service enterprises may have to be separated for two reasons. It may be desirable to know what the new tractor's expenses are compared to those of an older tractor. In addition, it would be valuable to know where and how the tractors are used; therefore, the two tractors should be treated as two enterprises. Moreover, if the rates at which the two tractors are charged to a production enterprise are significantly different, they should be treated as separate enterprises. For example, the hourly charge for using a 200-horsepower, four-wheel-drive tractor that is used 400 hours per year will be more than the hourly charge for a 100-horsepower tractor used 600 hours per year.

Holding enterprises are necessary to make the enterprise accounting system work. There are many expenses (utilities, insurance, fuel, fertilizer, interest) that cannot be charged to a particular enterprise when incurred. A noncash transfer is made later to the appropriate enterprise. Labor is normally considered a holding enterprise because all gross wages and other labor expenses will be charged to it and later transferred to other enterprises according to the hours worked. A crop storage enterprise allows for termination of the production enterprise at harvest time, rather than waiting until the crop is sold, and for separation of marketing activities from production activities. The crop should be valued at its harvest price when it is transferred to the storage enterprise.

Enterprise Accounting

Each enterprise should have a separate ledger, book, or file. There should be major column headings for each type of expense and each type of receipt. For income and expenses that may occur as both noncash transfers and taxable or tax-deductible transactions, the column should be split into two parts, one labeled "cash" and the other "transfers." The transfer column is used for noncash transfers from one

TABLE 5-5. A Sample Weekly Labor and Machine Time Card

Job performed	Enterprise	Propelling machine	Equipment	List hours by date						Total hrs.
Total hours worked										

Worker name _____   Month _____   Wage rate/hr. _____

enterprise to another. The cash column is used for tax-deductible expenses or taxable income.

A record of labor and machine time by enterprise provides the most accurate basis for allocating labor and machinery expenses to the production enterprises. This record is often kept on labor and machine time cards. There are many types available; Table 5-5 illustrates one example that has space for one week's information.

For some purposes, such as estimating the total expenses for producing a crop, it may be necessary to allocate such overhead expenses as taxes, insurance, and utilities to a production enterprise. There is no completely accurate procedure available, but one possibility is to allocate such expenses to each enterprise in proportion to the amount of revenue the enterprise contributes to the farm. This procedure requires some caution because an arbitrary allocation of fixed expenses may result in a less valid analysis.

An Example

A set of enterprise accounts indicating their interrelationships is illustrated in Table 5-6. In this example, the total farm has been divided into four enterprises: crops, pasture, livestock, and machinery. The first three enterprises are production enterprises, but the machinery enterprise is a service enterprise for accumulating the costs associated with machinery ownership and operation and allocating these costs to the three production enterprises.

The revenue for the total farm after adjusting for inventory changes is $180,000. In addition to crop and livestock sales, the individual enterprises show revenue based on credits from the other enterprises. For example, there is a $4000 credit

TABLE 5-6. Statement of Accounts for the Total Farm and Major Enterprises,
January 1 to December 31

	Total Farm	Enterprises			
		Crops	Pasture	Livestock	Machinery
Revenue					
Crops for sale	$154,000	$154,000			
Livestock	26,000			26,000	
Crops fed		4,000			
Pasture grazing			5,500		
Machinery use value					42,850
Total	$180,000	$158,000	$5,500	$26,000	$42,850
Expenses					
Purchased feed	4,000			4,000	
Labor hired	2,000	1,700		300	
Repairs	10,000		100	1,900	8,000
Fertilizer and chemicals	33,000	31,500	1,500		
Livestock expense	1,500			1,500	
Other cash expenses	48,900	41,900		2,000	5,000
Crops fed				4,000	
Pasture grazing				5,500	
Machinery use cost		41,650	200	1,000	
Depreciation	20,000		500	1,600	17,900
Interest on capital	40,290	23,600	4,000	2,000	10,690
Operator labor and management	18,000	13,800	200	3,000	1,000
Total	$177,690	$154,150	$ 6,500	$26,800	$42,590
Profit	$ 2,310	$+3,850	$−1,000	$ −800	$ +260

to the crop enterprise for the feed fed to livestock. Likewise, there is a $5500 credit to the pasture enterprise, which represents a return for grazing. The machinery enterprise provides service to the other enterprises for an estimated hourly cost based on typical machinery rental charges. The summation of the charges to the other enterprises generates a credit to the machinery enterprise.

The expenses in these accounts are more inclusive than those in the income statement (Table 4-4). The total interest on capital assets includes both debt and equity; a charge has also been added for the operator's labor and management input. The remainder, then, after subtracting expenses from revenue, is *profit*. The relationship between profit and net farm income is as follows:

Profit	$ 2,310
+ interest on average net worth	25,690
+ operator labor and management	18,000
= net farm income	$46,000

Note that the profit for the entire farm is equal to the sum of the profits indicated for each enterprise.

How should these profits be interpreted? Both the pasture and the livestock enterprises show negative profits, or losses. Does this mean that these enterprises should be discontinued? Before answering this question, the costs of these enterprises and their classifications should be examined.

Analysis of Enterprise Accounts

The classification of costs into fixed and variable categories was explained in Chapter 2. Costs can also be divided into cash and noncash categories. Combining these classifications provides a better framework for interpreting enterprise accounts. The records for a farm flock of sheep are used as an example (Table 5-7).

TABLE 5-7. Sheep Enterprise Costs for a Farm Flock of 100 Ewes

Category	Quantity	Cost
Variable cash costs		
Hay	37.5 tons	$ 1,950
Grain	13.0 tons	1,958
Other		100
Breeding		250
Marketing		150
Veterinary and medicine		300
Shearing		248
Other operating costs		680
		$ 5,636
Variable noncash costs		
Pasture grazing		$ 2,600
Interest on ewes		800
		$ 3,400
Fixed cash costs		
Taxes and insurance (buildings and equipment)		$ 200
Fixed noncash costs		
Depreciation (buildings and equipment)		$ 800
Interest (buildings and equipment)		1,200
Operator's labor	300 hours	1,500
		$ 3,500
Total Costs		$12,736

SOURCE: Based on Cost Study for 100-Ewe Sheep Enterprise, Willamette Valley, Oregon, prepared by the Oregon State University's Department of Agricultural and Resource Economics, October 1979.

Variable Cash Costs.   All out-of-pocket costs that must be incurred if the enterprise is to be carried on are included in the category of variable cash costs. An example of a variable cash cost is the veterinary expense incurred in connection with the sheep enterprise. In this example, hay and grain were purchased, so they are cash costs. If they were home grown, they would be included with variable noncash costs.

Variable Noncash Costs.   Variable noncash costs are those that have to be met if the enterprise is to be carried on, but they do not involve an outlay of money. Normally, this is not a large category and may be nonexistent for many enterprises. In the case of the sheep enterprise, the value of pasture grazing and interest on the investment in ewes are entered here. Both are variable, because they could be avoided if the sheep enterprise were discontinued. The cost of pasture grazing is based on what the land would return if used by another enterprise or leased to another farmer.[1] The interest on the ewes is noncash because the investment is financed with the operator's own capital, not borrowed capital. This is a variable cost because it can be avoided by selling the ewes.

Fixed Cash Costs.   Fixed cash costs are the costs that would be borne regardless of the existence or size of the enterprise. They are directly associated with the particular enterprise; in this case, the building is used exclusively for sheep. The taxes and insurance on this building are fixed cash costs. They may also be joint, shared by some other part of the farm business. If a building were used for machinery storage as well as for sheep, some method of allocation would be needed for complete enterprise accounting.[2]

Fixed Noncash Costs.   Examples of fixed noncash costs are depreciation, the value of the operator's labor, and interest on the investment. Again, some of these costs can be attributed directly to the enterprise, and others have to be allocated among enterprises or parts of the farm business.

Can enterprise accounts be used to indicate whether a given enterprise should be expanded or eliminated? The principle involved here can be illustrated by the use of the data in Table 5-5. These data are summarized as follows:

Variable cash costs	$ 5,636
Variable noncash costs	3,400
Fixed cash costs	200
Fixed noncash costs	3,500
Total costs	$12,736

[1] A problem of valuation is encountered with the variable noncash costs. How should the pasture grazing be valued? If the pasture did not exist, more hay would have to be fed, or pasture would have to be leased from a neighbor. Pricing the grazing at the cost of production is not workable, because this land is wasteland in small tracts between larger tracts of cropland, making costs difficult to estimate. Probably, the best basis for valuation is what the pasture would return if leased out for grazing.

[2] How should taxes be allocated for a building used for both sheep and machinery? Any procedure will necessarily be arbitrary, but one method is to allocate the taxes according to the space used by each enterprise.

Assume the following receipts from the sheep enterprise:

Lamb (13,255 pounds at $0.65/pound)	$8,616
Wool (960 pounds at $1.15/pound)	1,104
Cull ewes and rams	255
Total receipts	$9,975

The returns from the sheep enterprise do not cover all of its costs; there is a loss of $2761. What would be the effect of discontinuing the sheep enterprise? Suppose the net farm income for the entire farm for the previous year were $38,000. If the sheep enterprise were dropped, the receipts would be reduced by $9975. But what about the costs? Would they be reduced by $12,736? In Chapter 2, it was shown that variable costs are the relevant costs. By carefully examining Table 5-6, it is possible to determine the amount by which costs would be reduced. All of the variable cash cost items would be eliminated.

The pasture grazing cost can be avoided if another use can be found, and the opportunity cost of capital invested in the ewes would be eliminated when they are sold. The taxes and insurance would continue whether sheep are produced or not. The depreciation and interest on buildings and equipment would not be reduced, because the investment in these facilities will be continued. Finally, there is the operator's labor. This charge is eliminated, but unless the labor released can be used productively elsewhere on the farm, the charge has no effect. Because the value of the operator's labor is not deducted from net farm income, the reduction in costs would amount to $9036.

The effect on net farm income would be as follows:

Net farm income with sheep enterprise	$ 38,000
− reduced receipts	−9,975
+ reduced costs	+9,036
Net farm income without sheep enterprise	$ 37,061

Net farm income would be reduced by $939, or, to put it another way, this is the contribution of the sheep enterprise to the net farm income. The farmer, then, by spending 300 hours on the sheep enterprises, can command a return of $3.13 per hour.

The preceding example illustrates how the information provided by enterprise analysis can be useful in decision making. However, the analysis is incomplete because alternatives for using the resources in another enterprise were not considered. Budgets should be developed to test alternative uses of the resources. For example, budgets might test the effects of expanding, curtailing, eliminating, or substituting enterprises. The enterprise accounts would provide some of the information needed for these budgets, but not all that would be needed for decision making. The preparation of farm budgets is the subject of the next chapter.

Summary

Analysis of the information provided by the accounting system will identify the strong and weak aspects of the management of the farm business. No standard approach exists that is satisfactory for all farms. However, this chapter has presented several factors for measuring the profitability, capital position, cash flow adequacy, size, productivity, and efficiency of the farm business. Those factors most appropriate to the particular farm can be compared to standards or objectives to identify the symptoms of problems needing management attention. An analysis procedure is suggested as a guide to the process, emphasizing profitability, business size, marketing, production efficiency, and enterprise mix.

Enterprise accounting involves allocating revenue and expenses to the individual enterprises that make up the total farm business. Although enterprise accounting can be complex, the information derived can be useful in determining which enterprises are profitable and which ones should receive further analysis. Farm business and enterprise analysis is a continuing process directed toward identifying opportunities to increase profitability.

Recommended Readings

PENSON, JOHN B., JR., and CLAIR J. NIXON. *Understanding Financial Statements in Agriculture*. College Station, Tex.: Agri-Information Corporation, 1983.

SHULTIS, ARTHUR. *Farm Enterprise Accounting and Management*. California Agricultural Experiment Station and Extension Service Manual 31, August 1961.

CHAPTER 6

Planning and Budgeting Techniques

The preceding chapter was concerned with diagnosing the strengths and weaknesses of the farm business based on past performance. This chapter focuses on the use of budgeting techniques to prescribe future plans or courses of action based on the evaluation of the alternatives.

Budgeting involves a calculated assessment of the likely outcome(s) of a particular course of action that might be taken to improve the farm business's profitability. In other words, budgeting tests the alternatives on paper before they are actually implemented.

A *budget* is a logically consistent device for examining alternative plans for the farm business and estimating the profitability of each alternative. A budget indicates only whether a proposed change will be more or less profitable than the situation with which it is compared. It does not indicate if the change will be the most profitable use of resources. However, a series of budgets for various alternatives will allow the manager to approximate the most profitable alternative.

To evaluate alternative decisions, three questions must be answered regarding returns, repayment, and risk. The farmer needs to know (1) whether the proposed alternative will be profitable, (2) whether debt repayment obligations can be met when they become due, and (3) whether the business has the capacity to assume the risk. The management tools used to analyze the first two questions are (1) the partial budget or the total farm budget and (2) the cash flow budget. The consideration of risk, the third question, is described in Chapter 8.

Three types of budgets are discussed in this chapter. The *partial farm budget* is used to test the profitability of decision alternatives that may affect various parts of the farm business but that do not involve a major change in the farm's organization. The *total farm budget* is used to consider reorganization of the entire farm business. The *cash flow budget* is used to estimate the amount and timing of cash requirements and the availability of cash to meet these requirements.

The fourth type of budget is the *enterprise budget*, which summarizes the revenue and expenses for a specific enterprise. It includes estimates of production, revenue, quantities of inputs used, variable and fixed expenses, and measures of profitability. The preparation and application of enterprise budgets for crop and animal enterprises are discussed in chapters 15 and 17, respectively.

Partial budgeting is appropriate when the proposed change will have a relatively minor effect on the farm business or will produce benefits within a relatively short period of time, a year for instance. A total farm budget is more appropriately used when the change involves a comprehensive reorganization of the business. When the benefits are not immediate or vary over time, investment analysis techniques, as described in the next chapter, should be used.

Although decisions can be made without these budgeting tools, each tool serves a purpose, allowing a more careful evaluation of a proposed alternative. They can help farmers develop better information to make difficult decisions.

In the discussion that follows, considerable emphasis is placed on the mechanics of budgeting. However, reference is made to the appropriate use of economic principles and accounting information in budget preparation.

A detailed description of the use of computer programs to facilitate the application of planning and budgeting techniques is beyond the scope of this book. Even so, it should be emphasized that a computer can greatly increase the manager's capacity to do budgeting and other analyses (such as linear programming, discussed in Chapter 14) for planning and decision-making purposes. Computer programs are particularly advantageous for evaluating risk because of the ease with which a budget can be recalculated with different price, yield, and cost assumptions. However, the proper use and interpretation of computerized planning and budgeting aids begin with an understanding of the concepts in this chapter.

The Partial Budget

Many changes in the farm business can be made that do not involve a complete reorganization of the business. In other words, there are many parts of the farm business that remain constant, that need not be considered when analyzing the consequences of a particular decision. The partial budget is the appropriate tool for analyzing changes of this kind.

The distinction between fixed and variable expenses is important when considering such changes. Partial budgeting involves estimating the effects of a change on the expenses and revenue of the existing farm organization. Because only those expenses that change with the new way of doing business are estimated, the manager must separate expense items that will change from those that will not.

Partial Budget Outline

An outline of the partial budget is presented in Table 6-1. It is simple to prepare, and it eliminates the time needed to determine those revenue and expense items not affected by the decision. Only the affected items are listed. Adding a small beef cow herd, for example, will involve only extra cattle costs and returns; so, to test for profitability, the farmer does not need to include all of the crop costs and returns that will remain unchanged. The result of the partial budget is to show the change in net income above relevant (variable) expenses. Again, the important rule in partial budgeting is that only those revenue and expense items that will actually change as a result of the alternative action should be included.

TABLE 6-1. The Partial Budget Outline

1. *Added revenue*—expected increase in farm revenue
   from products sold and services rendered as a result
   of the proposed change. $ _____

2. *Reduced expenses*—estimated annual expenses that
   will be eliminated or decreased if the change is
   made.   _____

3. Added revenue plus reduced expenses (total
   credits). $ _____

4. *Added expenses*—estimated new expenses and
   increases in current expenses directly associated
   with the proposed change. These might include
   depreciation, interest, repairs, taxes, insurance,
   labor, materials, and marketing expenses. $ _____

5. *Reduced revenue*—expected value of revenue that
   will no longer be received if the change is made.   _____

6. Added expenses plus reduced revenue (total debits). $ _____
7. Difference (change in net farm income). $ _____

It may be helpful to examine the relationships among the partial budget components in the following example. Suppose a vegetable farmer is considering growing carrots rather than peas. The partial budget, presented in the usual format, is as follows:

Added revenue (carrots)	$16,000	
Reduced expense (peas)	13,300	
		$29,300
Added expense (carrots)	$13,450	
Reduced revenue (peas)	14,100	
		$27,550
Difference		$ 1,750

Another way to consider this decision is to compare the net income of the added enterprise with that of the discontinued enterprise. For example:

Carrots		Peas	
Added revenue	$16,000	Reduced revenue	$14,100
Added expense	13,450	Reduced expense	13,300
	$ 2,550		$ 800
Difference			$ 1,750

This comparison leads to the same conclusion as the previous format did, namely, that the enterprise to be added would contribute more to net farm income than the one to be discontinued.

The outline of the partial budget suggests that there are five ways to increase the profitability of the farm business. Changes in the farm's organization that have the following characteristics will result in a positive change in net farm income:

1. Changes that increase revenue and decrease expenses.
2. Changes that increase revenue by more than the increase in expenses.
3. Changes that increase one source of revenue by more than the decrease in other sources, with no change in expense.
4. Changes that decrease expenses by more than the decrease in revenue.
5. Changes that decrease one item of expense by more than the decrease in other items, with no change in revenue.

Although these five ways to increase profitability may be obvious, the manager's job is not easy. The challenge is to discover the changes that are possible and then to evaluate them within this partial budget framework.

When to Use Partial Budgets

Most decisions for an ongoing farm business involve only parts of the business; that is, only certain revenue and expense items will be affected if a particular action is taken. Examples of decisions that partial budgets can be used to evaluate are (1) adding a new enterprise with no effect on existing enterprises, (2) buying a machine rather than custom hiring, (3) applying a chemical herbicide, and (4) choosing the time to sell grain. On the basis of their effects, proposed changes or alternative actions for an ongoing farm business can take one of three forms:

1. *Input–Output.* This change involves increasing or decreasing the use of a resource with a resulting effect on output. Examples include adding fertilizer and applying herbicide.
2. *Input–Input.* This type of change involves substituting one resource for another without an effect on output. Examples include substituting a new crop tillage system for the current system and renting rather than purchasing machinery.
3. *Output–Output.* This change results from substituting one enterprise for another. Examples are replacing barley with wheat and seeding cropland to pasture.

Decisions frequently involve combinations of these three types of changes.

The term *partial* budgeting does not imply that less care or detail is required than that for a total farm budget. This is not the difference between the two methods of budgeting; the difference is the impact of the proposed change on the farm organization. If the proposed change will affect the entire farm business, a total farm budget is needed. This is obviously the case if a person is just starting to farm or is changing from one type of farming to another. The partial budget is appropriate when some of the expenses and revenue will remain constant; it involves identifying those expenses and revenue that will change and the degree or amount of change.

Partial Budgeting: An Example

Assume that a grain farmer has been paying $12,000 for custom harvesting but is considering buying a $50,000 combine to harvest the grain herself. She anticipates that revenue would come from having more grain to sell as a result of the reduction in field losses if she operates the combine herself. The reduced expense is the $12,000 cost of the custom harvester; the total credits (added revenue and reduced expense) amount to $14,500 (Table 6-2, item 3). To determine if the purchase of a combine will be profitable, the debits (added expense and reduced revenue) must also be estimated.

Added expenses include both operating and ownership expenses if a capital purchase is involved. The fuel used in operating the combine is an operating expense. Although it is not reasonable to include the initial purchase cost of the combine in the budget, the expenses associated with owning it are included. They include the following:

1. *Depreciation.*
2. *Interest* on the invested capital.
3. *Repairs.*
4. Property *Taxes.*
5. *Insurance* premiums.

As listed, these expenses form an easily remembered acronym—DIRTI.

TABLE 6-2. Partial Budget for Buying a Combine

Decision alternative: Whether to buy a $50,000 owner-operated combine to replace custom grain harvesting		
1. *Added revenue*		
Reduction of field loss (20 tons × $125)	$ 2,500	
2. *Reduced expense*		
Custom harvesting	12,000	
3. Added revenue plus reduced expenses		$14,500
4. *Added expenses*		
Depreciation [($50,000 − $20,000)/5 years]	$ 6,000	
Interest ($35,000 × 12%)	4,200	
Repairs	800	
Taxes	600	
Insurance	300	
Fuel	800	
Operator labor	1,200	
5. *Reduced revenue*		
None	0	
6. Added expenses plus reduced revenue		13,900
7. Difference (change in return to management)		$ 600

To estimate the average annual depreciation expenses, the salvage value ($20,000) at the end of the asset's useful life (five years) is subtracted from the original cost ($50,000). This difference is then divided by the useful life ($30,000/5) to give an estimate of the annual depreciation ($6000). The estimated salvage value of the asset should be consistent with its useful life. That is, the longer the expected life of the asset, the lower the anticipated salvage value.

The interest on the invested capital, whether borrowed or not, is an opportunity cost based on the return forgone by investing in this capital purchase. One rule of thumb is to use the prevailing interest rate on farm loans for similar assets. The average annual interest expense is then determined by multiplying the interest rate by the average value of the asset over its useful life. This average value is the average of the asset's original cost and salvage value; it is calculated by totaling the original cost and salvage value and dividing the sum by 2. The average value of the combine over its five-year life is [($50,000 + $20,000)/2], or $35,000. Using an interest rate of 12 percent gives an annual interest expense of $4200.

Estimated repair costs, which depend on the type of asset, conditions, amount of use, and maintenance requirements, can be based on personal experience and records. In this example, they are $800.

The asset's value determines the property tax. Tax rates vary widely; depending on the state in which the farm is located, some assets are not subject to property taxes. To estimate taxes for the example, the local tax rate is multiplied by the average value of the combine.

Insurance costs should be included to account for the risk of loss from fire, wind, and other hazards. The owner can bear this risk, or an insurance company can be paid to assume it. The annual cost is estimated by multiplying the insurance premium per $100 by the average value of the combine.

The fuel and labor costs depend on the number of hours required to complete the grain harvest. The fuel cost per hour is the product of the fuel use per hour multiplied by the cost per gallon. The labor cost per hour should reflect the opportunity cost of the operator's labor.

The total added expense for owning and operating the combine amounts to $13,900. No reduction in revenue is anticipated as a result of this decision. So, the total debits (added expense and reduced revenue) also amount to $13,900.

The difference between the credits and debits is $600. Note that this is the change in the return to the operator's management. This positive result indicates that the decision will be profitable; a negative result would have indicated that the decision will not be profitable. The difference between the credits and debits in this case is not the change in net farm income, as in the outline (Table 6-1). Recall that net farm income is the return to the operator's labor, capital, and management. In this partial budget example, costs for the operator's capital and labor have been included. Thus, the $600 difference refers only to the change in the return to the operator's management. It is important to note which costs have been included in the partial budget in order to know whether the computed difference refers to a change in net farm income, return to labor and management, return to capital and management, or return to management.

The farmer may be uncertain about the estimate of a reduced field loss of grain as a result of operating her own combine. The reduction in field loss that will permit her to just break even can be calculated from the information in the partial budget by means of the following formula:

$$\frac{\text{Item } 6 - \text{item } 2}{\text{Grain price/ton}} \quad \text{or} \quad \frac{\$13,900 - \$12,000}{\$125} = 15.2 \text{ tons}$$

If the reduction in field loss is less than 15.2 tons, the farmer would be better off continuing to have the grain custom harvested. The farmer estimates that the reduction in field loss will be an average of 20 tons per year over the five-year life of the combine. It is likely, however, that this result will not be achieved until the second year, when she has become proficient in adjusting and operating the combine.

It is possible to calculate other break-even points in addition to field loss reduction. The same could be done for grain prices, custom harvest costs, and repair costs. The variable chosen for break-even analysis is usually the one that has the most uncertainty associated with it. Then the break-even point is found by holding the other variables constant and solving for the break-even value that makes both sides of the equation equal—the value that makes the total credits equal to the total debits (see Table 6-1).

Although the preceding example illustrates the use of the partial budget when the purchase of a new machine is being contemplated, a similar analysis can be used when the manager is considering a new enterprise, the construction of buildings, or any other changes that do not call for a complete reorganization of the farming operation. Obviously, it is important to estimate systematically the added and reduced revenue and the added and reduced expenses. Although the nature of the revenue and expense items will vary, depending on the decision, the logic is the same in all cases. If the logic of the method is understood, the outline in Table 6-1 will allow an accurate estimation of profitability.

The Total Farm Budget

The total farm budget includes all the expenses and revenue of the farm business; it is used in a decision-making situation where the alternatives affect the use of all or most of the farm's resources. Although such a major reorganization usually occurs infrequently, the decisions have such a long-term effect that it is important to make a comprehensive analysis of their impact. Examples of decisions involving the use of the total farm budget are changing the combination of farm enterprises, increasing the size of the farm, and introducing a new practice such as irrigation. A young man returning home to take over the family farm and a manager purchasing a farm can use total farm budgeting to develop a farm plan.

The main purpose of budgeting is to compare the profitability of different plans for the farm business. More than one budget is needed to make a comparison. The present organization may constitute a base for this purpose. If adequate farm information and records have been kept, it may not be necessary to prepare a budget representing the current farm plan. However, if the manager is just starting, there may not be a current plan from which to work. In this case, the manager will probably want to compare budgets for alternative plans so that the one that best suits the available resources and the manager's objectives can be selected. Preparing the second and following budgets usually involves much less work than the preparation of the first. One reason is that many of the same data can often be used for subsequent budgets. Another is that the second and succeeding budgets are often only modifications of the first. Using a computer will greatly reduce the time and effort required to make the calculations. And, if the alternative plan will affect only a portion of the total operation, partial budgeting may be used to save time.

The total farm budget may be used for other purposes as well. Only by considering the whole farm business can *all* of the expenses and revenue be anticipated, and by coordinating the whole farm budget with the family living budget, the adequacy of the farm business to meet family living expenses can be checked. Total farm planning can be helpful when credit is being obtained and repayment ability is being determined. The budget can be used to negotiate credit and land leasing arrangements and to obtain counsel and advice from experts. Consequently, the budget should be capable of communicating the farm plan to someone other than the person who prepared it.

Before Starting

To see how total farm budgeting is used to develop, compare, and select farm plans, it is useful to consider the planning process. The farm planning process involves answering the following questions:

1. What do you want to accomplish? (Goals)
2. What do you have to work with? (Resources)
3. What has been done? (Analysis of past performance)
4. What might be done? (Alternative plans)
5. What will be done? (Decision)
6. How will it be done? (Implementation)
7. Is it working? (Monitoring)

The goals of the manager should be considered when preparing the budget; the budget and the plan it represents should serve as a guide for reaching those goals. One approach is to adjust the organization of the farm at various steps in the planning process according to the preferences and goals of the manager. Another approach is to budget a plan developed to maximize profits. This budget can be compared with other plans that incorporate personal preferences. The manager can then see how much it costs to indulge these preferences.

The next step in the preparation of a total farm budget is to make an inventory of the resources available. If adequate farm records have been kept, much of this information will be found there. Capital, land, labor, and management are the resources that the farmer must transform into marketable products. A rather detailed listing of the available capital assets should be made. Capital may be in the form of land, livestock, machinery, or cash and other liquid assets. The balance sheet illustrated in Table 4-1 might be used to list both assets and liabilities.

In the planning process, it is important to determine the fixed resources and to work around them. For many situations, the total acres of land available is the appropriate starting point. Such an approach is appropriate when there is little possibility of adding to the acreage either by purchase or by lease. It should be recognized that this is not always the correct approach. In some instances, the acreage of land may not be fixed, but other capital, available labor, or management (in terms of how large an operation the manager can handle adequately) may well be limiting factors.

The procedures for analyzing past performance, presented in the preceding chapter, provide a basis for evaluating the manager's skills and ability. What are the manager's strengths and weaknesses? Can poor past performance be corrected? This honest self-appraisal is an important step in planning.

Once the goals, resources, and past performance are evaluated, the manager is in a position to develop alternative crop and livestock enterprise plans, considering the land, labor, capital, and management available. Resources available include not only those on hand but also those that can be acquired by borrowing, renting, or hiring. Certain enterprises may be eliminated from consideration even before the plans are budgeted. For example, if the farmer believes that the most limiting factor is capital, enterprises with high capital requirements should probably not be considered. It is usually not feasible to budget all possible alternatives. Therefore, the selection of the alternatives to be budgeted is quite important. For this reason, the possible alternatives should be screened in light of the available resources, the manager's goals, and past performance.

This screening process should allow the manager to narrow the range of alternatives to those that are most promising. Asking the following four questions helps to narrow the selection:

1. Will the alternative action contribute to your goals? It is not realistic to assume that you will implement an alternative that does not agree with your preferences. It may require labor that you do not want to perform or risks that you are not willing to take.
2. Is the alternative action consistent with the financial and physical resources available? Is labor available to implement the decision? Does the alternative require capital that is not available?
3. Does the alternative action help solve a management problem? If the business has a heavy debt load, alternatives that require additional borrowing are not feasible.
4. Is the alternative action within your management capabilities? Because of the wide differences in manage ment skills and abilities among farmers, the range of feasible alternatives varies considerably from one farmer to another.

The final three steps are deciding on a plan, implementing it, and then monitoring it. Total farm budgeting will facilitate the selection of a feasible and profitable plan that is consistent with the manager's goals, abilities, and available resources. Implementation may involve obtaining additional resources, and the total farm budget will be useful in convincing the lender of the soundness of the business plan. Also, monitoring procedures must be developed to ensure the plan's success and to make adjustments as necessary.

Gathering the Information

Employing the budgeting procedures discussed here requires much information gathered from a variety of sources. These sources include farm records, the university extension office, lenders, agricultural publications, the experience of neighbors, agricultural research and experiment stations, and so forth. The good manager is constantly searching for needed information. By putting together information from a variety of sources, budgeting figures can be developed with greater accuracy. A margin of error must be allowed in estimating these figures, however, because they are still subject to some degree of uncertainty.

Managers should also consider the relative importance of the various expense and revenue items as they prepare budgets. Some cost items, such as fuel expenses, may be of only minor significance in terms of the final accuracy of the budget, whereas others are of major importance. Time spent in gathering information should be allocated accordingly: The greater the size of the item, the more influence it will have on the budget total, and the greater the need that it be accurately determined. For example, a 5 percent error in the quantity of feed consumed may change expenses by $1000, whereas a 50 percent error in the cost of insurance may affect the total by only $100.

Preparing the Total Farm Budget

The total farm budget relies on a net farm income budget (projected income statement) and the supplementary worksheets, which may vary in the amount of detail they include. At a minimum, they might consist of (1) a crop plan, (2) a livestock plan, (3) a livestock feeding plan, (4) an estimated expense schedule, and (5) the net farm income budget. This last component of the total farm budget is prepared in the same format as the income statement (Table 4-4), except that this net farm income budget contains projected figures, unlike the income statement, which is based on actual results.

Crop Plan.   The crop plan summarizes the crops to be grown, the total production, and its disposition during the year for which the budget is being prepared. Table 6-3 indicates the type of cropping system information that must be developed. To provide a basis for determining the cropping system, a detailed map of the farm should be drawn, different soil classes should be identified, and land of similar topography and productivity should be grouped together. This map provides a basis for determining the cropping system.

The technical relationships between the enterprises and economic principles (Chapter 2) form the basis for choosing the crops to be grown. Once this decision is made, the areas to be devoted to each crop must be specified and the expected yield of each crop must be estimated. These estimates can be made on the basis of existing farm records, county records, or experimental data. The disposition of the crop must also be specified. That is, what percentages of the crop will be fed to livestock, used for seed, sold, and so forth? If the crop is to be sold, a price estimate is necessary. Estimating prices is usually one of the more difficult tasks involved in farm budgeting. Because price and yield information is so important in budgeting, some of the problems involved are discussed later in this chapter. Once the price is estimated, crop sale receipts are determined.

TABLE 6-3. The Crop Plan

Crop	Production			Disposal				Expected Sales Receipts ($)
	Acres	Yield per Acre	Total Production	Livestock Feed	Seed	For Sale	Expected Price	
Total		X X X	X X X	X X X	X X X	X X X	X X X	

Livestock Plan.   Table 6-4 is a worksheet for recording comparable information for livestock. The selection of the number and size of livestock enterprises is obviously an important step in the planning process. Once the kind and number of livestock have been determined, additional questions call for still more information. The plan must indicate the number of livestock that will be purchased, born, sold, and kept, as well as the number likely to die of natural causes. Similar data must be recorded for livestock products.

It is a good idea to check the calculations on each line of Table 6-4. The numbers of each kind of livestock should balance as follows: Beginning inventory plus purchases plus births equals deaths plus sales plus ending inventory.

Livestock Feeding Plan.   Table 6-5 may be used to estimate the amount of feed that will be required by the livestock, as well as how much feed must be purchased and how much will be produced on the farm. The proper combination of feedstuffs is an economic question, which the principle of substitution can address. Again, farm records can be helpful in showing past requirements and providing a base from which to work.

Expense Schedule.   The estimated expenses for the year are summarized in Table 6-6 using the information from the crop and livestock production plans. Expense items for livestock and feed purchases are transferred from Tables 6-4 and 6-5. The other expenses can usually be estimated fairly accurately if the physical quantities of the inputs used are carefully planned and if the manager has kept records. These records of the past are extremely valuable in estimating future expenses. The amount of fertilizer to apply obviously should not be determined solely on the basis of what has been applied in the past, but the price of fertilizer can be checked, as can the cost of chemical applications or any custom work. The amount of fertilizer to be applied or the farm practice to be undertaken should be determined by using the principles outlined in Chapter 2.

Labor, fuel, and oil costs for both crops and livestock must be estimated. A labor budget showing each month's labor requirements can be used to estimate the amount of labor that must be hired to perform critical operations. This expense schedule illustrates the potential usefulness of farm records in general and enterprise records in particular. Past performance may be a good basis for estimating the expenses of crop and livestock enterprises.

Net Farm Income Budget.   The revenue and expenses for the farm plan are summarized in an outline like Table 6-7. This table corresponds to the income statement (Table 4-4) described in Chapter 4, but with some important exceptions. First, the net farm income budget indicates the projected, or estimated, profitability of a proposed farm plan. The income statement in Chapter 4 indicates the actual profitability of the farm business based on past performance. Second, the net farm income budget as presented in Table 6-7 does not include revenue adjustments or paid and payable expense adjustments, which are contained in the net income statement. These adjustments are needed in the income statement to account for the changes in beginning and ending inventory values of crops and livestock and for changes in paid and payable expenses between the beginning and end of the year. For the

TABLE 6-4. The Livestock Plan

Kind of Livestock or Products	Beginning Inventory		Animals to be Purchased		No. Born	Disposition of Animals and Products				
	No.	Value ($)	No.	Value ($)		No. Died	No. for Sale	Weight Per Head	Price Per Unit	Expected Sale Receipts ($)
Total								X X X	X X X	

TABLE 6-5. The Livestock Feeding Plan

Kinds of Animals	No.	Length of Feeding Period (Months)	On-Farm Grazing		Rented Grazing			Roughage		
			Pasture (AUMs)	Aftermath Grazing (AUMs)	Pasture (AUMs)	Aftermath Grazing (AUMs)	Rental Costs ($)	Home Raised (Tons)	Purchased Amount (Tons)	Purchased Cost ($)
Total required		X X X								
Amount to be raised		X X X								
Amount to be purchased		X X X								
Cost of purchases		X X X								

Kinds of Animals	No.	Length of Feeding Period (Months)	Grain			Concentrates		Salt, Minerals	
			Home Raised (Weight)	Purchased		Amount (Weight)	Cost ($)	Amount (Weight)	Cost ($)
				Amount (Weight)	Cost ($)				
Total required		X X X							
Amount to be raised		X X X							
Amount to be purchased		X X X							
Cost of purchases		X X X							

TABLE 6-6. Expense Schedule

Estimated Operating Expenses	Annual Amount ($)
Livestock purchased (from Table 6-4)	
Feed purchased (from Table 6-5)	
Land rent payments	
Fertilizer and lime	
Seed	
Chemicals	
Custom machine hire	
Labor hired	
Gasoline, fuel, oil	
Repairs	
Supplies	
Breeding fees	
Veterinary fees, medicine	
Storage fees	
Property taxes	
Insurance	
Utilities	
Marketing and transportation	
Other	
Total expenses (excluding depreciation and interest)	$

TABLE 6-7. The Net Farm Income Budget

Revenue		
Crop sales (Table 6-3)	$	
Livestock (Table 6-4)		
Breeding stock (Table 6-4)		
Value of production	X X X	$
Expenses		
Operating expenses (Table 6-6)	$	
Interest		
Depreciation		
Total expenses	X X X	$
Net farm income	X X X	$

total farm budget, these changes are normally zero, meaning that revenue and expenses are based on actual production during the year.

To prepare the net farm income budget (Table 6-7), the totals from the other worksheets must be transferred. Crop sales is the total of the expected sales receipts from the crop plan (Table 6-3). Revenue from livestock, livestock products, and breeding stock comes from the livestock plan (Table 6-4). Operating expenses is the total from the expense schedule (Table 6-6). Interest payments will have to be estimated based on current debt (see Table 3-5) and any anticipated changes in financing associated with the proposed plan. Depreciation likewise will include annual depreciation for assets currently owned (see Table 3-4), adjusted for new acquisitions (and sales) needed to implement the proposed plan.

An income comparison can be made of the net farm income budget for the current plan with the net farm income budget for the proposed plan. A general format for comparing the results of total farm budgeting in its simplest form is as follows:

	Current Plan	Proposed Plan
Farm revenue	$_____	$_____
Farm expense	−_____	−_____
Net farm income	$_____	$_____

Of course, there are numerous ways of arranging this information in order to reveal various relationships. It was pointed out in Chapter 4 that the allocation of net farm income to capital, on the one hand, and to labor and management, on the other, is somewhat arbitrary. However, if the same rate of interest is used to charge capital for the various plans, a comparison can be made of the return to labor and management.

Table 6-8 illustrates the relationship between total farm budgets and partial budgets. As can be seen, the two approaches are closely related, and, if used correctly, both will give the same result when used to compare decision alternatives. The approach, or budget, used depends on the situation and alternative being

TABLE 6-8. Relationship between Total Farm Budgets and Partial Budgets

	Total Farm Budget (Current)	Partial Budget		Total Farm Budget (Projected)
Farm revenue	Current revenue	+ added revenue	− reduced revenue	= projected revenue
Farm expenses	Current expenses	+ added expenses	− reduced expenses	= projected expenses
Net farm income	Current net farm income	Change in net farm income (+, −)		= projected net farm income

considered. If relatively few revenue and expense items are affected, the partial budget is much easier to use. If the alternative is more complex and involves more items, total farm budgeting is more practical.

General Budgeting Procedures

The total farm budget may or may not be detailed. The worksheets included here are merely illustrative. More appropriate worksheets can be prepared to fit the particular farming situation. The more familiar the manager is with the present and the proposed farm business, the less detail is required in the budget.

The preceding description of the budgeting process may seem to be more mechanical than it actually is. The usefulness of economic principles and farm records of previous years has been mentioned, as well as the interdependence of the various components of the farming business—for example, the relationship of the cropping program to the livestock program and livestock feeding to the selection of livestock. Here the cropping program was considered first. In practice, the cropping program must be planned with the livestock enterprise in mind. The order in which the two are planned is not of great importance and should be determined on the basis of the problem at hand. This statement is not meant to imply that all, or even a major part, of the feed for livestock should be grown on the farm, nor does it mean that all the feed grown on the farm should be fed. These are economic questions that should be decided after examining the budgets rather than imposed on the budgets in advance.

The Cash Flow Budget

In the preceding discussion, budgets were used to analyze the profitability of proposed changes in the farm business. However, managers of businesses, even profitable businesses, that cannot raise sufficient cash to meet their obligations must either reorganize or discontinue operation. Thus, the bottom line for business survival is a positive cash balance in the checking account after payment of obligations due. And this is the purpose of the cash flow budget—to project the cash inflows and outflows over some period of time. The cash flow budget is sometimes called a *whether forecast*, because it is used to determine a plan's financial feasibility.

Major changes in plans for the farm often involve borrowing to finance new investments, and there is often a transition period of a few months, or perhaps two or three years, before the new plan is fully implemented and generating the projected profit. The cash flow budget is used to evaluate the feasibility of repaying the required loan and to predict whether cash flow during the transition period will be sufficient to maintain liquidity until the plan is producing as projected by the total farm budget or the partial budget.

An advantage of the cash flow budget, compared to the total farm and partial budgets, is that the timing of projected revenue and expenses is considered. It shows when money will be needed and when it will be received. Timing is important, because the manager usually must pay for a new machine or building at a faster rate than it depreciates. Also, early in the life of the investment, expenses such as interest will be higher. These considerations are not included in total farm or partial budgets. Although these budgets play an important role in analyzing the profitability of a decision, a cash flow budget is also needed to analyze loan repayment and other obligations.

The cash flow budget includes the projected operating receipts, capital sales, operating expenses, capital expenditures, income taxes, family living expenses, money borrowed, and debt repayments. When such information is summarized by months (or quarters), the timing of cash flows is brought into focus. The budget shows the cash flow surplus or deficit for each month and the end-of-the-month balance. Thus, the availability of cash to meet commitments can be evaluated throughout the year.

The cash flow budget is like a calendar of cash inflows and outflows. The cash flow budget combines personal and business affairs into one report so that the amounts and timing of cash deficits and surpluses can be anticipated. This information makes it possible to arrange credit needs in advance, to plan repayment schedules, and to invest cash surpluses. And these projections make it easier to convince the lender to provide financing, because the cash flow budget indicates that it can be repaid.

Preparing a Cash Flow Budget

Preparing the cash flow budget is simply a matter of projecting the cash that will be deposited in the checking account during the year and the amounts that will be withdrawn. A special form is not needed; a large sheet of ledger paper from the local bookstore will suffice. It should have 13 columns, including one for the annual total and one for each month of the year, and enough lines for each category of cash inflows and outflows.

A simplified format for a cash flow budget is illustrated in Table 6-9. Note the headings at the beginning of each line. The major headings are "Cash Available" and "Cash Required." The cash available for the month is the total of the beginning balance, operating receipts (crop and livestock sales), capital sales, and nonfarm income. The cash required is the total of all uses of cash for operating expenses, capital purchases, nonfarm expenses, family living expenses, and income taxes. The "Available Minus Required" cash is found by subtracting the total uses from the total sources. To this difference is added any new borrowing required to maintain liquidity; then any scheduled or optional loan payments are subtracted. The result is the ending balance for the month being budgeted, which becomes the beginning balance for the next month.

There are two basic ways to prepare the cash flow budget. The first is to make estimates based on last year's records. Starting from last year's figure for a particular receipt or expense item, the first step is to adjust for any changes in price or quantity anticipated. Then this annual total is prorated to the months when the cash will be paid or received.

The operating receipts from crop, livestock, and other enterprises can be estimated based on last year's production and marketing patterns. Operating expenses can be estimated from previous livestock and crop records, and monthly allocations

TABLE 6-9. Simplified Cash Flow Budget

	Total Year	Months			
Beginning Balance					

Operating Receipts
  Crop sales
  Other					

Capital sales
  Machinery
  Breeding animals					

Nonfarm income					
CASH AVAILABLE					

Operating expenses
  Fertilizer
  Feed
  Other					

Capital purchases
  Machinery
  Breeding animals					

Nonfarm expenses Family living Income taxes					
CASH REQUIRED					
AVAILABLE MINUS REQUIRED					
Money borrowed Loan payments					
Ending Balance					

for family expenditures likewise can be estimated from last year's needs.

However, preparing the cash flow budget based on last year's records works well only if there will be no major changes in the farm plan, the number of acres for each crop, the number of livestock, and so forth.

If major changes are planned, then the second approach should be used—preparing a detailed budget. This procedure involves crop and livestock plans, complete with yield projections, livestock purchases, feed requirements, crop sales, livestock sales, and expenses, as described earlier for the total farm budget. The timing of income and expenses must also be anticipated, and capital expenditures, repairs, loan requirements, taxes, and family living expenses must be projected.

Preparing a detailed projected cash flow statement is a time-consuming process if done by hand. However, with the help of a microcomputer and a spreadsheet program, the process can be simplified.[1] Another advantage of these spreadsheet

[1] Examples of spreadsheet programs include VisiCalc, SuperCalc, Multiplan, and Lotus 1-2-3.

packages is that they allow the user to ask if–then questions. This is a particularly important aspect of cash flow management, because the outcome—the end-of-the-year cash balance—is dependent on price, yield, and cost assumptions made in the budget. Because the computer program can incorporate a wide range of assumed prices, yields, and costs, it is easy to look at several cash flow possibilities.

Table 6-9 illustrates a monthly breakdown of cash inflows and outflows. This breakdown may be more detailed than necessary for a farm business with one enterprise and a fairly constant cash flow. In this case, two-month or quarterly periods might be used.

Suppose a manager is considering investing in a new milking parlor for the dairy herd. This plan will take two or three years to implement—it will take perhaps six months to complete construction and to install the equipment, and then another year or two to expand the size of the herd in order to utilize the new facility to capacity. To analyze the cash flow for this plan, the cash flow budget might be divided into annual, or perhaps semiannual, periods to check the financial feasibility of the plan, particularly during the first two or three transitional years.

Calculating Borrowing Needs

It is important to understand the detailed procedures involved in estimating the amounts of money that must be borrowed to maintain liquidity. When the cash available minus cash required is projected to be negative for the month, the manager must borrow to cover the shortfall. On the other hand, when the projected difference is positive, the manager can use the surplus to repay loans. However, the calculations are somewhat more complicated. Table 6-10 presents an example with the bottom part of the cash flow budget shown in more detail. The projected cash flows represent the pattern of a typical farm operation. This example is organized by quarterly periods.

Examining one quarter at a time, it can be seen that "Available Minus Required" in the first quarter indicates a deficit of $2000. Rather than borrowing to cover this shortfall, the manager withdraws $3500 from savings. This withdrawal leaves an ending balance of $1500. The manager has estimated that this ending balance will be large enough to provide a cushion for larger than expected expenses or smaller than expected receipts.

In the second quarter, the "Available Minus Required" (including the $1500 first quarter ending balance) indicates a $54,000 deficit. This shortfall is too large to be covered by savings; the money must be borrowed. A total of $55,000 is borrowed, leaving an ending balance of $1000 and increasing the outstanding loan balance from $120,000 to $175,000.

The third quarter shows a small surplus. This money is used to make a small principal payment, reducing the loan balance to $169,200 and leaving a $1200 ending cash balance.

In the fourth quarter, crop and livestock sales are expected to produce a cash surplus of $94,400. The first use of this surplus will be to pay the accrued interest— $17,800. Then $71,200 will be applied to the principal, leaving an ending balance of $1400 after transferring $4000 to savings. This example has a happy ending.

TABLE 6-10. Cash Flow Budget with Borrowing and Repayment Detail

	Jan.–Mar.	Apr.–June	July–Sept.	Oct.–Dec.
AVAILABLE MINUS REQUIRED	$ −2,000	−54,000	7,000	94,400
Transfer from savings	3,500			0
Cash balance before borrowing	1,500	−54,000		
Money to be borrowed		55,000		0
Loan payments				
Interest				17,800
Principal			5,800	71,200
Transfer to savings				4,000
Ending cash balance	1,500	1,000	1,200	1,400
Total loan balance	$120,000	175,000	169,200	98,000

The farmer should be able to repay the current year's loan and reduce the carryover from previous years.

The successful operation of a farm business requires careful planning and good information for accurate cash flow budgeting, as for total farm budgeting and partial budgeting.

Price and Yield Information for Budgeting

Earlier in this chapter, reference was made to the necessity of having price and yield information for budgeting. Because poor data can invalidate budget results, it is appropriate to examine this problem in more detail.

Agricultural product prices are extremely difficult to predict. Some experts advocate using long-run (five-year) average prices for budgeting. The argument is that prices tend to have long-run relationships even though these relationships may not hold in a particular year. Unusual conditions may prevail in a given year that make product prices high or low relative to the costs of production. There is little doubt that using prices for an individual year without reference to the long-run situation is questionable. On the other hand, the use of long-run averages may lead to difficulty, because basic changes in the economy may occur. An example is a persistent rise in the general price level.

Unfortunately, there is no simple answer to this complex question. Managers should study long-run price relationships; both commodity price relationships and the relationship of prices to costs should be analyzed in order to provide a framework for evaluating current prices. For decisions that have short-run implications, such as the purchase of additional livestock or the expansion of an enterprise, using current prices adjusted for the best outlook information available is recommended. For long-run decisions, such as the purchase of land or the construction of a building, long-run price relationships should be given more weight. If current price levels indicate changes in the relationship of supply and demand in the long run, then some adjustment in long-run averages is indicated.

Of course, the manager is not limited to using only one set of price relationships. It usually takes very little additional work to revise the price assumptions and to rework the budget. By comparing the outcomes of these different assumptions, the manager is better equipped to make the decision.

Trends in the price level also may be important in budgeting long-run decisions. If the purchase of land is being considered, the general level of prices is obviously crucial. The consequences of inflation, deflation, and price stability can all be budgeted. The results must then be interpreted on the basis of what the manager believes to be the most probable outcome. Historically, the trend has been a rising price level and inflation. It must also be recognized that this trend has been broken by economic recessions and depressions of varying length and intensity.

As far as yields are concerned, it is usually possible to get fairly accurate estimates of average yields. Historical farm records are one source. The records of a neighboring farm may be useful. County average yields can often be adjusted and used. Moreover, cooperative trials by a university experiment station may provide yield data for a crop produced by the use of some new farm practice.

Farmers may not be satisfied with the average yield in the area and may believe they can exceed it. When preparing budgets, they should consider yields in relation to the practices they plan to use. Practices designed to raise yields above the average may come at a cost. If so, these costs should be included in the budget. If significantly higher than average yields are projected, the manager should have a clear picture of what will be done to get better results than those achieved by others.

In some parts of the country, yields vary greatly from one year to the next. Average yields may not be appropriate to use in budgeting certain decisions. This problem involves the subject of risk, which is examined in Chapter 8. A good question for the manager to ask is, how serious would it be if the yield or price information were in error? If the result is serious and if the manager lacks confidence in the information, the assumptions should be varied and the budget reworked. This task can be done easily by using a computer program.

Monitoring and Controlling the Plan

Once the plan is set into motion, events can occur that will draw the plan off target. A system for monitoring and controlling the plan is required to adjust the plan before the deviations become serious. This system consists of three components: the plan itself, a device for monitoring or comparing actual with planned results, and the control or corrective actions.

The cash flow budget is frequently used to summarize the important features of the plan. It integrates the various enterprise activities, as well as the nonfarm family activities.

The device used for monitoring the plan is called the *cash flow monitoring worksheet*. It compares the actual and budgeted cash flows to check the implementation of the manager's plan for the farm business. By identifying deviations from the plan, the worksheet provides early warning of potential problems.

Once the problems have been identified, they need to be corrected. Deviations can be controlled by adjusting the implementation of the plan or by changing the plan. In other cases, the event causing the deviation may be unavoidable and outside the manager's control. If this unavoidable event occurs, the manager should have a contingency plan to follow.

The Monitoring Worksheet

Monitoring is a continuing process of comparing the actual results to those planned. Monthly estimates of cash inflows and outflows are compared with actual results as soon as possible after the end of each month. This is done for the month just ended and also for the year-to-date totals.

The "Budgeted" column of the worksheet (Table 6-11) refers to the estimates from the cash flow budget. At the beginning of each month, the actual results of the preceding month are entered on the worksheet in the "Actual" column. After the monthly columns are completed, the year-to-date totals are calculated. The amounts in the "Budgeted Year-to-Date" column are determined by adding the

TABLE 6-11. The Cash Flow Monitoring Worksheet

	Month _____		Year-to-Date	
	Budgeted	Actual	Budgeted	Actual
Beginning Balance				
Operating receipts Crop sales Other				
Capital sales Machinery Breeding animals				
Nonfarm income				
CASH AVAILABLE				
Operating expenses Fertilizer Feed Other				
Capital purchases Machinery Breeding animals				
Nonfarm expenses Family living Income taxes				
CASH REQUIRED				
AVAILABLE MINUS REQUIRED				
Money borrowed Loan payments				
Ending Balance				

current month's amounts to the budgeted year-to-date amounts from the monitoring worksheet of the preceding month. Next, the actual year-to-date amounts from the previous month's worksheet are added to the current month's actual results to arrive at the cumulative totals in the "Actual Year-to-Date" column.

Using the Worksheet to Control

An examination of each cash flow item helps to determine the cause of any difference. If the difference is significant, the manager must decide whether an adjustment must be made in the implementation of the plan and/or whether the plan and the projected cash flows for the remainder of the year should be revised. For example, a change in expected prices either paid or received may lead the manager to change the original projections.

Of course, this monitoring process is much easier with a computerized accounting system. Monthly cash flow projections are stored for easy comparison with actual results after each month. In many nonfarm businesses, the ability to compare cash flow projections with actual outcomes is considered essential. As producers acquire computers to use in the farm business, accounting and other numerical operations should become easier.

The advantage of this monitoring and control process is that orderly borrowing and repayment adjustments can be negotiated as the business proceeds through the year, thereby avoiding crises. Because of the variability and uncertainty in agriculture, projected cash flows should be compared with actual receipts and expenses regularly throughout the year in order to monitor and control the implementation of farm plans. In essence, a system of monitoring and control means having a plan, timely information on actual performance, and a worksheet for identifying deviations. This procedure allows exploitation of the favorable deviations from the plan and correction, where possible, of the unfavorable deviations to bring the plan back on target.

Summary

This chapter has presented three different types of budgets—partial, total farm, and cash flow—that are useful in planning and controlling the future of the farm business. The budgeting techniques may seem to be quite formal and complex. Of course, the total farm budget is more elaborate and formal than the partial budget. Even so, the type of planning suggested by these budgets is more detailed than that used by many farmers. Some farmers do their budgeting mentally while performing other duties. Often they make rough partial budgets on paper before making a decision. All successful managers are adept at weighing alternatives. The budget is merely a method for testing alternatives by using arithmetic and an organized system. Farm businesses are becoming more complicated, and it is difficult to take into account all the ramifications of a decision without writing them down.

The most useful results can be obtained from budgets if a few simple rules are kept in mind. First, budget only those alternatives that are pertinent to the problem at hand. Second, establish clearly what the analysis is to show before the budgeting is begun. When the budgets are put into practice, it is important to understand clearly what was, and was not, changed during the budgeting process. Otherwise the budget cannot be expected to be a satisfactory predictor of future developments.

The third important consideration is the amount of information to gather. Although additional information may add to the accuracy of the budgets, it usually comes at a cost. The marginal principle can be used subjectively to decide when sufficient information has been collected. When the added accuracy just equals the added cost of data collection, adequate information has been obtained. Although complete information is never available, decisions must be made, and budgeting will usually yield better results than intuition.

Finally, once the budget has been completed and the plan decided, it should not be put aside and forgotten. The budget provides the basis for monitoring and controlling the plan. By comparing the actual outcomes with the budget, managers can learn from experience and improve their budgeting skills.

Recommended Readings

NELSON, A. G. and T. L. FREY. *You and Your Cash Flow*. Skokie, Ill.: Century Communications, Inc., 1983.

SCHMISSEUR, E., and D. LANDIS. *Spreadsheet Software for Farm Business Management*. Reston, Virginia: Reston Publishing Co., 1985.

SONKA, S. T. *Computers in Farming: Selection and Use*. New York: McGraw-Hill Book Co., 1983. Chapter 9.

THOMAS, K. H., R. O. HAWKINS, R. A. LUENING, and R. N. WEIGLE. *Managing Your Farm Financial Future*. University of Minnesota North Central Regional Extension Publication 34, no date.

CHAPTER 7

Investment Analysis Techniques

In the preceding chapter, various budgeting techniques, including partial budgets, total farm budgets, and cash flow budgets, were presented as methods for evaluating decision alternatives. These techniques work well for many of the decisions faced by farm managers, but major investment decisions should receive a more comprehensive analysis. That is, the analysis of investment decisions should allow for the explicit consideration of the time value of money, inflation, and income taxes. This chapter introduces these methods of analysis and illustrates their application to agricultural investment decisions.

Investment is defined as an addition of intermediate and fixed assets to a business. Typically, an agricultural investment falls into one of four major categories:

1. Replacement of obsolete or worn capital items.
2. Adoption of mechanized technology to reduce costs and increase profits.
3. Expansion of existing enterprises.
4. Addition of new enterprises.

Capital investment decisions are particularly important to farmers because the effects are long lasting, as opposed to the effects of operating decisions such as feeding and fertilization, which last for only a short period of time. For example, if a dairy farmer finds that a ration chosen earlier is not appropriate, it can be changed. However, if the building constructed last year is not working out, it cannot be changed easily or inexpensively. Capital investment decisions are usually costly to reverse.

Investment decisions are also important because of the dollar amounts involved: Tractors, equipment, buildings, and land are expensive. These decisions are further complicated by the fact that large investment expenditures made now generate returns over a long period of time, perhaps as many as 10 or even 20 years. It is difficult to predict what the returns to the investment will be that far into the future. Thus, the importance of these decisions justifies the use of a comprehensive method for analyzing their soundness. However, to paraphrase one of Parkinson's laws, the time spent on a decision is often in inverse proportion to the sum involved.[1] The reason for this inverse relationship is that managers often lack the tools to evaluate these more complex and significant decisions. The purpose of this chapter is to provide these tools so that the thoroughness of the analysis will be proportional to the importance of the decision.

The Time Value of Money

The methods for analyzing capital investments are based on the concept of the *time value of money*, which means that a dollar received today is valued more highly than a dollar received tomorrow or any time in the future. For example, in order to decide whether to buy a new tractor that will last for 10 years or a used tractor that will last for 5 years and will then be traded for another, a farmer must put a value on money to be spent and received in the future. Another example of a decision dependent on the time value of money is choosing to plant either an orchard that will not produce a crop for several years or crops that will be harvested annually.

Why is a dollar received today worth more than a dollar received a year from now? If you lend someone $100 for one year, would you expect to receive just $100 at the end of the year or would you expect to receive more? You would probably expect to receive the $100 plus another $10. The extra $10 is compensation for the fact that $100 received a year from now is not worth as much as $100 today. There are three reasons for the fact that a future payment is worth less than the same payment received today:

1. *Opportunity cost.* As a result of lending your money, you have forgone the possibility of earning returns from other investments. These other returns represent the opportunity cost, the dollar amount you could have earned if you had invested elsewhere.
2. *Risk.* Because of misfortune or dishonesty, your borrower may not pay you back. The extra charge represents a reward for taking this risk.
3. *Inflation.* Assuming that there is price inflation in the economy during the lending period, the dollars that you get back will not buy as much as those you loaned. This decrease in purchasing power also justifies an additional charge.

Because of these three factors, a dollar today is not equal to a dollar tomorrow, a dollar 1 year from now, or a dollar 10 years from now. So, comparing an investment that will yield $100 in a year to another that will yield $125 in three years is like comparing apples to oranges. To make these investment returns directly comparable, a common denominator is needed. Using the concept of the time value of money, the dollars from the two investments can be adjusted to their respective values at the same point in time.

There are two types of adjustments that can be made to determine the value of money at a given point in time: compounding and discounting. *Compounding* is

[1] C. N. Parkinson, *Parkinson's Law*, New York: Ballantine Books, 1957, p. 40.

used to find the value of the money at some future date when its present value is known. *Discounting* is used to find the value of the money today when its value at some future date is known.

Compounding and Interest

The time value of money is the major reason that banks and other lenders charge interest. Interest can be seen as the difference between the values of present and future dollars. Compounding is used to find the future value of a dollar amount when its present value is known. Compounding implies growth. If a sum of money is deposited in a savings account, it will grow until withdrawn. Suppose $1000 were invested in a savings account at 8 percent interest compounded annually. What would the value of this $1000 be after three years? The interest payments and balances for each year would be as follows:

Year	Balance at Beginning of Year	Annual Interest Earned (8%)	Balance at End of Year
1	$1000.00	$80.00	$1080.00
2	1080.00	86.40	1166.40
3	1166.40	93.31	1259.71

The interest earned in the first year is 8 percent of the $1000 ($80). The total value at the end of the first year is $1080 ($1000 + $80). In the second year, however, interest is also earned on the first year's interest; it compounds. The 8 percent multiplied by $1080 equals $86.40, which is added to the $1080 to give a value at the end of the second year of $1166.40. This compounding process continues in the third year to give an ending value of $1259.71.

As can be seen, the process of compounding to find future values is very tedious if many time periods are involved. The following formula can be used to simplify the process:

$$FV = PV(1 + i)^n$$

Here FV is the future value, or the balance in the savings account at the end of n periods; PV is the present value, or the amount originally deposited in the savings account; i is the interest rate per period; and n is the number of periods (years, months, or days). For this example:

$$FV = \$1000 (1 + 0.08)^3$$
$$= \$1000 (1.2597)$$
$$= \$1259.70$$

The future value of a dollar depends on the length of time to that future date and the rate of interest that can be earned. These are the key variables in the compounding formula for determining future value. The formula just presented can be easily executed on a pocket calculator, or the appropriate factor can be found in a table like Appendix Table 1. This table gives the future value of $1 at different rates of interest and for different numbers of periods. For example, the factor for three periods at 8 percent interest is 1.2597. Therefore:

$$FV = \$1000 \times 1.2597$$
$$= \$1259.70$$

As a simplified example, suppose a farmer is deciding whether to continue growing the annual crop now being grown or to grow a new perennial crop. The current crop returns $160 per acre per year. The projected returns for the new perennial crop and the current crop are as follows:

	Net Return/Acre	
Year	New Crop	Current Crop
1	$ – 100	$160
2	75	160
3	150	160
4	400	160
5	300	160
	$ 825	$800

The new crop has a life of five years, at the end of which the farmer could either shift back to the current crop or replant the new one. Although the returns on the new crop totaled over the five years are higher, they are distributed unevenly over the years. In fact, because of the costs of establishing the perennial crop and its low first-year yield, the net return is negative the first year.

To make a valid comparison, the time value of the returns must be evaluated. Suppose the farmer can invest the dollar value of the returns, earning 10 percent interest. To calculate the future value of the returns on each crop alternative, they are compounded to the end of the five-year period. One way to do this is to make the calculations illustrated in Table 7-1; the interest is accumulated as if the returns were deposited in a savings account. A simpler method is to multiply the factors from Appendix Table 1 by the net returns for each year and sum the products (see Table 7-2). Considering the time value of money and assuming a 10 percent interest rate, the current crop returns the higher future value over the five years and, therefore, is the more profitable crop. Regardless of the method used (Table 7-1 or Table 7-2), the future value of the current crop is $977, compared to $875 for the new crop.

Discounting

Discounting is the mathematical procedure for finding the present value of an amount of money when its future value is known; discounting is the inverse of compounding. Referring back to the savings account example, find the present value of the $1,259.70 to be received at the end of three years, assuming 8 percent

TABLE 7-1. Calculation of Future Values of Two Crop Alternatives
by Accumulating Interest Earned

Crop Alternative and Year	Value at Beginning of Year	Interest Earned (10%)	Net Return	Value at End of Year
New Crop				
Year 1	$ 0.00	$ 0.00	$ − 100.00	$ − 100.00
2	− 100.00	− 10.00	75.00	− 35.00
3	− 35.00	− 3.50	150.00	111.50
4	111.50	11.15	400.00	522.65
5	522.65	52.27	300.00	874.92
Current Crop				
Year 1	$ 0.00	$ 0.00	$ 160.00	$ 160.00
2	160.00	16.00	160.00	336.00
3	336.00	33.60	160.00	529.60
4	529.60	52.96	160.00	742.56
5	742.56	74.26	160.00	976.82

TABLE 7-2. Calculation of Future Values of Two Crop Alternatives
Using Compound Factors

Crop Alternative and Year	Net Return		Compound Factor (10%)		Future Value (After 5 Years)
New Crop					
Year 1	$ − 100.00	×	1.4641[a]	=	$ − 146.41
2	75.00	×	1.3310[b]	=	99.83
3	150.00	×	1.2100	=	181.50
4	400.00	×	1.1000	=	440.00
5	300.00	×	1.0000	=	300.00
Total					$ 874.92
Current Crop					
Year 1	$ 160.00	×	1.4641	=	$ 234.26
2	160.00	×	1.3310	=	212.96
3	160.00	×	1.2100	=	193.60
4	160.00	×	1.1000	=	176.00
5	160.00	×	1.0000	=	160.00
Total					$ 976.82

[a] This is the factor for four years and 10 percent from Appendix Table 1. The amount in the first year compounds for four years (five minus one).

[b] This is the factor for three years and 10 percent from Appendix Table 1. The amount in the second year compounds for three years (five minus two).

interest compounded annually. The present value is $1000; it is the quantity of money that would have to be deposited or invested today at 8 percent compound interest to realize $1259.70 after three years.

Now suppose a manager is interested in the present value of a crop of Christmas trees that will be ready for market in eight years at a value of $10,000. The current interest rate is 10 percent. To find the present value of money to be received or paid at some future time, the following formula is used:

$$PV = FV/(1 + i)^n$$

The definitions of these terms (PV, FV, i, and n) are the same as for compounding, except that here the manager is solving for the present value, PV. For this example, the calculations are as follows:

$$PV = \$10,000/(1 + 0.10)^8$$
$$= \$10,000/(2.1436)$$
$$= \$\ 4,665$$

If it will cost no more than $4665 to plant the Christmas trees today and there will be no other costs later, then this investment alternative will be profitable. It may not be the *most* profitable alternative, but it is profitable in providing at least a 10 percent rate of return.

Again, factors are available to simplify the calculations (see Appendix Table 2). The factor for eight years and 10 percent interest is 0.4665. Multiplying this factor by $10,000 gives the same result as calculated previously:

$$PV = \$10,000 \times 0.4665$$
$$= \$\ 4,665$$

Note that this factor is the inverse of the compounding factor for eight years and 10 percent as found in Appendix Table 1.

Now, referring back to the new crop/current crop decision and Table 7-2, instead of comparing these two alternatives in terms of their future values, their present values will be compared. The calculations are presented in Table 7-3 using the factors from Appendix Table 2. First, note that the present values for each of the five years' returns for each crop are smaller than their corresponding future values. This is true because a dollar amount five years from now has to be larger to be equivalent to a dollar amount today. Also, note that the total present value of the net returns for the current crop is larger than that of the new crop. Compounding to a given future date and discounting to the present provide the same results when comparing and ranking investment alternatives. Compounding may be easier to understand, but discounting, by reducing future dollar amounts to present-day values, makes the magnitude of the differences between investment alternatives easier to judge.

TABLE 7-3. Calculation of Present Values of Two Crop Alternatives Using Discount Factors

Crop Alternative and Year	Net Return		Discount Factor (10%)		Present Value
New Crop					
Year 1	$ − 100.00	×	0.9091	=	$ − 90.91
2	75.00	×	0.8264	=	61.98
3	150.00	×	0.7513	=	112.70
4	400.00	×	0.6830	=	273.20
5	300.00	×	0.6209	=	186.27
Total					$ 543.24
Current Crop[a]					
Year 1	$ 160.00	×	0.9091	=	$ 145.46
2	160.00	×	0.8264	=	132.22
3	160.00	×	0.7513	=	120.21
4	160.00	×	0.6830	=	109.28
5	160.00	×	0.6209	=	99.34
Total					$ 606.52

[a] A shortcut to calculate the present value of this uniform $160 payment for five years would be to use the factor for five years and 10 percent from Appendix Table 3. PV = 160 (3.7908) = 606.52.

Appendix Table 3 is used to find the present value of a uniform series of payments. For example, suppose a manager is considering the purchase of a $50,000 machine that will save $9000 per year during its expected 10-year life. If the interest rate is 12 percent, is this a good investment? First, look up the discount factor for 10 years and 12 percent interest in Appendix Table 3. It is 5.6502. Multiplying this factor by the $9,000 saved per year gives the present value of the cost savings:

$$PV = \$\ 9,000 \times 5.6502$$
$$= \$50,852$$

The present value of the cost savings from the machine is greater than the original cost of the machine, so the investment is profitable.

Farm managers make many investments involving large initial expenditures, but the returns from those investments are only realized over a number of years in the future. Examples include machinery purchases, building construction, and orchard establishment. To evaluate such investments properly, future returns and costs must be discounted back to their present values so that all values are considered on an equivalent basis. If the present value of the net returns is positive, then the investment is profitable. This method of analyzing investments is called *net present value* analysis.

Net Present Value Analysis

Although other methods can be used for analyzing capital investment decisions, the approach recommended and used here is net present value analysis. The advantages of the net present value method are as follows:

1. This measure properly accounts for the time value of money.
2. The net present value method is accepted by most financial experts as the superior approach.
3. This method will not result in misleading or multiple answers, which can happen with other methods in certain situations.
4. Compared to other approaches, the net present value is relatively easy to compute.
5. Explicit consideration can be given to the effects of income taxes and inflation on the profitability of the investment.

Following are the steps involved in applying the net present value method for analyzing capital investments.

Steps in Investment Analysis

1. *Identify the investment to be analyzed* (tractor, combine, truck, dairy cows, hog facility, land, and so forth) after making a thorough search for all investment possibilities. It is important not to limit the analysis to only one possibility; consider other alternatives that may be better investments.
2. *Determine the initial cash outlay required* to make the investment. This is the total amount of cash required, whether borrowed or not.
3. *Estimate the annual net cash flows that result from the investment for each year* of the planning period. These annual cash flows include only those changes in receipts or expenses resulting directly from the investment. Do not include those that would occur regardless of the investment. Include the salvage value of the investment, if any, as a positive cash flow in the last year.
4. *Choose the discount rate.* This discount, or interest, rate is the investor's minimum acceptable rate of return, and should reflect the interest paid on the debt and the opportunity cost of the equity invested. The opportunity cost is that rate of return the investor would expect to earn if the money were invested in the best alternative.
5. *Calculate the net present values* of the cash outlay to purchase the investment and the annual net cash flows resulting from the investment by multiplying the amount for each year by the appropriate discount factor and then summing the discounted amounts. The result is the net present value of the investment.
6. *Decide whether to accept the investment.* If the net present value is positive, then the returns to the proposed investment will more than cover the cost of the capital; the investment is profitable. If the net present value is negative, it

will return less than the investor's cost of capital. Therefore, the investment should not be made.

Projecting Cash Flows

Cash flow, sources of data, and budgeting concepts are discussed in the previous chapter. This section deals with those special aspects that are unique to the net present value analysis of capital investments. To analyze investments using the net present value method, the costs and benefits of the investment are measured as the changes in net cash flows for each year in the planning period.

Before estimating net cash flows, the length of the planning period for the investment must be determined. The length of the planning period should be based on the useful life of the most durable asset of the investment to be considered. For example, a grain handling facility might last for 25 years or more, but it may become obsolete or require major remodeling sooner; so 10 to 15 years would be a more reasonable planning period. The timing of the returns associated with the investment also affect the planning period; for example, five to seven years might be reasonable for a tractor but much too short for a new orchard, which would be barely in production at the end of such a short period. The salvage or terminal values should be reasonable and consistent with the length of the planning period used.

Determining the appropriate cash flows and the method for projecting or estimating them is considered here. The cash flows should be estimated as a partial analysis of the investment alternative, which means that only the changes in cash receipts and cash expenses resulting from the investment are considered. This is the same as the partial budgeting approach used in the preceding chapter, but in this case only cash flows are considered. The important point is that all changes in receipts and expenses that are likely to result from the investment should be included. Changes in either receipts or expenses that will occur regardless of this investment should not be included. Separate budgets should be prepared for each year, or perhaps for groups of years, in which different cash flows are expected. Projecting the cash flows is probably the hardest and most time-consuming part of investment analysis.

Cash Outflows. The most obvious cash outflow is that required to make the investment. This is the total of the investor's cash and the money borrowed. In most situations, this cash outflow is simply the purchase price or construction cost, but in others it may be more complicated. Take the example of a hog farmer considering a new farrowing house with a useful life of 15 years. It is likely that some of the equipment will need to be replaced before the end of the planning period. If so, the cash required for those replacements should be entered as outflows in the year(s) that they will be made.

Cash Inflows. The cash inflows are the benefits expected from the proposed investment. Most investments result, however, in a combination of outflows and inflows; that is, annual receipts and expenses will both be affected by the investment. Thus, these cash inflows are actually net changes resulting from the investment. In estimating these net cash flows, it is important to remember that they can be affected by four types of changes: (1) added receipts, (2) reduced receipts, (3) added expenses, and (4) reduced expenses.

Another cash inflow included at the end of the planning period is the expected salvage value or terminal value of the investment, whether it is to be sold or not. Income taxes (discussed later) should also be considered in estimating the net cash inflows.

Depreciation and Interest.   Depreciation and interest are not included in net cash flow projections for important reasons. Assume that an investment proposal involves a depreciable asset. Because the entire cash flow was included to purchase this asset at the beginning of the planning period, including the depreciation also in the annual net cash flows would be double counting. Depreciation, a legitimate accounting expense for computing net farm income, is not a cash flow item and is not included for purposes of investment analysis.

Now suppose that borrowed money financed this investment. The interest on this loan would be a cash expense, but it is not included. In net present value analysis, the interest charge is considered in the discount rate (as discussed in the next section). If interest paid were also included in the cash flow, interest would be charged against the investment twice.

Choosing the Discount Rate

The selection of the discount rate is a key step in net present value analysis. This is the interest rate used in the discounting process discussed earlier. A high discount rate lessens the present value of future receipts and expenses; a lower discount rate does not discount, or shrink, them as much.

In choosing the appropriate discount rate, it is first important to recognize that the investment is likely to be financed by both debt and equity capital. Thus, in arriving at the appropriate discount rate, both the debt and equity cost components should be considered. Debt and equity funds are usually intertwined in most businesses, but the relative amount of each used to finance investments is indicated by the debt to asset ratio calculated from the balance sheet. This ratio is used to calculate the weighted average cost of additional capital. This cost for additional capital is the investor's minimum acceptable rate of return and is used to discount the net cash flows.

Cost of Debt.   Placing a cost on debt capital is somewhat easier than placing a cost on equity capital. The former involves finding the rate of interest currently being charged on new loan obligations of the business, including any new loans associated with the investment being considered. Discovering the actual interest rate on loans can be complicated by service charges, minimum deposit requirements, and the like. (These details are discussed in Chapter 11.)

Cost of Equity.   The cost of equity capital represents an opportunity cost, the rate of return the equity capital for the investment being considered could earn if invested at comparable risk elsewhere—either another investment in the farm business or outside the business. Note that this is a marginal concept: For the cost of debt, it is the interest on new or additional loans that is relevant; for the cost of equity, it is the opportunity cost of the equity available for this investment. It is not the opportunity cost for the total equity.

This marginal cost of equity should usually be viewed as higher than the cost of debt capital, because the owner of a business takes more risks than its lenders. If a business becomes bankrupt, the lenders take their money first and the owner gets what is left. The other argument for the cost of equity to be higher than the cost of debt is that it is not rational to borrow money to invest in the business if the rate of return to equity capital is less than the cost of the borrowed money.

Although the guideline is to use an opportunity cost for equity capital that exceeds the cost of debt, it should not be set too high. By using an unrealistically high rate of return on equity, some investments that are indeed profitable and in the best interests of the business might be overlooked.

Computing the Discount Rate.   The discount rate (the minimum acceptable rate of return) is the weighted average of the marginal costs of debt and equity capital. The use of the foregoing information to calculate the discount rate is illustrated by an example: Suppose Ivan Investor is considering investing in an adjacent acreage of land. He will finance this investment with a new mortgage that has an interest rate of 12 percent and with equity capital at an opportunity cost of 15 percent (Table 7-4). After the investment, Ivan Investor's total outstanding debt, including an existing mortgage and a loan for machinery, will be $350,000. Subtracting the debt from the total value of the assets gives an equity, or net worth, of $150,000. Dividing each amount by the total value of assets gives the weights for debt and equity capital, as in column 2 of Table 7-4, that reflect the financing proportions that Ivan intends to use over the planning period of this investment. In other words, the weights reflect his long-term goal for the debt to asset ratio of his business.

In column 3 of the table, the cost (interest) for new debt and the opportunity cost for equity capital are recorded. Multiplying the weights by the costs in column 3 gives the products in column 4. These products are summed to calculate the weighted average cost of capital that is to be used as the discount rate. In this case, the discount rate is expressed before taxes. The following section deals with the way tax considerations are incorporated into investment analysis decisions.

TABLE 7-4. Calculation of the Discount Rate Based on the
Weighted Average Cost of Capital

Source of Capital	(1) Amount ($)	(2) Weight	(3) Interest or Cost (%)	(4) Weight × Cost (%)
Debt	350,000	0.70	12	8.4
Equity	150,000	0.30	15	4.5
Total	500,000	1.00		
			Discount rate =	12.9

Income Tax Considerations

One of the advantages of using net present value analysis of investment decisions is that income tax effects can be readily considered. This after-tax net present value analysis is consistent with the manager's primary objective: what will be available after taxes to reinvest or withdraw for family living expenses.

Tax considerations are also important because different assets may have different income tax implications. For example, some machinery and livestock facility investments are eligible for investment tax credits and depreciation deductions. As a result, tax considerations can affect the profitability of alternative investments as well as their relative desirability.

These income tax considerations affect the net present value analysis in two ways: (1) They affect the discount rate used to calculate the net present value; the weighted average cost of additional debt and equity capital should therefore be expressed on an after-tax basis. (2) Taxes also affect cash flows and therefore should be taken into account when making cash flow projections. If the investment increases the farm's income, for example, then income taxes must be paid on that additional income.

There are several income tax–related features that affect the net present value of investments: (1) the marginal tax rate, (2) the after-tax discount rate, (3) ordinary income, (4) capital gains income, (5) depreciation, (6) investment credit, and (7) taxes on terminal values.

The intent here is not to give all the details for adjusting cash flows to an after-tax basis. Some additional discussion of income tax principles is presented in a later chapter. Because regulations are subject to change, advisers familiar with these regulations should be consulted, or current Internal Revenue Service publications such as the *Farmer's Tax Guide* should be used.

Finding the Marginal Tax Rate.   The marginal tax rate is needed to adjust the discount rate to an after-tax basis and to calculate the after-tax cash flow. The easiest approach to determine this marginal tax rate is to refer to the most recent income tax return. The first step is to find the amount of taxable income, usually identified as a specific line item. If income was higher or lower than normal that year, the taxable income should be adjusted to a normal expected level.

The next step is to find the appropriate tax rate schedule based on filing status—single, married filing jointly, married filing separately, head of household, or corporation. Using the appropriate schedule and level of taxable income (assuming it now reflects a typical year), find the tax bracket for that taxable income. The percentage figure that is used to calculate the taxes on income falling into this bracket is the marginal tax rate.

Many farmers are also subject to social security self-employment taxes. If the taxable income is below the maximum subject to this tax, add the social security tax rate to the marginal income tax rate. State income taxes must also be considered. Finding the marginal tax rate for state income taxes involves the same general approach used for federal income taxes. This state tax rate is added to the others to arrive at the total marginal tax rate.

Calculating the After-Tax Discount Rate.   The discount rate has to be expressed consistently with the annual net cash flows. If the annual net cash flows are calculated on an after-tax basis, then the discount rate must be as well. Assuming that both debt and equity sources of capital are treated in the same way for tax purposes, the after-tax discount rate can be calculated from the weighted average cost of additional capital. The after-tax discount rate is less than the before-tax discount rate because (1) the interest paid on debt is a tax-deductible expense, and (2) the opportunity cost on equity capital represents income subject to taxation.

Using the example of Ivan Investor with a before-tax discount rate of 12.9 percent (Table 7-4) and a marginal tax rate of 37 percent, the after tax discount rate is calculated as follows:

$$\text{After-tax discount rate} = 12.9\,(1-0.37)$$
$$= 12.9\,(0.63)$$
$$= 8.1\%$$

The 12.9 percent before-tax discount rate becomes an after-tax discount rate of 8.1 percent.

Adjusting Cash Flows for Ordinary Income Taxes.   Taxes are a cash outflow and must be subtracted when determining net cash flows. In most cases, the annual before-tax net cash flow can be adjusted to an after-tax cash flow by subtracting the added income tax to be paid from the before-tax cash flow. For example, if the net cash flow from an investment is $1000 and the investor is in the 37 percent income tax bracket, the additional tax is $370 and the after-tax cash flow is $630. This approach works well when changes in before-tax cash flows fall within one bracket. Larger changes in cash flow that affect the marginal tax rate require income taxes to be calculated with and without the investment to arrive at the appropriate change in after-tax cash flow.

Adjusting Cash Flows for Capital Gains Taxes.   Capital gains income can result from the sale of assets during or at the end of the planning period. An example of capital gains income during the planning period is the sale of livestock raised and held for dairy or breeding purposes. An example of capital gains income at the end of the planning period is land that has appreciated in value. Capital gains are taxed at a rate equal to 40 percent of the marginal tax rate that applies to ordinary income. This tax law tends to favor investments in which the receipts qualify as capital gains, with everything else being equal.

Adjusting for Depreciation.   An earlier section explained why depreciation should not be included in annual net cash flows. However, depreciation is considered in calculating after-tax cash flows. Depreciation is a tax-deductible expense, so it indirectly affects the after-tax cash flow. Various methods for calculating depreciation are discussed in Chapter 3. The important point, however, is that because of the time value of money, the present value of the tax savings associated with these depreciation options can be significant.

Adjusting for Investment Credits.   An investment tax credit of up to 10 percent of many farm investments can be claimed on federal income taxes. Some states also allow investment credits, and sometimes credits are available for special investments. These credits, which can be claimed in the year the asset is purchased, directly reduce the investor's tax liability. Therefore, the amount of the credits is an after-tax cash flow added at the end of the first year of the planning period. However, there are two important points to keep in mind: Only certain types of assets are eligible for these credits, and the amount of the credit usually depends on the asset's life.

Adjusting for Taxes on Terminal Values.   If assets have a salvage or terminal value at the end of the planning period, the tax implications of this value should be considered. This terminal value is to be considered in the net present value analysis whether the asset is sold or not. Even if it is not sold, taxes should be estimated as if it had been. When the rate of inflation (or deflation) is high, these salvage values have greater importance when analyzing investments.

The first step is to estimate the amount of taxable income involved, usually the difference between the market value and the book value of the asset. In the case of raised livestock and crops, the book value may be zero. The second step is to determine whether this income is subject to taxation at ordinary rates or at capital gains rates. The Internal Revenue Service complicates this issue further by requiring that depreciation be recaptured for certain types of assets. This means that part of this taxable income might be taxed at one rate and the remainder at another rate. It may also be necessary to subtract investment credit recapture if the asset is sold or traded too soon.

Accounting for Inflation (or Deflation)

Inflation is one of the factors contributing to the fluctuation of the value of money over time. Because of inflation, a dollar received in the future will buy less than it does today. Some special precautions must be employed in investment analysis to ensure that inflation's effects have been taken into account properly.[2]

Inflation affects the analysis of investment decisions in two ways: (1) The interest rate and the opportunity cost used in calculating the discount rate include an inflation component. In setting interest rates, lenders consider the potential loss in purchasing power due to inflation. As a result, interest rates tend to increase with inflation. Also, when the investor considers the opportunity cost of equity capital, the marketplace has already factored inflation into the returns of the various investment opportunities. (2) Future annual cash flows are affected by inflation in the prices of both receipt and expense items.

Either of two approaches can be used to account properly for the impact of inflation on investment analysis:

1. Use a discount rate that includes the inflation premium and cash flows that reflect the expected future rates of inflation in the prices of receipt and expense items.

[2] This section is based on unpublished material by George L. Casler, Professor of Agricultural Economics, Cornell University, April 1979.

2. Use a discount rate that does not include the inflation premium. The real cash flows are calculated and expressed in dollars of constant purchasing power.

Both of these approaches are correct. The key point is that the handling of inflation in the discount rate and in the expected cash flows must be consistent. The first approach is recommended because the discount rate, as it is calculated, includes the inflation premium, and it is easier to think in terms of inflated, or actual, cash flows than in terms of real cash flows. Past price trends for commodities and production inputs provide a guide for estimating the future effects of inflation on the annual net cash flows.

Haying Equipment Investment Example

This example illustrates the projection of cash flows, the estimation of after-tax cash flows, the handling of inflation, and discounting to find the net present value.[3] Although more complicated examples can be found, this one illustrates the major points.

Alfred "Alf" Alpha is a hay grower in Wyoming. He has 390 acres under irrigation. The average yield is 4 tons per acre, for a total of 1560 tons annually. All the hay is sold from stacks at the edge of the field. Alf swaths the hay with his own equipment and hires custom operators to bale and stack it at the edge of the fields. The total cost for this custom operation is currently $18,720.

Because of the increasing costs expected for these custom services, Alf is considering investing in a large baler with a bale accumulator and a front-end loader and doing the work himself with hired labor rather than with custom operators.

Alf has tractors, a truck, and other equipment not being used to capacity that can be used in conjunction with his new haying equipment. The total investment required to purchase the large baler with an accumulator and the front-end loader is $47,300.

The first step in analyzing this investment is to project the annual net cash flows for Alf's five-year planning period. Table 7-5 summarizes the results on a before-tax basis. As a result of purchasing the equipment, the custom harvest expense would not be incurred; thus, it is shown as a reduced expense. In any event, it is expected that this expense would continue to increase at a rate of about 6 percent per year.

The added expenses associated with the investment include the repairs and lubrication of the new equipment and of the tractors, truck, and other equipment to be used in the haying operation. Baler twine would be needed for the new baler, and diesel fuel and gasoline would be required for the tractors and truck. There would also be insurance and property tax to pay on the new equipment and hired labor to operate the baler and to stack bales. All of these added expenses are expected to inflate over time, except taxes and insurance, which decrease as the

---

[3] The data for this example are adopted from D. E. Agee, *Analyzing Alternative Haying Systems, Big Baler versus Custom Baling and Stacking*, University of Wyoming Agricultural Extension Service AE 80-20, August 1980.

TABLE 7-5. Projected Change in Annual Before-Tax Net Cash
Flows After the Purchase of Haying Equipment

Item	Year				
	1	2	3	4	5
Reduced Expense					
Custom harvesting	$19,843	$21,034	$22,296	$23,634	$25,052
Added Expense					
Repairs and lubrication	1,257[a]	2,032	2,154	2,284	2,421
Baler twine	1,030	1,092	1,157	1,227	1,300
Diesel and gas	1,701	1,803	1,911	2,026	2,147
Insurance and property tax	671	498	412	326	236
Hired labor	2,300	2,438	2,584	2,739	2,904
	$ 6,959	$ 7,863	$ 8,218	$ 8,602	$ 9,008
Change in before-tax cash flows	$12,884	$13,171	$14,078	$15,032	$16,044

[a] Baler and accumulator under warranty in the first year.

value of the equipment declines. The bottom line of Table 7-5 is the change in before-tax cash flows resulting from this investment.

The next step is to estimate the effects of this investment on income taxes (Table 7-6). The increases in before-tax cash flows, estimated in Table 7-5, reduce the deductible expenses. Offsetting these reduced deductions is the added deduction for depreciation on the new equipment. Alf's marginal tax rate is 44 percent. Multiplying this rate by the taxable income gives the change in income taxes. The haying equipment will be eligible for an investment credit. Subtracting this credit from the change in income taxes gives the bottom line of Table 7-6, the change in income taxes including the investment credit. Except for the first year, the income taxes will increase as a result of purchasing the haying equipment.

To arrive at Alf's after-tax discount rate, the calculations illustrated in Table 7-7 are used. The weights of 0.45 and 0.55 are used to represent the proportions of debt and equity financing for this investment. Loans to finance new investments include a 14.0 percent interest rate, and Alf sets the opportunity cost for his equity capital at 14.6 percent. Multiplying the weights by the costs and summing the products gives a before-tax weighted average cost of capital of 14.3 percent. To arrive at the after-tax discount rate, Alf multiplies 14.3 by 0.56 (1−0.44), 44 percent being his marginal tax rate. The result is 8.0 percent.

In Table 7-8, the net present value of this investment is calculated. In the second column the purchase cost is represented as a negative cash flow, and the terminal value is indicated at the end of the five-year planning period. In the next column, the changes in before-tax cash flow (from Table 7-5) are entered, and from these, the income taxes from Table 7-6 are subtracted. These differences are then multiplied by the discount factors based on the 8 percent discount rate to arrive at the present values in the last column. The total net present value of the investment is

TABLE 7-6. Projected Change in Income Taxes After the Purchase of
Haying Equipment

Item	Year 1	2	3	4	5
Reduced deductible expenses					
Change in before-tax cash flows	$ 12,884	$13,171	$14,078	$15,032	$16,044
Added deductible expenses					
Depreciation[a]	7,095	10,406	9,933	9,933	9,933
Change in taxable income	$ 5,789	$ 2,765	$ 4,145	$ 5,099	$ 6,111
Multiplied by tax rate	(0.44)	(0.44)	(0.44)	(0.44)	(0.44)
Equals change in taxes before credit	$ 2,547	$ 1,217	$ 1,824	$ 2,244	$ 2,689
Minus investment credit	− 3,784	—	—	—	—
Equals change in taxes after credit	$ − 1,237[b]	$ 1,217	$ 1,824	$ 2,244	$ 2,689

[a] The depreciation is calculated using the accelerated cost recovery system (Chapter 3).
[b] This negative change in taxes means that, as a result of the equipment purchase, taxes will be reduced in the first year, compared to those incurred with custom hiring.

TABLE 7-7. Calculation of After-Tax Discount Rate for Analysis
of Purchase of Haying Equipment

Source	Weight	Interest or Cost	Weight × Cost
Debt	0.45	14.0%	6.3%
Equity	0.55	14.6%	8.0%
		Before-tax discount rate	14.3%
		Multiplied by (1 − marginal tax rate)	× 0.56
		Equals after-tax discount rate	8.0%

$8730. Based on the assumptions in this example, the purchase of the haying equipment would be a profitable investment for Alf Alpha.

As can be seen from this example, the application of net present value analysis can be complicated, involving many calculations. However, the mechanics of this approach can be programmed on a microcomputer to facilitate the analysis.[4] One of the important advantages of such a program is the ease with which various assumptions about inflation in prices, tax rates, and discount rates can be analyzed to see how they affect the analysis.

[4] For an example, see H. R. Hinman, G. S. Willett, and M. E. Wirth, *Analyzing Agricultural Investments: Microcomputer Program*, Washington State University Western Regional Extension Publication 0073, January 1986.

TABLE 7-8. Summary of Annual Net Cash Flows and Calculation of the Net Present
Value of the Purchase of Haying Equipment

Year	Purchase Cost and Salvage Value	Before-Tax Cash Flow	Income Taxes	After-Tax Cash Flow	Discount Factor[a]	Present Value
0	$ – 47,300	$ 0	$ 0	$ 0	$1.0000	$ – 47,300
1		12,884	– 1,237	14,121	0.9259	13,075
2		13,171	1,217	11,954	0.8573	10,248
3		14,078	1,824	12,254	0.7938	9,727
4		15,032	2,244	12,788	0.7350	9,399
5	6,600[b]	16,044	2,689	19,955	0.6806	13,581
					Net present value	$ 8,730

[a] Factors for 8 percent interest from Appendix Table 2.
[b] The terminal value is the expected sale value of the equipment less the income taxes that would be due on
depreciation recapture.

Interpreting the Net Present Value

The following points and cautions are important when interpreting the results of
a net present value analysis of an investment:

1. If the results indicate a positive net present value, then the investment would
be profitable given the weighted average cost of capital. However, this
investment may not be the most profitable alternative. All other investment
possibilities must be considered.
2. The net present values of two investments cannot be compared directly if
they involve different planning periods. An easy way to put the two net
present values on a comparable basis is to compute the annual payments (see
Appendix Table 4), which would be equivalent to these net present values,
using the same number of years for both investments. The interest rate to use
for computing the annual payments is the same as that used for computing
net present values.
3. The size of the initial investment should be considered in relation to the total
availability of funds both owned and borrowed. One small investment with a
high net present value that uses only part of these funds may be less
desirable than one that uses all available capital but has a lower net present
value. The larger investment may allow the investor to utilize capital and
managerial skills more effectively and to obtain a larger total return,
considering all the resources available.
4. Assumptions made regarding prices, yields, and costs should be reconsidered.
It may not be necessary to look at all the possible levels and combinations
of these prices, yields, and costs, but it is important to do enough analyses to
be sure that the investment will be profitable given a reasonable range of
possibilities.

5. Before choosing an investment, there is one more important consideration other than profitability that should be analyzed. This is the financial feasibility of the investment. Will sufficient cash be generated to meet all the obligations, including principal and interest payments, as they become due? Some types of investments require large outlays of cash before any returns are realized. So, before the investment is made, it is important to determine if there are sufficient sources of cash to meet debt service obligations in the interim.

Determining Financial Feasibility

Although an investment may be profitable over the planning period, annual financial consequences can create difficulties for the investor. Annual payments and other cash requirements may exceed cash receipts, particularly in the early years of the planning period. The investor faces the risk that adequate cash will not be available to meet all the obligations as they come due.

When the cash flows were developed for the net present value analysis, only certain items were included. To analyze financial feasibility, all cash inflows and outflows must be considered. In particular, the following requirements for cash should be included in the analysis: down payments, principal and interest payments (less income tax savings due to interest deductions), and withdrawals for family living expenses.[5]

Cash flow feasibility can be evaluated as a partial analysis, or the complete cash flow for the business can be budgeted over the planning period. This complete cash flow analysis is particularly important in analyzing major investments. Whenever an investor must rely on other sources within the business for cash in order to meet the cash requirements of the investment, the entire business should be analyzed to determine the adequacy of these sources of cash.

To illustrate, the financial feasibility of Alf Alpha's purchase is presented as a partial analysis in Table 7-9. The sources of cash in this partial analysis are the reduced expenses from Table 7-5 and the income tax savings resulting from the deduction of interest.

Alf will be financing the purchase of the equipment with a five-year, 14 percent loan for $42,300. A $5000 down payment is required. The down payment, principal, and interest are listed as additional cash requirements in Table 7-9. The changes in income taxes, estimated in Table 7-6, are also included. The annual net cash flows are positive except when the down payment is made. However, because of the income tax savings, this $5000 will be offset by the end of the second year. By the end of the fifth year, this investment will have contributed an accumulated balance of $6361 to Alf Alpha's cash flow.

---

[5] The income tax savings resulting from interest deductions must be explicitly considered in the financial feasibility analysis. In the net present value analysis, they were implicitly considered by using the after-tax discount rate.

TABLE 7-9. Projected Change in Total Net Cash Flows After the Purchase
of Haying Equipment

	Year					
	0	1	2	3	4	5
Reduced outflows						
Operating expenses (Table 7-5)	$ 0	$12,884	$13,171	$14,078	$15,032	$16,044
Tax savings from interest[a]	0	2,606	2,211	1,762	1,250	666
Subtotal	$ 0	$15,490	$15,382	$15,840	$16,282	$16,710
Added outflows						
Down payment	$ 5,000	$ 0	$ 0	$ 0	$ 0	$ 0
Principal	0	6,400	7,296	8,317	9,482	10,805
Interest	0	5,922	5,026	4,005	2,840	1,513
Income taxes (Table 7-6)	0	−1,237	1,217	1,824	2,244	2,689
Subtotal	$ 5,000	$11,085	$13,539	$14,146	$14,566	$15,007
Net cash flow	−5,000	4,405	1,843	1,694	1,716	1,703
Accumulated balance	$ −5,000	$ − 595	$ 1,248	$ 2,942	$ 4,658	$ 6,361

[a] Estimated as 44 percent multiplied by the interest payment.

In this example, there are no apparent financial feasibility problems, but if cash deficits should occur, funds to cover them must come from other sources. One possibility is the loan payment itself, by renegotiating the terms to finance the investment. A longer loan repayment period would reduce the size of the annual principal and interest commitment. Investment decisions often carry important implications, not only for the profitability of the business, which can be checked using the net present value method, but also for the business's liquidity. Before the investment is made, future cash flows should be projected to determine whether adequate cash will be available to cover the cash requirements throughout the planning period.

Summary

When decisions will have income and expense effects over several years, conventional budgeting approaches are not as accurate as investment analysis techniques that consider the time value of money. These more powerful techniques are helpful when alternatives differ in the timing and amount of receipts and expenses. Net present value analysis, the technique presented in this chapter, is complex, and some may find it difficult to use. However, the net present value method is widely accepted and allows for explicit consideration of the effects of income taxes and inflation on the profitability of investments.

Applying the net present value method involves estimating the after-tax cash outflows and inflows associated with the investment for each year of the planning period. Each annual net cash flow is then discounted to its present value to account for the time value of money. The discount rate is based on the weighted average cost of capital for the farm business. The net present value is the sum of the discounted net cash flows over the planning period. A positive net present value means that the alternative is profitable, but it does not mean that this is the most profitable investment alternative. Careful consideration should also be given to the financial feasibility and risk of the investment.

Recommended Readings

CASLER, G. L., B. ANDERSON, and R. D. APLIN. *Capital Investment Analysis Using Discounted Cash Flows*, 3rd ed. Columbus, Ohio: Grid, Inc., 1984.

FREY, THOMAS L. *Time Value of Money and Investment Analysis: Explanation with Application to Agriculture.* University of Illinois Department of Agricultural Economics AET-15-76, June 1976.

APPENDIX TO CHAPTER 7

Reference Tables for Investment Analysis

Farm management decisions frequently involve compounding and discounting. Appendix Tables 1 to 4 are useful for reference purposes. Formulas and examples are provided for each table.

Future Value of a Present Sum

Appendix Table 1 is used if you have an amount of money now (PV) and choose to invest it for a number of periods (*n*) at an interest rate (*i*). Its future value (*FV*) is calculated as follows:

$$FV = PV(1 + i)^n$$

Examples

What is the FV of $8000 invested for five years at an interest rate of 12 percent compounded annually?

$$FV = \$\ 8,000(1 + i)^n \qquad \text{Appendix Table 1}$$
$$= \$\ 8,000(1.7623) \qquad n = 5$$
$$= \$14,098 \qquad\qquad i = 12$$

What is the FV of $8000 invested for five years at an interest rate of 12 percent compounded semiannually?

$$FV = \$\ 8,000(1 + i)^n \qquad \text{Appendix Table 1}$$
$$= \$\ 8,000(1.7909) \qquad n = 10$$
$$= \$14,237 \qquad\qquad i = 6$$

A friend borrows $3500 from you and agrees to pay it back in two years, including 10 percent interest. How much will you receive?

$$FV = \$3500(1 + i)^n \qquad \text{Appendix Table 1}$$
$$= \$3500(1.2100) \qquad n = 2$$
$$= \$4235 \qquad\qquad i = 10$$

APPENDIX TABLE 1. Future Value of a Present Sum of One Dollar
$$FV = PV(1 + i)^n$$

Period (n)	Interest Rate (i)						
	6.0	8.0	10.0	12.0	14.0	16.0	18.0
1	1.0600	1.0800	1.1000	1.1200	1.1400	1.1600	1.1800
2	1.1236	1.1664	1.2100	1.2544	1.2996	1.3456	1.3924
3	1.1910	1.2597	1.3310	1.4049	1.4815	1.5609	1.6430
4	1.2625	1.3605	1.4641	1.5735	1.6889	1.8106	1.9388
5	1.3382	1.4693	1.6105	1.7623	1.9254	2.1003	1.2878
6	1.4185	1.5869	1.7716	1.9738	2.1949	2.4364	2.6996
7	1.5036	1.7138	1.9487	2.2107	2.5023	2.8262	3.1855
8	1.5939	1.8509	2.1436	2.4759	2.8526	3.2784	3.7589
9	1.6895	1.9990	2.3579	2.7731	3.2519	3.8029	4.4355
10	1.7909	2.1589	2.5937	3.1059	3.7072	4.4114	5.2338
11	1.8983	2.3316	2.8531	3.4786	4.2262	5.1173	6.1759
12	2.0122	2.5182	3.1384	3.8959	4.8179	5.9360	7.2876
13	2.1329	2.7196	3.4523	4.3635	5.4924	6.8858	8.5994
14	2.2609	2.9371	3.7975	4.8871	6.2614	7.9875	10.1472
15	2.3966	3.1722	4.1773	5.4736	7.1379	9.2655	11.9737
20	3.2071	4.6609	6.7275	9.6463	13.7435	19.4608	27.3930
25	4.2919	6.8485	10.8347	17.0001	26.4619	40.8742	62.6686
30	5.7435	10.0627	17.4494	29.9599	50.9501	85.8498	143.3705
35	7.6860	14.7853	28.1024	52.7996	98.1001	180.3139	327.9969
40	10.2857	21.7245	45.2592	93.0509	188.8833	378.7207	750.3774

Present Value of a Future Sum

Appendix Table 2 is used to determine the PV of a sum of money when its FV is known. The FV is discounted for a number of periods (n) at discount rate (i). The PV is calculated as follows:

$$PV = FV(1 + i)^{-n}$$

Examples

What is the PV of $10,000 you will receive five years from today if the discount rate is 12 percent?

$$PV = \$10,000(1 + i)^{-n}$$
$$= \$10,000(0.5674)$$
$$= \$ 5,674$$

Appendix Table 2
$n = 5$
$i = 12$

You estimate that a piece of equipment will last for three more years and have a salvage value of $3000 when you sell it. Assuming a 14 percent discount rate, what is the PV of the salvage price?

APPENDIX TABLE 2. Present Value of a Future Sum of One Dollar
$$PV = FV(1 + i)^{-n}$$

Period (n)	Interest Rate (i)						
	6.0	8.0	10.0	12.0	14.0	16.0	18.0
1	0.9434	0.9259	0.9091	0.8929	0.8772	0.8621	0.8475
2	0.8900	0.8573	0.8264	0.7972	0.7695	0.7432	0.7182
3	0.8396	0.7938	0.7513	0.7118	0.6749	0.6407	0.6806
4	0.7921	0.7350	0.6830	0.6355	0.5921	0.5532	0.5158
5	0.7473	0.6806	0.6209	0.5674	0.5194	0.4761	0.4371
6	0.7049	0.6302	0.5645	0.5066	0.4556	0.4104	0.3704
7	0.6651	0.5835	0.5132	0.4524	0.3996	0.3538	0.3139
8	0.6274	0.5403	0.4665	0.4039	0.3506	0.3050	0.2660
9	0.5919	0.5003	0.4241	0.3606	0.3075	0.2629	0.2255
10	0.5584	0.4632	0.3855	0.3219	0.2697	0.2267	0.1911
11	0.5268	0.4289	0.3505	0.2875	0.2366	0.1954	0.1619
12	0.4969	0.3971	0.3186	0.2567	0.2076	0.1685	0.1372
13	0.4688	0.3677	0.2897	0.2292	0.1821	0.1452	0.1163
14	0.4423	0.3405	0.2633	0.2046	0.1597	0.1252	0.0986
15	0.4173	0.3152	0.2394	0.1827	0.1401	0.1079	0.0835
20	0.3118	0.2146	0.1486	0.1037	0.0728	0.0514	0.0365
25	0.2330	0.1460	0.0923	0.0588	0.0378	0.0245	0.0159
30	0.1741	0.0994	0.0573	0.0334	0.0196	0.0117	0.0069
35	0.1301	0.0676	0.0356	0.0189	0.0102	0.0056	0.0031
40	0.0972	0.0460	0.0221	0.0108	0.0053	0.0026	0.0013

$$PV = \$3000(1 + i)^{-n} \qquad \text{Appendix Table 2}$$
$$ = \$3000(0.6749) \qquad n = 3$$
$$ = \$2025 \qquad\qquad i = 14$$

You are offered $700 for your woodlot timber. Your estimate is that the timber will be worth about $1200 in another three years. With a discount rate of 8 percent, should you take the offer? If the *PV* of the $1200 is greater than the $700 offer, you should not accept the offer.

$$PV = \$1200(1 + i)^{-n} \qquad \text{Appendix Table 2}$$
$$ = \$1200(0.7938) \qquad n = 3$$
$$ = \$\ 952 \qquad\qquad i = 8$$

Present Value of a Uniform Series

Appendix Table 3 is used if you are to receive or pay a series of equal value payments (*PMT*) for a number of periods (*n*). At a discount rate (*i*), the total PV of all payments is calculated as follows:

APPENDIX TABLE 3. Present Value of a Uniform Series of One Dollar Payments

$$PV = PMT\left(\frac{1 - (1 + i)^{-n}}{i}\right)$$

Period (n)	Interest Rate (i)						
	6.0	8.0	10.0	12.0	14.0	16.0	18.0
1	0.9434	0.9259	0.9091	0.8929	0.8772	0.8621	0.8475
2	1.8334	1.7833	1.7355	1.6901	1.6467	1.6052	1.5656
3	2.6730	2.5771	2.4869	2.4018	2.3216	2.2459	2.1743
4	3.4651	3.3121	3.1699	3.0374	2.9137	2.7982	2.6901
5	4.2123	3.9927	3.7908	3.6048	3.4331	3.2743	3.1272
6	4.9173	4.6229	4.3553	4.1114	3.8887	3.6847	3.4976
7	5.5824	5.2064	4.8684	4.5638	4.2883	4.0386	3.8115
8	6.2098	5.7466	5.3349	4.9676	4.6387	4.3436	4.0776
9	6.8017	6.2469	5.7590	5.3283	4.9464	4.6065	4.3030
10	7.3601	6.7101	6.1446	5.6502	5.2161	4.8332	4.4941
11	7.8869	7.1389	6.4951	5.9377	5.4527	5.0287	4.6560
12	8.3838	7.5361	6.8137	6.1944	5.6603	5.1971	4.7932
13	8.8527	7.9038	7.1034	6.4236	5.8424	5.3423	4.9095
14	9.2949	8.2442	7.3667	6.6282	6.0021	5.4675	5.0081
15	9.7123	8.5595	7.6061	6.8109	6.1422	5.5755	5.0916
20	11.4699	9.8182	8.5136	7.4694	6.6231	5.9288	5.3528
25	12.7834	10.6748	9.0770	7.8431	6.8729	6.0971	5.4669
30	13.7648	11.2578	9.4269	8.0552	7.0027	6.1772	5.5168
35	14.4983	11.6546	9.6442	8.1755	7.0701	6.2153	5.5386
40	15.0463	11.9246	9.7791	8.2438	7.1050	6.2335	5.5482

$$PV = PMT\left(\frac{1 - (1 + i)^{-n}}{i}\right)$$

Examples

What is the total PV if you are to receive $1000 each year for five years and the discount rate is 10 percent?

$$PV = \$1000\left(\frac{1 - (1 + i)^{-n}}{i}\right)$$ Appendix Table 3

$$= \$1000(3.7908)$$ $n = 5$

$$= \$3791$$ $i = 10$

A repaired machine will save $2500 in operating costs each year it is used. Assume it lasts for eight years and the discount rate is 12 percent. What is the PV of these savings?

$$PV = \$\ 2,500\left(\frac{1 - (1 + i)^{-n}}{i}\right)$$ Appendix Table 3

$$= \$\ 2,500(4.9676)$$ $n = 8$

$$= \$12,419$$ $i = 12$

Capital Recovery or Loan Amortization

Appendix Table 4 is used if you are to pay or receive a series of equal value payments (PMT) for a number of periods (n). At an interest rate (i) and a known PV, the size of each payment is calculated as follows:

$$PMT = PV\left(\frac{i(1 + i)^n}{(1 + i)^n - 1}\right)$$

Examples

What size payments must you make each year to repay $10,000 borrowed for nine years at 12 percent interest?

$$PMT = \$10,000\left(\frac{i(1 + i)^n}{(1 + i)^n - 1}\right)$$ Appendix Table 4

$$= \$10,000(0.1877)$$ $n = 9$

$$= \$\ 1,877$$ $i = 12$

You borrow $7500 to be repaid in 10 equal semiannual payments with 12 percent interest. What size payments must you make?

$$PMT = \$7500\left(\frac{i(1 + i)^n}{(1 + i)^n - 1}\right)$$ Appendix Table 4

$$= \$7500(0.1359)$$ $n = 10$

$$= \$1019$$ $i = 6$

A farmer borrows $25,000 to buy some additional land. He plans to repay the loan over a 10-year period with 14 percent interest. How much additional income would the farmer need each year to amortize the loan?

$$PMT = \$25,000\left(\frac{i(1 + i)^n}{(1 + i)^n - 1}\right) \quad \text{Appendix Table 4}$$
$$= \$25,000(0.1917) \qquad n = 10$$
$$= \$ 4,793 \qquad\qquad i = 14$$

How much additional income would he need each year to amortize a 20-year loan?

$$PMT = \$25,000(0.1509) \qquad\qquad n = 20$$
$$= \$ 3,773$$

By amortizing the loan over a longer period, the farmer would need less additional income per year ($1020 less); however, he would pay more total interest for the loan ($27,530 more).

APPENDIX TABLE 4. Amount of Periodic Payment to Recover or Amortize a Present Value of One Dollar

$$PMT = PV\left(\frac{i(1 + i)^n}{(1 + i)^n - 1}\right)$$

Period (n)	Interest Rate (i)						
	6.0	8.0	10.0	12.0	14.0	16.0	18.0
1	1.0600	1.0800	1.1000	1.1200	1.1400	1.1600	1.1800
2	0.5454	0.5608	0.5762	0.5917	0.6073	0.6229	0.6387
3	0.3741	0.3880	0.4021	0.4164	0.4307	0.4453	0.4599
4	0.2886	0.3019	0.3155	0.3292	0.3432	0.3574	0.3717
5	0.2374	0.2505	0.2638	0.2774	0.2913	0.3054	0.3198
6	0.2034	0.2163	0.2296	0.2432	0.2572	0.2714	0.2859
7	0.1791	0.1921	0.2054	0.2191	0.2332	0.2476	0.2624
8	0.1610	0.1740	0.1874	0.2013	0.2156	0.2302	0.2452
9	0.1470	0.1601	0.1736	0.1877	0.2022	0.2171	0.2324
10	0.1359	0.1490	0.1628	0.1769	0.1917	0.2069	0.2225
11	0.1268	0.1401	0.1539	0.1684	0.1834	0.1989	0.2148
12	0.1193	0.1327	0.1468	0.1614	0.1767	0.1924	0.2086
13	0.1129	0.1265	0.1408	0.1557	0.1712	0.1872	0.2037
14	0.1076	0.1213	0.1358	0.1509	0.1666	0.1829	0.1997
15	0.1029	0.1168	0.1315	0.1468	0.1628	0.1794	0.1964
20	0.0872	0.0998	0.1175	0.1339	0.1509	0.1687	0.1868
25	0.0782	0.0937	0.1102	0.1275	0.1455	0.1640	0.1829
30	0.0727	0.0888	0.1061	0.1241	0.1428	0.1619	0.1813
35	0.0689	0.0858	0.1037	0.1223	0.1414	0.1609	0.1806
40	0.0665	0.0839	0.1023	0.1213	0.1408	0.1604	0.1802

CHAPTER 8

Evaluating and Managing Risk

To operate a farm business sucessfully in today's complex and rapidly changing world, the manager must confront the risks and uncertainties of the future. The ever-changing conditions of the farm's business environment are an important challenge to farm managers, who must consider the risks associated with these changes in every decision they make. Some examples of the risks farmers face are as follows:

1. It will not rain at the right time.
2. The old tractor will break down.
3. Prices will go up after the grain has been sold.
4. Government regulations will change.
5. The employee will quit.

When farmers make decisions, they do not and cannot know what will happen in the future. This is the definition of decision making under *uncertainty*. *Risk*, as commonly defined, refers to the chance or probability of adverse outcomes associated with an action. The greater the uncertainty, the greater the risk.

Uncertainty is a problem for several reasons. For instance, not knowing what to expect in crop prices and weather, farmers sometimes use less fertilizer than is most profitable under normal circumstances. They realize full well that with normal prices and weather, additional fertilizer would be more profitable. Yet they are willing to sacrifice the possible gain in order to avoid the possible loss in case of unfavorable prices or weather. Because of uncertainty, a farmer may not buy the additional 80 acres that would allow machinery and labor to be used more efficiently. Uncertainty affects not only the quantities of resources used but also the combination of products produced.

Uncertainty makes it more difficult for farmers to borrow money. If lenders view certain farm loans as too risky, they are likely to (1) reduce the amount of money loaned, (2) increase interest rates, (3) increase security requirements, and/or (4) decrease the repayment period for the loan. And as uncertainty increases in agriculture, it becomes more difficult for farmers to make decisions that will prove to be consistent with their goals.

Many economic principles are based on the assumption that the future can be predicted with some accuracy. In discussing economic principles in Chapter 2, the assumption was made that future yields, prices, and costs are predictable; in order to draw up a budget, a certain amount of forecasting is required. Yet obviously the future cannot be predicted with complete accuracy. This chapter discusses ways to manage the uncertainty and risk of future events.

Business managers must make risky decisions; they should remember, however, that a good decision does not guarantee a good outcome. A good decision is simply one that is consistent with the manager's information about the risks involved and with certain goals. Thus, a good decision is a carefully considered choice based on all available information. Whether the outcome of a good decision will prove to be favorable or not is partly a matter of luck.

Sources of Risk

Identifying the sources of risk is an important aspect of the decision-making process. However, the relative importance of these sources of risk differs among enterprises and changes over time. The following checklist provides an overview of the sources of risk:

1. *Production risk.*   Production risk is due to the variability in production caused by such unpredictable factors as weather, disease, pests, genetic variations, and timing. Examples include variations in crop yields, animal weaning weights, product quality, animal rate of gain, pasture-carrying capacity, feed conversion, death loss, labor required, and machinery breakdown.

2. *Market risk.*   Market risk involves the variability and unpredictability of prices that farmers receive for their products and must pay for production inputs. Price variability reflects changes in supply and demand.

3. *Financial risk.*   Financial risk relates to the financing of business assets. The increased use of borrowed capital and unpredictable cash flows create the risk of not having enough cash to meet all obligations, which could ultimately mean disaster. There is also the risk of losing the lease on the farmland.

4. *Obsolescence risk.*   Currently used production methods requiring large investments can be made obsolete by new technology. Adopting new technologies too soon or too late is a risk farmers must face.

5. *Casualty loss risk.*   Casualty loss, a traditional source of risk, is the loss of assets to fire, wind, hail, flood, and theft. Inflation increases the possibility that potential losses are underinsured.

6. *Legal risk.*   Laws and governmental programs that reflect society's changing attitudes are a growing source of uncertainty for farmers. Examples include environmental protection; controls on the use of feed additives, insecticides, and herbicides; and land use planning. In addition, there is the risk of lawsuits stemming from farm accidents.

7. *Human risk*.   The character, health, and behavior of individuals are unpredictable. The possibility of losing a key employee during a critical production period is one example of this risk. Dishonesty and undependability of business associates are others. If the farm manager is disabled, the operation's efficiency can be seriously disrupted. Also, family needs and goals change, sometimes unpredictably.

Psychological studies have shown that many business managers, farmers included, tend to suppress or disregard risk when making decisions. In order to maintain psychological equilibrium, ignoring risk may be perfectly natural. For example, suppose you have decided to drive to town. You know there is always a small chance that you will be injured in an automobile accident. By ignoring this risk you do not have to anguish over the probabilities and consequences. But past good luck does not guarantee continued success either in driving a car or in managing a farm. Good decision making requires an explicit consideration of the sources of risk.

Making Risky Decisions

Risk complicates the decision-making process. Some of these complications were mentioned in preceding chapters. In this chapter, they are treated in a more systematic fashion. The steps of the decision-making process described in Chapter 1 still apply, but following them is a somewhat less orderly procedure when the consequence of an action is uncertain. Decisions must be made, however, even with less than perfect knowledge.

As an example, assume that a farmer is trying to determine the best way to market grain. It is November 1, and the grain has been harvested. It must be sold by next March 1 to meet a loan payment. It can either be sold now or stored until March 1. If the grain is stored, there are several possible outcomes; for example:

Alternative	Outcome (Return on March 1)	Probability
Store grain	A_1 $150,000—high price	0.2
	A_2 $100,000—most likely price	0.5
	A_3 $ 75,000—low price	0.3

The farmer believes that the most likely return after subtracting costs and interest charges is $100,000. But higher and lower returns are also possible. Based on the market outlook, the farmer assigns probabilities to each outcome.

The other alternative, to sell now, can be analyzed similarly:

Alternative B	Outcome (Return on November 1)	Probability
Sell now	B_1 $90,000	1.0

In this case, based on the prevailing price, the farmer knows that the outcome will be a $90,000 return for the grain.

Now the weighted average can be calculated for each alternative:

Alternative A:	A_1 $150,000 \times 0.2$ =	$ 30,000
	A_2 $100,000 \times 0.5$ =	50,000
	A_3 $ 75,000 \times 0.3$ =	22,500
	Weighted average	= $102,500
Alternative B:	B_1 $ 90,000 \times 1.0$ =	$ 90,000
	Weighted average	= $ 90,000

The most profitable alternative, on the basis of the weighted average outcome, is alternative A. However, there are 3 chances out of 10 that it is not the best choice and that the farmer will lose $15,000 by storing the grain rather than selling it now. Because of the risk of not meeting the loan payment, alternative A is not necessarily the better choice. No general recommendations can be made in this risky situation because individuals differ both in their willingness to take chances and in their financial ability to bear the consequences of unfavorable outcomes. The financial position of the farmer is obviously quite important. In this example, the farmer will have to decide whether to forfeit the difference of $12,500 (based on the weighted average) in order to avoid a possible loss of $15,000. If the decision is to sell now, an opportunity to "make a killing" on the possible high return of $150,000 is also forgone.

This simplified example illustrates the considerations involved in making risky decisions. There are no easy solutions, but the following four steps provide guidelines:

1. Analyze decisions in terms of alternative actions, possible events, and payoffs.
2. Estimate the probabilities of the events that will affect the payoffs.
3. Consider your attitudes toward risk and your financial ability to take risk.
4. Adopt management strategies to control risk.

This management approach is logical, and, most important, it recognizes the personal elements in decision making: (1) the knowledge and beliefs of managers and (2) their goals and attitudes toward risk taking.

These steps simply specify explicitly the processes managers already employ intuitively. However, many decisions are too complex and important to be left to intuition alone. This more formalized approach provides the discipline to ensure that all the information available is fully utilized and that the decision does not conflict with the manager's goals. The first step is to organize the components of the decision using a payoff matrix.

The Payoff Matrix

When making risky decisions, a manager must choose from among *alternative actions*, the outcomes of which depend on *events*. These events are beyond the control of the manager; their occurrence is uncertain. The outcome of each action and event combination is termed a *payoff*; a table summarizing the actions, events, and payoffs is called a *payoff matrix*.

This section outlines some guidelines and procedures for constructing the payoff matrix. The process and format require an explicit consideration of risk. After completion, the matrix provides a convenient summary of the components of the decision under consideration. As an example, suppose you are faced with three alternatives involving the toss of a fair coin. Heads you win, tails you lose. The first alternative is to bet $100, the second to bet $1000, and the third to decline both bets (Table 8-1). These alternative actions are the first component of the decision. The second component is the events that determine the outcomes for each of the actions; in this case, the only possible events are heads and tails. The third component is the payoff, or outcome, associated with each action–event combination. In the case of the "no bet" action, the payoffs are zero dollars for each of the two events. For the "small bet" alternatives the possible payoffs are +$100 and −$100, and for the "large bet" alternatives the payoffs are +$1000 and −$1000.

The simplicity of the payoff matrix in Table 8-1 is deceptive. Usually several types of events affect payoffs. For example, both prices and yields are often uncertain, and the combined effect of the price event and the yield event determines the payoff. Also, more than three alternative actions may be possible. Finally, estimating payoffs may require very complex budgeting procedures.

It is impractical to assess every possible action and event. There are too many for practical consideration. Following are examples of just a few possible actions and events related to farm management:

Alternative actions

1. To rent or not to rent additional land.
2. To use more fertilizer or not and, if so, how much.
3. To sell corn at harvest or store and sell later on one of several dates.
4. To buy or not to buy crop insurance.

TABLE 8-1. Payoff Matrix for the Coin Toss Decision

	Alternative Action		
Event	No Bet	Small Bet	Large Bet
Heads	$0	$ 100	$ 1000
Tails	0	− 100	− 1000

Possible events

1. The monthly average wheat price may be $4.00, $4.50, or $5.00.
2. Rainfall may be abundant, moderate, or scarce.
3. There may be hail or no hail.
4. Cattle may gain 0.75, 1.25, or 1.75 pounds per day.

The key is to limit the matrix to the actions and events most likely to affect payoffs without losing the essentials. To illustrate further how the payoff matrix is used to summarize the components of decision making, refer to the store-or-sell grain example at the beginning of this chapter and to Table 8-2. If the manager sells now, the payoff is $90,000. If the grain is stored for sale on March 1, three different payoffs can be projected. Actually, there would be several more alternatives: The farmer could sell in December, January, or February; part of the crop could be sold now and part later—one-sixth in November, one-sixth in January, and so on. But the important point is that the payoff matrix provides a framework for organizing the components of the decision.

Considerable time and, perhaps, access to a computer would be necessary to study every possible alternative action and to project payoffs. Instead, a manager must eliminate those actions that are not feasible, leaving only a relevant few for consideration. The process of selecting the most promising alternatives was discussed in the preceding chapter.

Payoffs are affected by rainfall, temperature, the employee's work, prices of inputs, supply and demand, and so forth. In determining which uncertain events will significantly affect the payoffs, it is important to consider both the unfavorable and the favorable possibilities. Using a checklist such as the one used at the beginning of this chapter provides the discipline to analyze the sources of risk, to learn more about their origins, and thus develop a better understanding of how they can affect potential payoffs.

Past experience and a thorough knowledge of biological, physical, and economic phenomena help in determining the events that are most critical. It is also helpful to ask the following two questions:

1. What possible impact could this event have on the payoff?
2. Is there a significant chance that this will happen?

Obviously, if the event will have little impact and there is a low chance that it will occur, it can be ignored. On the other hand, an event with a large impact

TABLE 8-2. Payoff Matrix of Net Returns on Selling Grain Now Versus
Storing and Selling Later

Events	Alternative Actions	
	Sell Grain Now	Store and Sell Grain Later
High price	$90,000	$150,000
Most likely price	90,000	100,000
Low price	90,000	75,000

and a high chance of occurring should definitely be considered. But what about an event with a large impact and a low chance of occurring? Or one with little impact but a high chance of occurring? Whether these latter two events are included in the payoff will depend on the manager's judgment and experience. However, the manager might do some preliminary budgeting to estimate the possible extent of the impact and to try to learn more about its chance of occurring.

Once alternative actions and possible events have been specified, the next step is to budget the payoffs for each action–event combination. Budgeting can be a complex process. The procedures for preparing a total farm budget and a partial budget, both of which are appropriate for projecting payoffs, were presented in Chapter 6. The future may not be as planned, but after careful budgeting of all of the possible payoffs in the matrix, the actual outcome should be no surprise. This outcome will have been considered before making the decision.

Usually the payoffs are measured monetarily, but they can represent other values consistent with the manager's goals, such as production or leisure. If multiple goals are being considered, more than one payoff measure can be put in each cell of the matrix. For example, the payoff might be measured in terms of net income, return to labor, return to capital, or return to management. How the alternatives will affect the business's liquidity might also be of interest and can be measured as net cash flow. The appropriate measure or measures depend on the particular decision to be made and the manager's goals.

As in any other budgeting situation, when computing payoffs it is important to separate the relevant costs and returns from the irrelevant ones. Relevant costs are those variable costs that are affected by the action under consideration; conversely, irrelevant costs are those fixed, or sunk, costs not affected by the decision. Recall the grain sale example: Included in the cost of holding grain for sale in March are only those costs that would be incurred as a result of choosing this alternative. The fixed costs of the grower's storage capacity, such as depreciation, taxes, and interest, remain the same whether the grain is stored or not and, therefore, are irrelevant.

The payoff matrix approach offers several advantages: It provides a framework for specifying the various components of a decision; it divides elements according to what *can* be controlled (the alternative actions) and what *cannot* be controlled (the possible events). Organizing a payoff matrix helps the manager to find the most promising alternative actions and the events that will significantly affect results. The difficulty is in narrowing down the alternative actions and possible events to those that are most important to the decision to be made. However, in the end, a payoff matrix helps ensure that the risk inherent in the decision is not ignored but explicitly considered.

Probabilities

Probabilities provide a means for summarizing what is known or believed about the future. A *probability* is a number that measures the likelihood or chance that a particular event will occur. (Again, an event is something that might happen and over which one has no control.) This probability can be any number from 0 through 1. The number 0 means that there is no chance that the event will happen, and 1 means that it is certain to happen. Also, the sum of the probabilities of all the possible events that might occur relative to an action must equal 1.

An Example of Probability Information

Assume that a farmer in north central Oklahoma wants to estimate the chances of various rainfall levels for next September and October. In Table 8-3, past rainfall totals for each of the last 20 years are grouped into ranges (or intervals) to calculate the frequency and probability of each rainfall level. Using this table, the farmer can see that the chance of getting a disastrous 0 to 1.5 inches of rain is 4/20 (or 0.20). The range of 0 to 3 inches has a probability of 0.35 (0.20 + 0.15).

Probabilities based on past data can be quite useful to managers in some instances but less useful in others. For example, past rainfall data can provide a good guide to future rainfall, but soybean price probabilities based on historical data may be a poor guide to future prices because of changing supply and demand relationships.

Estimating Personal Probabilities

Probabilities based on a manager's personal beliefs are called *personal probabilities*. In estimating personal probabilities, managers should examine their own experiences and available data. Several farmers may assign different probabilities to the same event; however, this does not mean that personal probabilities are arbitrary. It means that the farmers have either different information or different experiences and that they interpret the same information differently. However, two reasonable

TABLE 8-3. Rainfall Amounts for September Through October, North Central Oklahoma

Rainfall Amount (inches)	Number of Years	Probability Based on Frequency
0 to 1.5	4	0.20
1.5 to 3.0	3	0.15
3.0 to 4.5	2	0.10
4.5 to 6.0	1	0.05
6.0 to 7.5	4	0.20
7.5 to 9.0	3	0.15
9.0 to 10.5	1	0.05
10.5 to 12.0	1	0.05
12.0 to 13.5	0	0
13.5 to 15.0	1	0.05
Total	20	1.00

SOURCE: From data provided by Odell L. Walker, Professor of Agricultural Economics, Oklahoma State University.

people having roughly the same experience and information regarding a particular event will assign it roughly the same probability.

In developing personal probability estimates, it is important to use all of the relevant information available and then to apply intuitive judgment. Note that personal probabilities are not constant or fixed. They should be revised and changed as new information becomes available.

An example will now be used to illustrate how personal probabilities can be estimated. Suppose a farm manager is considering next August's price of choice slaughter steers. To estimate the personal probabilities of particular prices (or of any uncertain event), a numerical weight is assigned to that event reflecting the manager's conviction that it will occur. To use this method, the first step is to divide the total range of possibilities into logical intervals. Intervals of $5 per hundredweight are used here, as shown in Table 8-4.

In this example, the manager has entered numbers from 0 through 100 in the second column to reflect the strength of belief that the actual choice steer price will fall in the corresponding price intervals. It is believed that there is no chance that the price will be either below $60 or above $95. In assigning the weight, the manager started by assigning a weight of 100 to the interval in which the price is thought most likely to occur; then, based on the manager's beliefs, weights are assigned to the remaining intervals.

Personal probabilities are then calculated by dividing the number (weight) in each range by the sum of the numbers (335). The quotients in the third column of Table 8-4 are the personal probabilities. The sum of the probabilities should be, and is, 1.

The following steps summarize this approach for estimating the probabilities of a particular set of unknown events:

1. Determine the lowest and highest possible events.
2. Divide the range into 10 or 12 convenient intervals—for example, $0.20 for wheat prices or four bushels for corn yields.

TABLE 8-4. An Example of the Estimation of Personal Probabilities for Cattle Prices

Price Interval ($/cwt.)	Conviction That Price Will Be in Given Range (weights 0–100)	Conviction Converted to Personal Probability
Less than $60	0	0
$60 to $65	15	0.04
$65 to $70	40	0.12
$70 to $75	70	0.21
$75 to $80	100	0.30
$80 to $85	65	0.20
$85 to $90	35	0.10
$90 to $95	10	0.03
$95 and over	0	0
Total	335	1.00

3. Consider what shape the probability distribution should be: Will it be flat, symmetrical, or skewed to the right or left?
4. Assign weights of 1 through 100 to each interval, assigning 100 to the most likely interval. Then, based on the general shape of the distribution, assign weights to the remaining intervals.
5. Total the weights.
6. Calculate the personal probabilities for each interval by dividing the weight for each interval by the total of all the weights.
7. Check to be sure that the sum of all probabilities is equal to 1.

Figure 8-1 is a *histogram* of the probability distribution of cattle prices; it helps the manager visualize the relationships among the probabilities. For example, it is easy to see the price interval with the highest probability of occurrence. It can also be seen that there is a 71 percent probability that the price will be in the range of $70 to $85 and that there is a 16 percent chance that it will be below $70.

The following cautions should be kept in mind in determining personal probabilities:

1. Be as objective as possible. Do not be influenced by what you *hope* will happen.
2. Consider the full range of possible events. Do not overlook the extremes— low and high.
3. The distribution should be consistent with the level of uncertainty. If you are highly uncertain, the shape of the distribution should be wide and flat.
4. Be ready to revise your probabilities as soon as more information is available. Aggressively seek out additional information.

These personal probabilities can be combined with the information from the payoff matrix to assess the overall risk. The result must be carefully interpreted in the context of your attitudes toward risk taking. Next, following a discussion of these attitudes, an example is used to illustrate how probabilities and a payoff matrix, when analyzed in the context of your attitudes, generate important management information.

Attitudes Toward Risk

So far, evaluating risks using the payoff matrix and personal probabilities has been discussed. The next step is to consider the manager's attitude toward assuming these risks. Attitudes toward risk are dependent on (1) *goals*, (2) *financial position*, and (3) the *potential gains and losses*.

How Attitudes Toward Risk Are Influenced

Because farmers have different goals regarding risk and income, they do not make the same decisions. What may be a good plan for your neighbor may not be appropriate for you. Setting goals is an important step in making decisions (Chapter 1). Although farmers may desire increased income to provide a higher standard

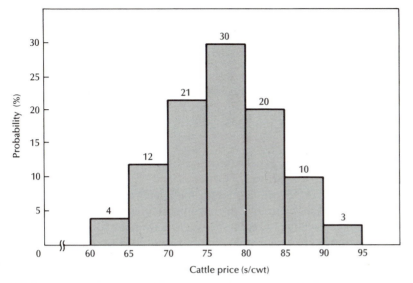

FIGURE 8-1.   Histogram of Personal Probabilities for Cattle Prices.

of living, they are also concerned about risk and the security and survival of the farm business.

Establishing priorities is difficult, because income and security goals often conflict with one another. For example, farmers might like to increase their net income, but to take action to do so might jeopardize the survival of the business. The combination of risk and income chosen depends on the priority placed on these two goals—increased income versus security. These priorities vary among farmers, which explains why, when faced with the same risky situation, they respond differently.

In addition to goals and the potential gains and losses, a third consideration influences how managers respond to risk: their financial position. Financial position, measured by solvency ratios and cash flow requirements (as discussed in Chapter 5), determines risk-taking ability or, put another way, the vulnerability of the business to risk. Solvency can be measured by the *debt to net worth ratio*, that is, the adequacy of equity relative to debt. The higher this ratio, the more precarious the business's financial position. Annual cash flow obligations include income taxes, loan repayments, and family living expenses. The higher the financial requirements, the less risk the business can safely assume.

The following hypothetical situations illustrate how goals, potential gains or losses, and financial position affect attitudes toward the risk associated with a particular decision. Put yourself in each of the following situations and decide whether to accept the bet or not:

1. You are offered a wager in which you will gain $20 if a fair coin falls heads or lose $10 if it falls tails.
2. You have accumulated a fortune worth $5 million. You are now offered the opportunity to triple this fortune (a gain of $10 million) if a coin falls heads or lose the entire $5 million if it falls tails.

3. With sacrifice, you have accumulated $5000 that you plan to use for a long-planned vacation this month. The vacation will require the entire $5000. You are offered a bet that will yield a profit of $5000 (you would have twice as much to spend on the vacation) if the coin falls heads or a loss of $5000 (which would mean postponing the vacation) if it falls tails.

4. You are desperate to take a camping and fishing trip. You have $5000 in cash, but expenses will amount to $10,000. You are offered a chance to make a $5000 profit (enough to finance the trip) if the coin falls heads or a loss of the $5000 you now have if it falls tails.

Most people would accept the bets in situations 1 and 4, but not those in 2 and 3. Why is there a difference between situations 1 and 2? The two are similar in that the possible gains are twice as large as the possible losses. The difference is in the potential gains and losses relative to your financial resources. In situation 1, you would probably be willing to accept the consequences of a $10 loss relative to a potential gain of $20. In situation 2, however, your entire fortune is at stake. Ten million dollars is not likely to increase your satisfaction enough to offset the considerable dissatisfaction resulting from the loss of all your money.

In situations 3 and 4, the dollar amounts of the gains and losses are exactly the same. The difference is in your goal. In situation 3, accepting the bet would jeopardize the accomplishment of a goal already in sight; in situation 4, the objective could be accomplished now only by winning the bet. Thus, it is not the dollar amount of the gain or loss alone that affects the decision, but also what the decision maker would do with the money and the satisfaction that would be realized.

Risk Attitude Profiles

Just as people can be classified as optimistic or pessimistic and conservative or progressive, they can also be classified as *risk takers* or *risk avoiders*. It is important to recognize that attitudes toward risk do not reflect managerial ability. There are successful managers who tend to be risk takers, and there are successful managers who usually avoid risk.

The risk avoiders are the more conservative types who prefer less risky investments and are willing to sacrifice the small chance of higher income for reduced risk. They are more likely to market their production by forward contracting or hedging (techniques discussed later in this chapter).

At the other extreme are the risk takers. These more adventurous types are willing to accept the risk for the small chance of increased income. For instance, they tend to speculate in marketing their products.

A manager's attitude toward risk depends on personal feelings, experience, and temperament. To quote one farmer describing a neighboring cattle feeder, "His experience as a paratrooper in the army was good training for feeding cattle. Both take lots of guts."

Managers who cannot sleep at night because they are worrying about the future of their business are assuming too much risk. They should explore ways to control or reduce these risks, which may involve compromising on income and other goals. They should assess their aptitude for managing under the pressures of risk. Some managers are mentally capable of handling more risk than others. To be happy with their decisions, managers should be sure that these decisions are consistent with their goals and attitudes toward risk.

Evaluating Risk: An Example

An example involving a choice between two investment alternatives applies the foregoing approaches to risky decision making. Suppose a young farmer has the opportunity and desire to invest $100,000, as well as time and effort, in either a Montana wheat-cattle ranch or an Indiana corn-hog farm. For each of these ventures, the farmer has estimated the payoffs and the probabilities associated with them. Table 8-5 shows that investing in the corn-hog farm gives a 22 percent probability of a $12,000 net income, whereas the wheat-cattle ranch has a 15 percent probability for the same net income.

Next, the *weighted averages* of the net income for both alternatives are calculated. The weighted average involves weighing each net income according to its probability of occurring. Each probability is multiplied by the corresponding annual net income, and the results are totaled (see Table 8-6). This dollar figure represents the average net income the farmer can expect to receive over many years, although the net income received in any one year may be higher or lower. This weighted average is additional information to be used with the already calculated probabilities and payoff information.

If the farmer is risk neutral on this investment, the logical choice is the Montana wheat-cattle ranch because it has the highest weighted-average annual net income. But if the manager is concerned about the safety of these two ventures and tends to avoid risk, priority might be given to safety over the weighted average net income. The "safety first" rule might apply. It states that a manager should do the following:

1. Choose the alternative that maximizes the weighted average net income, subject to
2. An acceptably low probability that the net income will be below a certain minimum.

For example, any alternative that has a greater than 10 percent chance of yielding a net income below $5000 would not be considered. However, another manager, with a greater desire to avoid risk, might have a minimum income objective of $10,000, subject to only a 5 percent probability that it will not be achieved.

When applying this rule, the first priority is to be able to achieve a minimum level of income with an acceptable probability. The second priority, once the first has been satisfied, is to maximize average net income. The minimum income level usually reflects the manager's needs, that is, cash expenses, debt payments, and family living expenses.

The probability that the income will be less than a given level is called the *cumulative probability*. For example, in Table 8-7 the probability that the annual net income on the Montana wheat-cattle ranch will be $8000 or less is 25 percent, which is found by adding all of the probabilities associated with income levels at

TABLE 8-5. Example Payoffs and Probabilities for Two Agricultural
Investment Opportunities

Annual Net Income ($)	Probabilities	
	Indiana Corn-Hog Farm (%)	Montana Wheat-Cattle Ranch (%)
4,000	0	5
6,000	8	8
8,000	12	12
10,000	18	14
12,000	22	15
14,000	24	16
16,000	13	17
18,000	3	9
20,000	0	4

TABLE 8-6. Calculation of Weighted Average Annual Net Income for the
Indiana Corn-Hog Farm and the Montana Wheat-Cattle Ranch

Annual Net Income		Probability		Results
Indiana Corn-Hog Farm				
$ 4,000	×	0	=	$ 0
6,000	×	0.08	=	480
8,000	×	0.12	=	960
10,000	×	0.18	=	1,800
12,000	×	0.22	=	2,640
14,000	×	0.24	=	3,360
16,000	×	0.13	=	2,080
18,000	×	0.03	=	540
20,000	×	0	=	0
		Weighted average	=	$11,860
Montana Wheat-Cattle Ranch				
$ 4,000	×	0.05	=	$ 200
6,000	×	0.08	=	480
8,000	×	0.12	=	960
10,000	×	0.14	=	1,400
12,000	×	0.15	=	1,800
14,000	×	0.16	=	2,240
16,000	×	0.17	=	2,720
18,000	×	0.09	=	1,620
20,000	×	0.04	=	800
		Weighted average	=	$12,220

$8000 and below. Successively adding the net income probabilities in Table 8-5 determines the cumulative probabilities, as listed in Table 8-7.

After analyzing the financial situation, this prospective farmer or rancher finds that a net income of $9000 per year is needed to survive and maintain the business. As can be seen in Table 8-7, there is a 25 percent chance of not getting this income from the Montana wheat-cattle ranch, compared to a 20 percent chance of failing on the Indiana corn-hog farm. These probabilities may seem high and severe, but they are less so for an established farmer who has accumulated adequate surplus to weather these low-income years.

The risk-avoiding manager, after comparing the alternatives in Table 8-8, can easily see that the Montana wheat-cattle ranch has the higher profit potential and the higher risk. Depending on how the added risk is traded off against the higher income—that is, if the manager is willing to accept a 5 percent higher probability of failure for $360 more in income on average—the choice would be the Montana ranch. If not, the Indiana farm would be taken.

Managers who prefer to take risks will look at the same information differently. For example, suppose a goal of a risk-taking manager is to achieve a net income greater than $16,000. This manager can see that there is a 3 percent chance of achieving this goal in Indiana and a 13 percent chance in Montana.

TABLE 8-7. Cumulative Probabilities for Two Agricultural Investments

Annual Net Income ($)	Cumulative Probabilities	
	Indiana Corn-Hog Farm (%)	Montana Wheat-Cattle Ranch (%)
$ 4,000	0	5
6,000	8	13
8,000	20	25
10,000	38	39
12,000	60	54
14,000	84	70
16,000	97	87
18,000	100	96
20,000	100	100

TABLE 8-8. Comparison of Two Agricultural Investments

Decision Information	Indiana Corn-Hog Farm	Montana Wheat-Cattle Ranch
Weighted average annual net income	$11,860	$12,220
Cumulative probability of not exceeding $9000 net income	20%	25%

As this example illustrates, payoff and probability information, weighted averages, and cumulative probabilities are viewed from different perspectives by managers with different attitudes toward risk.

Risk Management Strategies

The fourth technique for dealing with risk is the adoption of strategies to manage or control it, that is, precautionary measures to reduce risk, mitigate its impact, or to increase the business's ability to survive unfavorable consequences. Not all of these strategies are appropriate for all farmers. Their use depends on the farm manager's situation, financial position, goals, and attitudes toward taking risk.

The first step in developing and adopting a set of strategies for managing risk is to consider systematically the various sources of risk affecting the farm business. (Refer to the checklist at the beginning of the chapter.)

The following sections discuss various strategies to control these risks; each has advantages and disadvantages. Risks cannot be eliminated completely, but they can be controlled.

Flexibility

Flexibility refers to the ability to make adjustments in the farm operation in response to changing conditions in order to reduce disruptive fluctuations in income. However, this flexibility may involve a cost. If so, the cost must be weighed against the advantage of flexibility.

One area emphasizing flexibility is agricultural engineering. Multipurpose structures are available to accommodate machinery, grain, or livestock, depending on the farmer's need. Farm machinery has been designed for use with several different crops by making only a few adjustments.

Another example of flexibility is the custom harvesting of grain rather than owning and operating a combine. If there is a complete crop failure, as can happen in some farm regions, no harvesting expense is incurred. But in the long run, hiring a custom operator may be more expensive than owning a combine.

When deciding whether to sell now, a farmer with on-farm storage facilities has more flexibility because off-farm storage facilities may not always be available. However, on-farm storage can be expensive.

Grazing wheat with stocker cattle is a common example of flexibility in states such as Oklahoma and Texas. Here the farmer has the flexibility to adjust the production of wheat or cattle in response to market changes. For example, when wheat prices are relatively high, the cattle might be sold early to allow for a higher wheat yield. Conversely, when cattle prices are relatively high, the cattle might be grazed for a longer period of time—a sacrifice of some wheat for a higher net income.

Equipment leasing may allow a flexible response to the risk of obsolescence. Sometimes investing in shorter-lived assets also makes it easier to make changes.

Flexibility should be considered when selecting farm enterprises. The number of *turning points* within an enterprise determine its flexibility; a turning point is any opportunity to continue or liquidate the enterprise. For example, a cattle producer who buys calves rather than yearlings enjoys the advantage of more turning points. The yearlings would go directly into the feedlot. After wintering the calves, however, there is a turning point in the spring, before they are put on pasture. If prices are good and pasture conditions seem unfavorable, they might be sold. At the end of the pasture season, the farmer can again decide whether to sell or to put them in the feedlot. Clearly, once the decision has been made to feed yearlings, little opportunity exists to change direction.

Certain enterprises are more inflexible than others in that considerable time must elapse before a return is realized. Examples are orchards, alfalfa, other perennial crops, and purebred livestock with the offspring being raised for sale as breeding animals.

Some enterprises require shorter production periods than others, so the manager can shift enterprises if necessary. Annual crops that do not require specialized equipment typify enterprises that can be started and stopped quite easily.

Enterprises also vary with respect to the proportion of fixed and variable costs as perceived at the beginning of the production period. If weaner pigs, for example, are purchased for feeding to market weight, most of the production costs—the pigs, feed, and labor—will be variable. This situation is in contrast to a farrow-to-finish hog operation, in which a higher percentage of production costs are fixed. The feeder pig program may be discontinued simply by not buying the pigs. When the farrow-to-finish operation becomes unprofitable, however, the breeding herd must be sold, probably at a loss.

Enterprise Selection and Diversification

Selecting enterprises to control risk involves studying price and production variability and choosing those with relatively stable incomes. In areas where crops are especially susceptible to weather, farmers should select crops that adapt well to drought or extreme weather.

Diversification, another precaution against risk, is exemplified by the adage "Don't put all your eggs in one basket." Because prices and yields for various enterprises do not fluctuate together, a combination of enterprises produces a more stable income than one enterprise does.

However, the results of an Indiana study suggest that diversification does not reduce variations in gross income very much.[1] Diversification as a means of reducing risk may be limited for two reasons:

1. All agricultural product prices are affected by the same basic supply and demand relationships, so fluctuations in various agricultural commodity prices over time tend to be positively associated.
2. Because the same factors tend to limit crop production in a given agricultural area, crop yields tend to be positively associated. In the Great Plains, for

[1] George F. Patrick, *Risk and Variability in Indiana Agriculture.* Purdue University Agriculture Experiment Station Bulletin No. 234, July 1979.

example, moisture is normally the limiting factor for all crops. In northern areas with short growing seasons, low temperatures affect most crops.

Diversification, therefore, is more effective in reducing the production risk when enterprises can be combined that are not equally affected by such factors as temperature and rainfall. Thus, when the yield for one crop is down, the other crops will not be reduced to the same extent. Diversifying to control market risk requires selecting a combination of enterprises with opposing price cycles, so that when the price for one commodity is down, a normal or higher price for the other offsets the loss. Two such enterprises might be cow-calf and feedgrain production.

Diversification does have disadvantages. Sometimes small enterprises are inefficient, and too many different enterprises reduce the efficiency of the manager. On the other hand, additional advantages of diversification are better utilization of labor and greater timeliness in operations. These issues are discussed further in Chapter 14.

Management of Production

Good management practices and production techniques eliminate some sources of production risk. For instance, purchasing more machinery than necessary for a normal crop year protects the farmer against years of bad weather with fewer days available for field work. This extra capacity can reduce the risk of yield losses due to late planting or late harvesting or the untimely performance of other operations. To determine how much extra machine capacity to purchase, potential yield losses due to late operations should be balanced against the cost of the extra capacity.

Weather and rainfall can vary considerably over only a few miles of farmland. Dispersing crop production over a larger area averages out these variations; however, there are increased costs for moving machinery over greater distances.

Certain production practices that reduce risk can be more costly, and some may actually lower yields, but they stabilize income over time. Some examples are irrigation systems, fallow rotations, tillage practices to conserve moisture, and the use of insecticides, herbicides, and fertilizers.

Another production risk control measure is to maintain reserves of hay, feedgrains, and other supplies necessary for the operation of the farm, which can be used if shortages occur as a result of drought or other conditions. The disadvantages are the carrying cost of maintenance, spoilage, and loss of quality.

Land-Lease Arrangements

Leasing arrangements are discussed more fully in Chapter 12. Here the emphasis is on controlling risk when negotiating land leases. With a share lease, for example, the landowner shares in the risk of production and market fluctuations, taking a specified percentage of production, but the actual quantity received is less or greater depending on the yield. Although tenants may end up paying more over time for this lease arrangement, their risk is reduced.

A variation of the cash lease arrangement can be negotiated to help control price risk. A flexible cash lease provides that the landowner shares the price risk and the tenant assumes all of the yield risk. To illustrate, if the lease rate is based on 35 bushels of grain per acre and the price at harvest is $3.00 per bushel, the cash lease payment is $105 per acre; if the price is $3.60 per bushel, the payment is $126 per acre.

To control financial risk, negotiating a long-term lease allows the tenant to make long-run plans without the fear of losing the land base. Another way to protect a land base is to lease smaller tracts from several landowners. However, production costs and the problems involved in negotiating a number of leases must be considered.

Marketing Strategies

Market risk can be reduced or controlled by a variety of measures. Forward contracting has become popular. Before the end of the growing season, the farmer signs a contract to sell a specified quantity at a specified price, which provides protection against unfavorable price movements. On the other hand, any chance for a higher return has been forfeited. Forward contracting with suppliers can also be used to fix the cost of inputs such as fertilizer and feed.

Another common strategy to control market risk is spreading sales over time. For example, one-twelfth of the crop might be sold each month or one-fourth each quarter of the year.

A more complex alternative is *hedging* on the futures market. This technique shifts the risk of price fluctuations from the farmer to the futures market by way of a futures contract. Before using this technique, the manager is advised to study the market carefully.

Financial Management Strategies

The inherent risk in agriculture has a profound effect on financial management. To be safe, farmers may hold a higher percentage of their assets in liquid form. Liquidity is the ability to readily convert assets to cash. Examples of liquid assets include government bonds, savings certificates, crop and livestock inventories, and cash. This extra liquidity gives farm managers greater flexibility in covering unforeseen expenses or taking advantage of unanticipated investment opportunities.

Some farmers may maintain an unused credit reserve as a precaution. If the lender has established a maximum credit level and the farmer has borrowed less than that amount to meet the year's operating expenses, then the balance is available to meet contingency requirements. Other possibilities that limit risk are maintaining a low solvency ratio, as discussed earlier, and negotiating flexible loan payment schedules.

Forming a corporation to hold all or part of the farm business's assets is another financial management strategy for isolating risk, particularly that related to more risky enterprises. The farmer's liability, then, can be limited to only those assets transferred to the corporation.

Backup Labor and Management

Many farm businesses are vitally dependent on one key person. If this person becomes disabled, dies, or retires, with no one qualified to take over, the business will suffer. Providing for backup management might include training other family members or bringing in a partner. In a large operation dependent on hired labor, losing a key employee during a critical time can be disastrous. To provide for continuity of the operation, the farmer could train other workers to take over, hire custom workers, bring in temporary employees, or simply work longer hours.

Obtaining More Information

More and better information helps to control risk. The following are some examples:

1. Crop production and animal performance records.
2. On farm fertilizer and variety trials.
3. Charts of past prices used for price forecasting.
4. Reliable sources of market outlook.
5. Current financial records comparing actual with projected cash flow in order to anticipate difficulties.
6. Research reports to anticipate technological changes.
7. Current regulations and legal interpretations of compliance requirements.

Gathering information and learning from experience are management processes that distinguish the farmer with 25 years of experience from the one with 1 year's experience repeated 25 times.

Insurance

Farmers must decide whether they will pay to transfer the risk to another business or carry it themselves. Insurance can be defined as the substitution of a known small cost for the possibility of a large loss. Following is a classification of the kinds of insurance available to farmers.

Property Insurance.   Farm buildings are usually insured against fire and wind damage. Motor vehicles and expensive machinery are generally insured against accidental damage and theft. Animals may also be insurable against certain losses.

Liability Insurance.   Liability insurance protects the farmer against losses due to bodily injury or damage to the property of others as a result of negligence in operating the farm. Workmen's compensation insurance is a type of liability insurance that protects employers in case of employee accidents or injuries incurred while working.

Crop Insurance.   Crop-hail insurance is available in nearly all parts of the country where the risk of hail damage is significant.
    Another type of crop insurance is all-risk insurance, a federal program made available through private agencies. This type of insurance covers losses caused by virtually all natural hazards: drought, flood, hail, wind, frost, insect damage, disease, and other "unavoidable causes."

Life Insurance.   Life insurance protects against loss of income due to death. The need for this protection is generally greatest when the family is young. Several types of life insurance are available. They vary in cost, amount of protection, and accumulation of savings. If the farmer's primary need is protection, consideration should be given first to term life insurance.

Health Insurance. Health and accident insurance can provide protection against catastrophic illnesses and accidents that cause major disruptions in farm management. Some protection against such disruptions may be worth the cost.

In choosing an insurance plan, it is important to determine just how large the loss will be if the worst event occurs and how likely it is to occur. If the insurance covering an automobile worth only $600, for example, has a $200 deductible and a $90 annual premium, the owner may decide to carry the $400 risk and keep the $90. On the other hand, if the car were wrecked, causing damage to another party in the process, the liability could run into thousands of dollars. In this instance, the annual premium, of perhaps $180, is protection against a much larger loss.

Summary

Effective management involves making risky decisions, a process that includes formulating alternative courses of action, evaluating the returns and risks as a result of these actions, and then choosing the best from among them. Management also involves prioritizing goals, carrying out the plans, and adjusting them to changing conditions.

The payoff matrix provides a convenient framework for summarizing the components of the decision guiding the budgeting process, for considering the sources of risk, and for determining the range of possible outcomes.

Carefully estimated personal probabilities based on available information combined with the payoff matrix help to evaluate the chances of both favorable and unfavorable outcomes and the expected return and risk associated with each alternative plan.

Attitudes toward risk taking are influenced by the financial position of the business—its net worth, solvency, and cash flow needs—and by personal disposition. Thus, a manager considers goals regarding income and business survival, the budgeted payoffs, and the estimated probabilities in choosing the best course of action.

Various strategies might be adopted to manage risk, such as flexibility, diversification, land-leasing arrangements, marketing options, financial management alternatives, and insurance.

These techniques for evaluating and controlling farming risks do not remove the agony of making the decisions. They do not eliminate risk, but they do help in selecting the right risks to take in the often uncertain world of the farm manager.

Recommended Readings

BOEHLJE, M. D., and V. R. EIDMAN. *Farm Management.* New York: John Wiley & Sons, Inc., 1984. Chapter 11.

NELSON, A. G., G. L. CASLER, and O. L. WALKER. *Making Farm Decisions in a Risky World: A Guidebook.* Oregon State University Extension Service, July 1978.

PART III

Acquisition of Farm Resources

One of the major problems facing farmers is acquiring control of
a sufficient quantity of resources so that the advantages of
modern technology can be realized. Part III deals with business
arrangements and farm size as well as the acquisition of
capital, land, and labor.

CHAPTER 9

Farm Business Arrangements

The family farm is the basic component of agriculture in the United States; the majority of U.S. farm businesses are operated by one individual or family. However, joint-venture business arrangements are becoming more common in agriculture. They pool resources through formalized business arrangements such as partnerships and corporations and through simpler employee, lease, and operating arrangements. A *joint venture* is defined as any business arrangement whereby two or more parties contribute resources to, and engage in, a specific business undertaking.

There can be several advantages to joint farming ventures. By working together, the parties can combine their knowledge, managerial abilities, farm assets, and financial resources. Increased efficiency may result from greater specialization of labor and management and from the economies of size (discussed in the next chapter) associated with a larger operation. By pooling resources, farmers may find that the investment per acre in machinery and equipment can be reduced. Another benefit of joint ventures is the chance for a day off, because another capable person with a stake in the business is available to supervise farm operations.

Also important, joint ventures can ease the transition of control of the business from older farmers seeking retirement to younger people who want to manage their own farm or ranch businesses. Because of the high capital and management requirements of agriculture, the best and most realistic way to enter farming and gain access to farm resources is to piggyback onto an existing farming operation. Joint ventures may be formed between relatives, such as a son or daughter and father or mother-in-law, or between unrelated parties to accomplish this transition.

There are disadvantages to joint venture arrangements. Unfortunately, people often think they are in agreement when, in actuality, they are not. Disagreements may result in the postponement of important decisions. The establishment and successful operation of a joint business venture require that all parties have a clear understanding of its goals and implications.

Before delving into the role of joint ventures in agriculture and the considerations involved in organizing workable arrangements, it is necessary to understand the alternative business arrangements that are used for farm and ranch businesses. This chapter is divided into two major sections. The first section compares the major features, advantages, and disadvantages of the business arrangements. The second presents the rationale and guidelines for selecting and designing farm business arrangements.

Comparing Business Arrangements

The three basic types of business arrangements are the proprietorship, partnership, and corporation.

The *proprietorship* is the most common business arrangement. The farm is owned and operated by one individual or family until the owner retires or dies. The owner establishes the business, provides or borrows all the capital, makes all decisions, takes all risks, and pays individual income taxes on the income of the business. The owner may hire employees and delegate managerial responsibility through contractual agreements. Proprietorships can be modified to accommodate joint ventures through various types of contractual agreements.

A *general partnership* consists of a business controlled by two or more owners contributing their resources to the business and sharing management responsibilities, profits, and losses. A partnership requires an agreement specifying each partner's contributions of capital, labor, and management and each partner's share of net income. The partnership itself does not pay income taxes; the partners are taxed on their individual shares. Each partner, however, is liable for the actions of the partnership for as long as it exists, usually until the death of one of the partners.

A *limited partnership*, when permitted by state law, limits the liability of the limited partners to their investments in the business. A limited partner is an investor who does not participate in management. A limited partnership must have at least one general partner who manages the business and is fully liable for all partnership debts and obligations.

A *regular corporation* is a legal entity, or artificial person, created by state law. It is a separate business entity, distinct from its owners for legal purposes. It can acquire, hold, and transfer property and conduct business in its own name; it is not affected by the death of its owners; it pays income taxes at corporate rates. The owners of the corporation, called shareholders because they own shares of stock that represent their capital contribution to the business, select a board of directors and officers to manage the business.

A *subchapter S corporation* is a type of corporation that receives different treatment under federal tax law. It has the same attributes as a regular corporation except that it pays no federal income tax. Instead, each shareholder individually reports a share of the corporate income. This type of corporation can have only one class of stock and is limited to no more than 35 shareholders.

The primary features of the proprietorship, partnership, and corporation are summarized and compared in Table 9-1. The following material discusses the features of each in more detail.

Proprietorships

Proprietorships are modified by various types of contractual agreements. Such a modified proprietorship is established when another party becomes a participant

TABLE 9-1. Comparison of the Major Types of Business Arrangements

	Proprietorship	Partnership	Corporation
Nature of entity	Single individual	Association of two or more individuals	Legal entity separate from shareholders
Source of capital	Personal funds or loans	Partners' contributions or loans	Contribution of shareholders for stock, sale of stock, or loans
Management decisions	Proprietor	Agreement of partners	Elected directors manage business through officers
Liability	Personally liable	General partner liable for all partnership obligations	Shareholder not liable for corporate obligations
Limits of business	Proprietor's discretion	Partnership agreement	Articles of incorporation and state law
Life of business	Terminates on death	Agreed term, but terminates on death of partner	Perpetual in most cases
Effect of death	Liquidation	Liquidation or sale to surviving partner	No effect; stock passes by will or inheritance
Transfer of interest	Terminates proprietorship	Dissolves partnership; but new partnership may be formed	No effect on continuity; stock transferable to anyone if not restricted

in the business's labor, ownership, and management. Examples include lease, employee, and operating arrangements. (Lease arrangements are discussed in Chapter 12.)

Employee Arrangements.   These arrangements (sometimes called incentive or profit-sharing plans) allow a capable and interested young person with no capital resources to furnish labor and to work into management under the guidance of the present owner of the farm business. The incentive for the younger party, of course, is a salary or share of the business's net income and an opportunity to accumulate capital and experience. Owners who want to stay actively involved in the management of the farm business may prefer the employee arrangement.

There are several variations of the employee arrangement. The simplest version is to pay the employee a specific salary. Some arrangements combine a profit-sharing or incentive plan with a minimum salary to encourage greater managerial involvement and reward extra effort. The incentive payment should be based on performance that is within the employee's control. It might be paid in the form of livestock or equipment to build the employee's capital investment in the business. The profit-sharing or incentive plan should be in writing, detailing the method of calculation, time of payment, and method of settling disputes.

As long as the arrangement provides for income sharing and not loss sharing, an employee relationship exists. This arrangement implies that the junior party, as an employee, is not responsible for the liabilities and losses of the business. It also implies that the employee's authority must be limited to those things authorized by the owner.

Operating Arrangements.   An operating arrangement is almost a partnership because the junior party supplies, and is reimbursed for, some capital as well as labor and management contributions. Operating arrangements are appropriate when the young person does not have the necessary capital to lease the farm but desires greater involvement than that provided by an employee arrangement. The junior party might begin by providing only labor and management, with the understanding that the operating arrangement will be modified as the young operator's capital investment increases. The income shares can be adjusted relative to the contributions of the parties. (Procedures and guidelines for sharing the joint venture's net income are presented in the section on "Partnerships.")

Operating arrangements require mutual trust between the two parties. Records must be kept, and the terms of the agreement must be well understood by both parties. Not all people can work harmoniously under such an arrangement. The essential idea is that the contributions of each party are valued in an agreed-upon manner, and the income is divided in the same way.

Because of its similarity to a partnership, the farm operating arrangement should be in writing, clearly specifying that the intent is not to create a partnership. An attorney can be helpful in drafting the actual operating agreement. Operating arrangements should be of limited duration, to test joint ventures until a more permanent arrangement can be developed.

Partnerships

Some of the most successful farm businesses are two-person or two-family operations. A partnership is a commonly used farm business arrangement for such joint ventures. Some advantages of a partnership are as follows:

1. A general partnership may be organized or dissolved with little or no expense and few legal restrictions.
2. A partnership allows for equal sharing of management responsibilities. Unless the agreement is otherwise, all partners have equal rights regardless of their capital contributions.
3. A partnership arrangement has greater flexibility because outside approval is not needed to change it.

Some disadvantages of a partnership are as follows:

1. All general partners have unlimited liability that extends to their personal assets. Because each general partner has the power to bind the partnership legally in other agreements, the partners must trust one another.
2. The death of a general partner, or a partner who wants to withdraw, can dissolve the general partnership and force a sale of the assets unless otherwise provided for in the partnership agreement.
3. Because unanimous agreement is usually required for major decisions, decisions may be postponed or made inconsistently.

Forming a Partnership.   Although a partnership can be formed with an oral agreement, it is advisable to have a written contract prepared by an attorney. The actual partnership agreement depends on the situation but, in general, should include the following:

1. Name and location of the business. Many states require the name to be registered and filed with a state agency.
2. Names and addresses of the partners.
3. Description of the business. It should be general and not restrict the business unless the partners intend to do so.
4. Provision for future expansion and inclusion of new partners.
5. Duration. A time should be specified for reviewing and revising the agreement if necessary.
6. Contributions of each partner. It should include how the contributions of money, property, and services will be valued; the identification of property that is loaned, rather than contributed to, the partnership; and when and how withdrawals can be made.
7. The accounting system to be used for computing net income. It is also desirable to specify who is to keep the records and exercise financial control.
8. Procedures for sharing partnership income. The method by which income shares will be calculated should be specified.
9. Management responsibilities. The agreement should indicate who is responsible for what decisions, and should specify the limitations on each partner's authority to commit the partnership.
10. The methods for settling disputes. It may include provision for a neutral party to arbitrate disagreements.
11. Procedures to be followed in case of the death, withdrawal, disability, or bankruptcy of a partner. How these events will affect the partnership should be specified. If one partner dies, the heirs can become partners and the business continued if the agreement so specifies. The agreement can include a procedure for one or more of the partners to buy out another.

A partnership agreement serves as a blueprint for the joint venture and should be specific enough so that the partners clearly understand their individual rights and obligations. One of the advantages of a partnership is its informality and flexibility; as conditions change, the agreement can be amended to reflect these changes.

Sharing of Income.   Net income (and losses) may be divided among partners in any manner they agree to, but generally should be proportionate to their contributions. If their contributions are equal, they should receive equal shares. The partners' contributions include labor, capital, and management, which are measured in different ways.

The labor contribution of each partner might be based on the actual number of hours worked during the year; in this case, each partner would be required to keep a daily time record. Alternatively, the partners might agree in advance regarding the labor to be contributed. For example, the contribution of one party might be the equivalent of 12 months, whereas for the other party it might be six months.

The value of the capital contributions of each partner should be determined when the partnership is first established. Partners may also contribute liabilities to the partnership. All asset and liability contributions are compiled in the balance sheet for the new partnership arrangement, and the percentage of the partnership's net worth contributed by each party is calculated. These percentages are then carried forward and multiplied by the new net worth at the end of the year to determine the amount of capital (net worth) each party has invested in the partnership at that time. A partner's capital contribution for any year, then, is the average of the partner's share of the beginning and ending net worth.

A commonly used procedure for distributing the partnership's net income is first to pay a previously agreed-upon salary for labor and then to allocate the remaining net income in the same proportion as capital is contributed. Conversely, the partners might each be paid, say, a 9 percent return on their respective capital contributions and the remaining net income divided on the basis of the amount of labor contributed by each partner.

There are two problems with these procedures, however. The factor (capital or labor) used to distribute the net income remaining after the first disbursement will allocate windfall gains or losses to the partner who is the primary contributor of that factor. The other problem is determining how to reimburse each partner for management, which has an important effect on the business's net income. The foregoing procedure includes management with the residual factor (capital or labor). So, if a salary is paid for labor and the remaining net income is divided on the basis of capital contributions, this approach would assume that the partners are contributing management in the same proportion as capital.

To overcome these problems, the partners can be reimbursed at preagreed-upon rates for both their capital and labor contributions, and then any remaining net income (or loss) can be allocated equally among the partners. This arrangement assumes that management is provided equally by all partners and that they should be equally entitled to any windfall gain or loss.

Another approach is to reimburse the partners for their capital and labor contributions; then, rather than sharing the remaining net income (or loss) equally, they share it in proportion to the total value of their respective labor and capital contributions. This approach assumes that management is provided in proportion to the total value of each partner's contribution of labor and capital and that windfall gains or losses should be shared accordingly.

Table 9-2 illustrates how the two approaches work if a partnership consists of two partners contributing labor and capital (partnership net worth) for the year as follows:

	Partner A	Partner B
Capital (average net worth)	$625,000	$62,500
Labor	40 weeks	50 weeks

The example assumes that the partners agreed at the beginning of the year to reimburse themselves for capital contributions at a rate of 8 percent and for labor at $500 per week.

In part A of Table 9-2, the partnership is assumed to have earned a $110,000 net income. The two approaches illustrate how this income would be shared if the partners share the management return equally or in proportion to their total contributions of labor and capital. Depending on the approach used, partner A's share of net income would be $75,000 or $77,000, and partner B, with the smaller total contribution, would receive $35,000 or $33,000.

TABLE 9-2. Two Approaches for Sharing Partnership Net Incomes of
$110,000 and $90,000

		Partner A	Partner B
Part A: net income is $110,000			
1. Management return is shared equally	Capital return at 8%	$50,000	$ 5,000
	Labor return at $500 per week	20,000	25,000
	Management return	5,000	5,000
	Net Income	$75,000	$35,000
2. Management return is shared in proportion to contributions	Capital return at 8%	$50,000	$ 5,000
	Labor return at $500 per week	20,000	25,000
	Management return	7,000	3,000
	Net Income	$77,000	$33,000
Part B: net income is $90,000			
1. Management return is shared equally	Capital return at 8%	$50,000	$ 5,000
	Labor return at $500 per week	20,000	25,000
	Management return	− 5,000	− 5,000
	Net Income	$65,000	$25,000
2. Management return is shared in proportion to contributions	Capital return at 8%	$50,000	$ 5,000
	Labor return at $500 per week	20,000	25,000
	Management return	− 7,000	− 3,000
	Net Income	$63,000	$27,000

Part B of Table 9-2 shows what will happen if the net income is lower than
expected and the partners must distribute a negative return to management. The
process is the same. It can be shared equally or in proportion to the partners' total
contributions. The choice between the two approaches depends on the preferences
of the partners and the extent to which each of them participates in the management of the farm. (Note that by sharing the management return in proportion to
total contributions, the junior partner's share of net income will be less variable
and, therefore, less risky.)

Capital and Credit Implications.   Capital for operating the partnership can be contributed
by the partners, obtained by borrowing, or increased by investing net income
derived from the farm business if the partners agreed to do so. The pooling of
capital and resources may increase the opportunities to obtain credit; however,
lenders may place some restrictions on such partnership loans. Another source of
capital is new partners. It is important to remember that each partner is personally
liable for all debts and obligations of the partnership.

Continuity of Management and Ownership.   Proper planning can provide a reasonable degree of continuity of a partnership. For example, the partnership agreement might permit surviving partners to purchase the interest or assets of a deceased partner. Partners might be beneficiaries of a life insurance policy carried on the other partners' lives, with premiums paid by the partnership. Policy proceeds would then be used to purchase the deceased partner's interest in the partnership. The important point is that if continuity is an important goal of the partners, effective provisions must be included in the partnership agreement.

Tax Considerations.   A partnership does not pay federal income tax but is required to file a return reporting the income of the partnership and the amount of each partner's share. The partnership's net income must be determined each year and divided among the partners. These amounts are income to the partners, whether or not they actually receive them. If the income is left in the partnership, the partners are still required to pay taxes on it. Because the partnership's taxable income is divided among the partners, the individual tax rates of the partners will determine the incidence of income taxes.

Limited Partnership.   A limited partnership can provide a greater degree of continuity for the business. It is used when one or more of the partners is supplying only capital—no labor or management—to the business. Most limited partnership agreements specify that a limited partner's capital contribution cannot be withdrawn until an agreed-upon date.

The limited partnership agreement must be signed by the partners, and in some states it must be recorded in a state or local government office. It insulates the limited partners against unlimited liability for the firm's obligations; their liabilities can be no greater than the amount of their investment in the business. They may not participate in the management of the business in any way, and their death, bankruptcy, or withdrawal does not ordinarily dissolve the business.

A limited partnership offers flexibility. Active managers can control the business, and other individuals can maintain an ownership interest with limited liability. It is simpler and more informal than a corporation, although the initial structuring of the limited partnership must be carefully planned with the help of an attorney so that there are no disadvantages in income tax treatment.

Corporations

As farm businesses grow, involving additional family members and possibly non-family participants, incorporation should be considered. Forming a corporation is a complex process that is not easily reversed. It is the most permanent of the farm business arrangements. Some advantages of a corporation are as follows:

1. A corporation limits the liability of the owners. Shareholders are liable for the actions and obligations of the corporation only to the extent of their investment.
2. A corporation provides continuity of the business. The life of the corporation is not terminated by the death of a shareholder or by the transfer of stock.

3. Ownership transfer is easier because shares of corporate stock can be immediately divided and transferred.

Some disadvantages of a corporation are as follows:

1. There are initial and annual corporation fees that must be paid. Other expenses include attorney's fees, accountant's fees, and appraisals.
2. Corporations have strict requirements for organization and record keeping. Reports and minutes of meetings must be prepared, corporate tax returns filed, and records maintained indicating the shares of stock held by each shareholder.
3. A corporation may have less flexibility to change its structure and operation. It has only the powers provided by law and those stated or implied in the articles of incorporation. Dissolving a corporation is more complicated than dissolving other business arrangements.

Forming a Corporation.   Usually a farm corporation will be organized in the state in which its property is located. Because of the numerous technical and legal questions involved and because state laws are not uniform, legal advice should be sought when planning and organizing a corporation.

The following steps outline the procedures involved in forming a corporation:[1]

1. The incorporators file a preliminary application with the proper state office or official, usually the secretary of state.
2. The incorporators may make a preincorporation agreement regarding the major rights and duties of the parties after the corporation has been established.
3. Articles of incorporation are filed with the state office. The state issues a certificate recognizing the organization of the corporation.
4. The shareholders or incorporators hold an organizational meeting and elect directors. In some states, the initial directors must be named in the articles of incorporation.
5. The directors meet to elect officers, adopt bylaws, select a depository bank, issue stock (or stock and debt securities) in exchange for property or cash or both, and begin business in the name of the corporation.

The articles of incorporation establish the powers and limitations of the corporation and its shareholders. They constitute the basic charter, or governing instrument, of the corporation. Some states provide forms for the articles of incorporation. Articles generally include the following;

1. Names and addresses of the incorporators.
2. Name of the corporation. The name may not be identical to or deceptively similar to the name of an existing corporation. Most states require that the corporate name include the word *corporation, incorporated, company,* or *limited* or an abbreviation.

---

[1] Based on Neil E. Harl and John C. O'Byrne, *The Farm Corporation: What It Is, How It Works, How It Is Taxed,* Iowa State University North Central Regional Extension Publication No. 11, 1983, pp. 18–19.

3. The registered office or agent of the corporation.
4. The duration of the corporation. A corporation continues until dissolved by the shareholders, by operation of law, or by termination of its stated life. Usually the life is made perpetual.
5. The powers and purposes of the corporation. A corporation has only the powers to carry out the general purposes provided by law plus those stated or implied in the articles of incorporation. The general purpose is to engage in a farming business. If necessary, a "catch-all" clause might be included to authorize all business necessary or reasonably related to the primary purpose of farming.
6. Class, number, and value of shares of authorized stock, voting rights, and any preferences or restrictions.
7. Directors and officers. State laws usually require a minimum number of directors. In farm corporations, the directors will probably be the shareholders. In addition to the office of president, some states require a vice-president, and most states require a treasurer and secretary.

The corporate bylaws are the rules governing the internal conduct of the corporation. Included are such items as the time and place for shareholders' and directors' meetings, quorum requirements, a listing of officers and their duties, and miscellaneous provisions such as the fiscal year, the bank to be used, kinds of insurance to be carried, and special limitations on the authority of corporate officers to borrow money and enter into contracts.

Limited Liability.   Limited liability means that the corporation is responsible only for business done in its name, not for the obligations of stockholders and employees acting outside the scope of the corporation. In other words, losses are limited to the value of the stock of the corporation. The advantage of limited liability may be offset if officers are required to endorse the actions of the corporation, particularly its contractual obligations. Legal action can also be taken against the stockholders rather than, or in addition to, the corporation.

Capital and Credit Implications.   Corporations may be able to attract capital by selling stock to outside investors. However, the acquisition of equity capital through the sale of securities on the open market has limited possibilities. Most farm corporations are not large enough to create the necessary confidence of investors who are not personally familiar with them.

A corporation has both advantages and disadvantages in obtaining credit. Lenders may be more willing to lend money to a corporation because its continuity is more assured than that of a proprietorship or partnership, both of which are tied to the lives of individuals. Also, the opportunity of a corporation to be more adequately capitalized than a single proprietorship or partnership may enhance a corporation's credit rating. However, lenders may view the limited liability provision as a disadvantage because it limits the security for a loan.

Continuity of Management and Ownership.   One of the principal difficulties of transferring the family farm as a unit to the next generation is its lack of divisibility. Incorporation of the farm or ranch unit is one way to solve this problem. Shares of stock can be transferred by sale, gift, will, or inheritance. Because these shares of stock can be transferred, it is possible to divide ownership of the farm unit among a number of people without breaking up the farm business. It is possible, too, for stockholders to form agreements to keep stock within a group such as a family, if this is desired. For example, an agreement can be made that stock will not be sold unless it is first offered to the corporation or to existing stockholders.

Tax Considerations.   A corporation may choose to pay federal income taxes as a regular corporation or as a subchapter S corporation. A regular, or subchapter C, corporation pays federal income tax on its net earnings at applicable corporate rates; the subchapter S, or tax option, corporation does not pay federal income tax directly but distributes its earnings to the shareholders, who pay taxes individually on their respective shares of income.

A regular corporation may provide a savings in federal income taxes if the regular corporate tax rate is lower than the individual tax rates of the owners, provided earnings are not distributed to shareholders. With a regular corporation, however, there is the possibility of double taxation: Dividends distributed to shareholders are taxed as part of the corporation's income and again as taxable income of the individual shareholders. One way to minimize this problem is to pay shareholders a salary for labor and management services performed. These salary payments are tax deductible expenses for the corporation. The individual must report the income, but double taxation is avoided.

A subchapter S corporation, by meeting certain requirements, can be taxed as a partnership. The requirements are that the number of stockholders not exceed 35, that all stock be of one class, and that all stockholders agree to this classification for tax purposes. If the corporation decides to follow this procedure, the income of the corporation is passed to the stockholders, who report this taxable income individually.

Selecting and Designing Arrangements

The most appropriate business arrangement depends on the circumstances of the managers and potential managers and on the life-cycle stages of the farm and family.

The Farm and Family Life Cycle

The typical farm business has a life cycle that parallels that of the farm family. A typical family farm business and the family that operates it progress through three stages during a lifetime: entry, growth, and exit (Figure 9-1).

At the beginning of the *entry stage*, the farm family evaluates the opportunities and decides whether to continue. If this initial testing phase is favorable and capital resources and managerial skills are sufficient, a viable economic unit is established.

During the *growth stage*, the business (and perhaps the family as well) expands. This business expansion involves leasing or purchasing additional resources, which

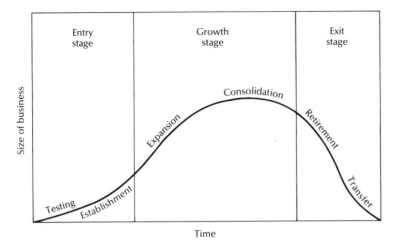

FIGURE 9-1.   The Farm and Family Life Cycle.

usually means borrowing more money. Later in the growth stage, the farm family shifts its emphasis from expansion to consolidation, with less emphasis on increasing the size of the business and more on stabilizing its income.

In the *exit stage*, consideration shifts to retirement and transferal of farm assets to the next generation. The family seeks to reduce labor and management responsibilities while maintaining control of the assets in order to generate retirement income. The transfer phase involves selling, gifting, or willing the farm assets to the next generation. This transfer might involve simply distributing the wealth among the heirs. Alternatively, it might involve more careful planning to provide for the continuation of the farm business, including the designation of managerial responsibility.

Table 9-3 illustrates the farm and family life cycle with a comparison of farm sizes (measured by the value of farm sales and acres per farm) for farm operators in various age categories. The relationship is similar to the life-cycle graph in Figure 9-1. As the age of the operator increases, the size of the farm increases until the operator is approximately 45 years of age. Then the growth in farm size levels off and begins to decrease as the operator begins to plan for retirement.

Each family must decide whether to provide for the continuation of the business after the death of the owners. The alternative is to sell the farm assets so that they can be recombined with those of other farm businesses. Inefficiencies result from the lack of continuity between generations, that is, the birth and death of a farm within one generation. If several heirs exist, an efficient farm business may be fragmented. Even if the land is controlled by one heir, there may not be adequate capital to operate the farm. Alternatively, the young farmer may have to spend several years buying out the other heirs. Through the use of various arrangements, it is possible to provide for ownership and management succession so that the farm business can continue to operate efficiently.

Many established farmers are interested in having one of their children take over the business. Others, who do not have children interested in farming, may nevertheless want to ensure the perpetuation of the business. Many young people are

TABLE 9-3. Size of Farm by Age of Operator Whose Principal
Occupation Is Farming (1982)

Age of Operator (years)	Average Value of Sales ($)	Land Area per Farm (acres)
Under 25	48,600	289
25–34	95,800	534
35–44	130,500	714
45–54	125,800	772
55–64	91,600	653
Over 65	41,800	475

SOURCE: U.S. Department of Commerce, Bureau of the Census, *1982 Census of Agriculture*, Vol. 1, Part 51, Washington, D.C., October 1984, pp. 44–48.

interested in pursuing farming and ranching careers, but not all of them have sufficient financial resources to establish a farm business of adequate size to get them through the entry stage. The careful selection of business arrangements can help these parties achieve their goals. The following sections present guidelines to help solve these problems of entry and exit.

A Decision Flowchart

In this section, a decision flowchart for selecting and progressing through various arrangements is presented. Before entering into a farm business arrangement, all the parties to the arrangement should carefully analyze their current situations and identify their goals. Then they should do a feasibility analysis (Figure 9-2) including compatibility of goals, potential profitability, fairness of the arrangement, and interpersonal relationships.

If the feasibility analysis suggests that a successful joint venture can be developed, then the parties should consider a preliminary arrangement, which allows the young farmer to decide whether farming is the right career choice. This preliminary arrangement also allows each party to test the personal and business abilities of all parties to make the venture work.

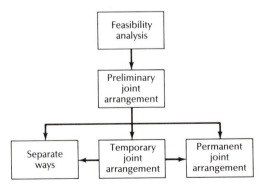

FIGURE 9-2.   A Flowchart for Selecting Business Arrangements.

In some cases, the parties may decide to skip this preliminary stage and develop a more permanent, complex arrangement. However, the failure of such an arrangement may be costly; a preliminary arrangement, therefore, allows for a less expensive trial period.

After the preliminary arrangement has been tested, which may take one to three years, the parties are ready to evaluate the arrangement and decide how to proceed. One possibility is that they will decide to go their separate ways. Another is that they will enter into a long-term or permanent business arrangement. The third alternative is the development of a temporary or short-term joint venture arrangement that will provide flexibility. A temporary arrangement can be adjusted for changes in goals, for impending retirement, and for the inclusion of other parties. Once the joint venture has stabilized, the temporary arrangement can be converted to a more permanent one.

Conducting the Feasibility Analysis

Before engaging in any type of joint venture, the parties involved should carefully examine their current situations and analyze the potential for a successful joint business arrangement. Several factors that should be included in this feasibility analysis: compatibility of goals, potential profitability, fairness of the arrangement, and interpersonal relationships.

Compatibility of Goals.   Each party in the joint venture should know and appreciate the goals of the other party. The parties must decide if their goals are sufficiently compatible so that the joint arrangement will be mutually beneficial. They may be willing to compromise on some aspects of ownership and management in order to obtain other advantages. On the other hand, some conflicts in goals cannot be accommodated in a joint venture. This possibility should be considered before going any further in organizing the venture.

The first step in assessing this feasibility factor is for all of the parties to write down and frankly discuss their respective goals. (See the section in Chapter 1 on goal setting.) Setting and sharing goals establish a good basis for communication and planning.

Potential Profitability.   The potential profitability of the plan, which should be projected before the parties enter into any type of arrangement, encourages the parties to consider all phases of the proposed joint venture. No matter how well the parties get along, the arrangement will not be successful if the business's net income is not sufficient to provide all parties with an acceptable standard of living and compensation for their labor, capital, and management contributions.

Is the business large enough, or must it be expanded, to provide adequate financial support to all the parties? To answer this question, a total farm budget might be developed (Chapter 6) based on the actual records of the farm business for the past few years. These historical figures may have to be adjusted if the business will be expanded to accommodate the additional resources that will be available in the joint venture.

Fairness of the Arrangement.   To be successful over any length of time, a joint venture arrangement must be fair to all parties. This means that they should share in the income from the operation in the same proportion as they contribute labor, capital, and management. The idea is basically simple, but different kinds and amounts of resource contributions complicate the actual process. Therefore, annual contributions should be valued so that the percentage provided by each party can be calculated and used as a basis for dividing net income.

   To maintain its fairness, the arrangement must be flexible to allow for adjustments in shares of income over time. The junior party may want the opportunity to invest additional capital in order to gain increased equity and control of the business. The distribution of annual income, then, must reflect changes over time in relative capital, labor, and management contributions. Consideration of fairness may also extend to off-farm heirs of the senior parties. If one child is to acquire part or total ownership of the farm, an estate transfer plan should be developed that clearly provides for fair treatment of all heirs.

Interpersonal Relationships.   As part of the feasibility analysis, the parties to the joint venture should honestly and frankly evaluate their ability to work and live together. In a farm business, it is difficult to separate business and personal matters. Not only the business participants but also their families must be compatible. More business arrangements are dissolved because of trivial disagreements than matters of real consequence.

   Interpersonal relationships are complicated by differences in age and maturity. This is particularly true with father–son arrangements, which are common in agriculture and unfortunately have a high potential for conflict. The father who established the farm business is often reluctant to delegate the prestige and authority of managing it. The son, on the other hand, is seeking increased responsibility and freedom but is frustrated by his father's reluctance to give up control.

   The greatest potential for conflict between fathers and sons occurs when the sons are approaching 40 years of age and their fathers are in their sixties. At that point, sons are interested in adding new enterprises and expanding the business, whereas fathers wish to protect their retirement security by consolidating the business.

   There are several ways to minimize potential conflicts in business arrangements. Because of the predominance of father–son arrangements in agriculture, the approaches are described in relation to this arrangement, but they apply equally to other relationships.

1. Sons should acknowledge the importance of their fathers' desire to maintain an active role in managing the farm business. Fathers' experience and judgment should be recognized and respected.
2. Fathers should acknowledge their sons' maturity, increasing ability, and desire to progress. It is important for sons to be able to play a greater role in management. For example, sons might assume full responsibility for separate new ventures within the business. This opportunity gives them the freedom to make their own mistakes while developing skills and experience.
3. Before entering into a family business joint venture, sons might prove

themselves first in another farm or a nonfarm business. Once they have done so, they have a better opportunity to participate in the management of the joint family venture.

4. Farm-labor and decision-making responsibilities should be divided between the parties: Each party should have certain tasks to perform and final decision-making responsibilities, or authority, in certain areas. Even with this division of responsibilities, it is important that father and son maintain frequent communication to be sure that they are not working at cross purposes.

5. Joint participation in certain major managerial decisions is necessary. Those decisions that affect the entire business are crucial to the success of a joint venture, and communication is the key to making them. Deadlines should be established for making these decisions, and then time should be set aside to do so by going through the steps of the decision-making process (Chapter 1) together. This process requires tolerance for debate and disagreement; the parties should be honest with each other and willing to work out differences. Productive debate results in better decisions, because they are based on the combined knowledge and experience of the parties.

6. There will be disagreements regarding the management of the business, but this does not mean that the parties need to be disagreeable. It is important to remember that the objective is to find a workable solution to the disagreement rather than to impress or intimidate the other party. The following tips may be helpful in resolving disagreements:
   a. Identify the areas of agreement before proceeding to the points of disagreement.
   b. Stick to the issue.
   c. Do not bring up past grievances.
   d. Listen and attempt to understand the other party's position.
   e. Present the argument so that the other person can agree without losing face. Use phrases like "Have you considered," "Don't you think," and so forth.
   f. Be prepared to change your mind if the other person's points are valid.
   g. Be willing to compromise.

7. Sometimes it will not be possible to come to an acceptable compromise. It is important to anticipate this possibility and to use a predetermined procedure to arrive at a decision, which might involve consulting with a neutral third party, such as the county extension agent, vocational agriculture teacher, lender, or any other mutually respected individual. This procedure might provide for binding arbitration. Discussing problems with another party helps to prevent a compounding of the problem. The mere existence of such a procedure for arbitrating differences may be sufficient to reduce the likelihood that such differences will occur.

It is wise to test the compatibility of the parties with a preliminary arrangement (lease, employee, or operating arrangement) before committing to a more permanent arrangement (partnership or corporation).

General Recommendations About Business Arrangements

Each of the business arrangements—proprietorship, partnership, and corporation—has a place in the flowchart in Figure 9-2. Family farm businesses are typically arranged as proprietorships, so this is the usual starting point. Following are some general guidelines for choosing the appropriate arrangement.

Preliminary Arrangements.   The purpose of the preliminary arrangement is to allow the parties to try out the joint venture and their personal compatibility before making a more permanent commitment. There are several types of farm operating arrangements that can be used in this preliminary phase. If the parties decide to develop a preliminary agreement, a modified proprietorship involving a lease, employee, or operating arrangement will probably be most appropriate. The choice depends on the circumstances of the parties, that is, the labor, capital, and management they can contribute. These are relatively simple yet flexible arrangements that can be developed at low cost and with a relatively minor commitment. Thus, they provide the opportunity to test the desires and compatibility of the parties involved.

The preliminary arrangement should be of specific, limited duration. For example, the parties might agree that after two years they will evaluate the arrangement and decide whether to modify, continue, or terminate it. One to three years should be enough time to test the compatibility of the parties and the financial feasibility of the venture. Agreeing to a specific date in advance sets a deadline for making a decision that might otherwise be postponed. It also reduces any sense of failure associated with terminating the arrangement.

Temporary and Permanent Arrangements.   At the end of the preliminary phase, the situation should be reappraised. At this point, the employee or junior member of the joint venture may decide to leave farming, or the junior party may have accumulated sufficient experience and capital to start a separate farm business. Alternatively, the joint venture may have proved successful, and all parties may choose to continue it.

If so, the next step is to consider a temporary joint venture that might include a more complex operating agreement or a partnership. The choice will depend on the parties' circumstances, available resources, and long-range goals. These arrangements allow for flexibility: They can be dissolved within a few years, or the joint venture might progress to a more permanent arrangement.

When more formally organized joint venture business arrangements are contemplated, the long-range goals of the participating parties become more important. Partnership and corporation arrangements require more complex procedures for sharing management and capital; the detrimental effects that can result from discontinuing the joint venture should also be considered.

When choosing the type of arrangement best suited to a particular farm situation, the size of the business, the number of individuals supplying capital and management, and the outside interests of these individuals are some of the factors that should be considered. In fact, a combination of business arrangements might be used to accomplish the goals of the owners and managers. For example, a corporation may own the land, a partnership may operate the farm, and some other type of business arrangement may supply the machinery or handle marketing.

A permanent joint venture usually involves creating a partnership, a corporation, or a combination of the two. A corporation particularly reduces the flexibility of the business to adjust to changing circumstances, and dissolving it may have significant tax implications. For this reason, it is important to follow the process outlined in Figure 9-2: conducting a feasibility analysis, going through a preliminary testing phase, and then progressing to a temporary or permanent joint business arrangement.

Recommendations for Designing Business Arrangements

Farm business arrangements have important legal, tax, and financing implications. It is important that they be developed with expert advice and put in writing. Advisers should be involved in the early stages of the planning process, including the attorney, the accountant, and the lender. Close teamwork among the advisers is important. Even with this expert advice, it is necessary that the parties involved have a general understanding of the legal, tax, and financial implications so that they can evaluate the advice and feel confident that the arrangement will be consistent with their goals.

Many unfortunate, frustrating conflicts arise because of different recollections about an agreement made some time ago; all parties want to abide by the agreement, but they cannot agree on what it was. An agreement should be in writing for the following reasons:[2]

1. The process of writing the agreement provides an opportunity to think through the questions, "Is this really what we want to do?", " Will it work?", and "What could go wrong?"
2. The writing process also provides for clear understanding. As one person writes, the other may say, "No, that's not what I thought we agreed to." The difference can be resolved then and there, not a year later.
3. A written agreement facilitates the settlement of the estate following the death of either or both parties.
4. Finally, in the event that legal action must be taken, the written document provides evidence to enforce performance of the terms of the agreement.

The agreement should allow for changes in goals, capital invested, labor input, and management responsibilities and should anticipate the possibility of termination so that it can proceed as painlessly as possible. For example, the agreement might provide for equal division of the physical assets or for one party to buy assets from the other party at a price determined by appraisal.

[2] J. H. Atkinson, Robert W. Taylor, George F. Patrick, and John F. Martin, *Farming Together: A Farm Management Perspective*, Purdue University Cooperative Extension Service EC 473, 1978.

Summary

Through joint ventures, farmers and ranchers can pool their resources to organize larger, more efficient units. Joint ventures also offer opportunities to young people seeking to enter agriculture, and can facilitate the transfer of farm assets between generations while providing for the continuation of the farm business. Continuity prevents inefficiencies associated with the entry and exit stages of the typical farm and family life cycle.

Before establishing a joint venture, the feasibility of the proposed arrangement should be studied, including the compatibility of the parties' goals, potential profitability, fairness of the arrangement, and interpersonal relationships. Developing the skills and tolerance to resolve disagreements is crucial to the success of the joint venture. It is also advisable to start with a preliminary arrangement in order to allow the parties an opportunity to test their ability to work together. This preliminary testing phase might involve an employee, lease, or operating arrangement.

If the preliminary arrangement proves successful, the joint venture might evolve to a more permanent arrangement such as a partnership or corporation. These two business arrangements have different advantages and disadvantages, which should be considered in deciding which one will best facilitate the continuation and growth of the farm business as well as its transfer to the next generation. Regardless of the business arrangement used, it is crucial that the agreement among the parties be in writing and that legal advice be sought in its formation.

A successful joint venture requires a deliberate approach to planning and negotiation so that the arrangement results in the achievement of goals important to all parties. Also essential are communication, compromise, profits, positive interpersonal relationships, and luck.

Recommended Readings

HARL, NEIL E. *Farm Estate and Business Planning*, 9th ed. Skokie, Ill.: Century Communications, Inc., 1984.

HARL, NEIL E., and JOHN C. O'BYRNE. *The Farm Corporation: What It Is, How It Works, How It Is Taxed.* Iowa State University North Central Regional Extension Publication No. 11, 1983.

THOMAS, KENNETH, ROBERT LUENING, and RALPH HEPP. *Planning Your General Farm Partnership Arrangement.* University of Minnesota North Central Regional Extension Publication 224, 1985.

CHAPTER 10

Farm Size and Growth

The rapid adoption of new techniques of production in recent years has resulted in a continuing increase in the average size of U.S. farms, which in turn has been associated with a decrease in the number of farms. There is no evidence that this trend has run its course. Size is, therefore, an important consideration when organizing a farm business. The farm should be sufficiently large to provide not only an efficient operation but also a net income adequate to support the farm family. The process of getting the farm business to the desired size is referred to as *growth*. The economic principles related to size and growth should be applied early in the farm planning process.

This chapter presents the principles and relationships related to the size and growth of the farm business. The relationship of costs to farm size is discussed and illustrated; then the processes of determining the optimal farm size and planning for growth to reach that size are presented.

How Costs Relate to Size

Changes in farm size refer to increases or decreases in one or more factors of production, that is, land, labor, capital, or management or any combination of them. Various measures of size, along with their advantages and disadvantages, were discussed in Chapter 5. The possibilities for changing farm size depend on the period of time being considered. In the short run, say one year, it is usually not possible to vary all the factors of production. The amount of land will probably be fixed, and it will not be possible to vary the amount of machinery greatly. Within a year, it is possible to hire more labor, use more fertilizer, or intensify production in some other way.

In the long run, perhaps three to five years, it is possible to vary more factors of production in changing the size of the farm business. Because each farm and each manager are different, there is no such thing as an optimal or ideal size appropriate for all situations. For planning purposes, it is important to understand how changes in farm size affect the profitability of the business. Using this information, the manager can establish a goal regarding the ideal farm size. For most farms, progress toward this ideal size will be made in piecemeal fashion, a step at a time. Of course, new techniques of production and other changes will affect this ideal size, and the goal may change even before it is achieved. Even so, it is important to use this long-run perspective when planning.

In this chapter, the discussions of size and growth imply that the planning period is long enough so that none of the factors of production is considered fixed. Management decisions for the short run are analyzed in other chapters.

General Cost–Size Relationships

There are several considerations that influence farmers' goals regarding the ideal farm size. One of the most important considerations is the relationship between costs and size.

For most agricultural commodities, the number of producing farmers is so large and their individual production so small relative to total output that the output of any one farmer generally has no influence on price. Therefore, the price received usually remains constant regardless of the amount produced by an individual farmer. Total returns, then, vary directly with output, because price remains constant. Costs, in turn, depend upon the inputs required to obtain different quantities of output.

To examine the relationship between inputs and output, all inputs, or factors of production, must be reduced to some common denominator. This is done by converting them to dollars and relating the costs of the inputs to output. If output changes relative to inputs, costs will also change in the same proportion. This will be true if the price or cost per unit of input remains the same regardless of the quantity purchased.

Costs are constant relative to size if the percentage change in total cost is equal to the percentage change in output. Decreasing costs are evident when the percentage change in cost is less than the percentage change in output. Conversely, costs are increasing when the percentage change in cost is greater than the percentage change in output. These relationships may be expressed as a ratio:

$$\text{Constant costs:} \quad \frac{\text{percent change in costs}}{\text{percent change in output}} = 1$$

$$\text{Decreasing costs:} \quad \frac{\text{percent change in costs}}{\text{percent change in output}} < 1$$

$$\text{Increasing costs:} \quad \frac{\text{percent change in costs}}{\text{percent change in output}} > 1$$

Decreasing costs indicate economies of size, and increasing costs indicate diseconomies of size. Specifically, *economies of size* refers to reductions in the total cost per unit of output resulting from increases in the total size (output) of the business. Farm managers consider economies of size when making long-run plans that affect the size of their operations. Information on economies of size indicates the relative efficiency of different farm sizes and how changes in size may affect efficiency.

Consider the example presented in Table 10-1. The percentage increase in costs is less than the percentage increase in output when output is increased from 10,000 to 15,000 units.

TABLE 10-1. Output–Cost Relationships

Output	Cost	Percentage Change in Output	Percentage Change in Cost	Cost per Unit of Output
10,000	$100,000			$10.00
		+50	+30	
15,000	130,000			8.67
		+33.3	+20	
20,000	156,000			7.80
		+25	+25	
25,000	195,000			7.80
		+20	+20	
30,000	234,000			7.80
		+16.7	+20	
35,000	280,800			8.02
		+14.3	+15	
40,000	322,920			8.07

NOTE: The data are hypothetical.

$$\frac{(130,000 - 100,000)/100,000}{(15,000 - 10,000)/10,000} = \frac{30\%}{50\%} = 0.6$$

In this decreasing cost situation (0.6 is less than 1), economies of size exists. The cost per unit of output declines from $10 to $8.67. Constant costs describe the situation from 20,000 to 30,000 units, with the cost per unit of output remaining constant at $7.80. After 30,000 units, costs increase at a more rapid rate than output (diseconomies of size), and the cost per unit increases. This cost–size relationship is shown graphically in Figure 10-1.

Obviously, the farmer producing in the AB range will increase net income by increasing output. If the farmer is operating at B, more total net income can be earned by producing at C even though the net income per unit remains the same. In fact, it may be profitable to go beyond C, again assuming that all prices remain constant. The marginal principle explains this fact: Production should be increased

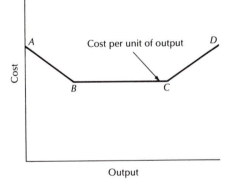

AB = decreasing costs (economies of size)
BC = constant costs
CD = increasing costs (diseconomies of size)

FIGURE 10-1. A Cost–Size Relationship with Decreasing Costs, Constant Costs, and Increasing Costs.

as long as the additional costs are less than the additional returns. If the price is higher than the average cost of production, the marginal cost will be equal to the marginal revenue at some output greater than C.

Factors Affecting Costs

Average total costs tend to decline at first as fixed costs are spread over more units of output. The total annual depreciation, interest, taxes, and insurance costs for a tractor will be the same if only 1 acre is farmed or if 100 acres are farmed. In other words, the total cost per unit of output will decrease until the fixed inputs are fully utilized.

Quantity discount buying may also result in economies of size. In other instances, a farmer may become a more efficient manager by concentrating on one particular enterprise and expanding its size.

There may be several farm sizes that are equally efficient. That is, a 480-acre farm may be just as efficient as a 640-acre farm. Constant costs may prevail over rather wide ranges for certain types of farms. In Figure 10-1, this range is labeled BC. Net income is increased when moving through this range by producing more units rather than by increasing the amount of net income per unit.

If output is increased sufficiently, a point will be reached where some factor will become limiting. It may be that the farmer is unable to buy or lease more land. To be sure, it may be possible to increase production by irrigating or by applying additional fertilizer, but if the amount of land is fixed, costs will eventually rise because of diminishing returns.

Even if the farmer can acquire all the land, labor, and capital desired, the management factor will ultimately become limiting; for example, diseconomies of size may result from the increased burden of supervising additional workers. One of the reasons extremely large-scale farming has not become prevalent in the United States is increasing costs. Many family farms are large enough to fall within the BC range.

The preceding considerations raise fundamental questions regarding U.S. agriculture. Why do small farms persist year after year? How do they remain in business if larger farms are more efficient? Larger farms tend to be more profitable, but this does not mean that if small farms became larger they would generate a higher net farm income. Even so, some farms could be more efficient if they were larger. Following are some other reasons that small farms persist:

1. Many farmers are not motivated entirely by profit considerations. For reasons of health or age, they may not be interested in operating a larger unit that would return the greatest profits.
2. Others who are more conservative may want to avoid the additional risk associated with increased size.
3. Some may be unable to obtain the capital necessary to make the adjustments that would increase farm size and profits.
4. The life cycle of the family farm suggests that some farms are smaller because they are still expanding in size, whereas others have decreased in size because the farmer is anticipating retirement.

5. Farmers are continuously adjusting to new economic conditions. New technology increases the volume of production necessary for efficient operation, and just as some farmers adjust by increasing the size of their farms, others are forced out of agriculture.

6. The opportunity cost of the farmer's labor is particularly important to part-time farmers. If these farmers reduced their amount of off-farm work, their income might suffer even though they could use the labor saved to expand the farm and produce more efficiently. That is, their opportunity costs would rise sharply. Many part-time farmers have off-farm jobs that allow for farming without disrupting their off-farm work, but an increase in the size of the farming operation may cause the off-farm work to suffer. There are many part-time farmers in the United States, and their off-farm work opportunities must be considered when they make their farming decisions.

Technical Economies of Size

To illustrate the relationships between cost and size, actual data are presented for grain farms and dairy farms. These studies examine the technical economies of size, that is, the physical relationship between inputs and outputs, assuming that the prices paid and received are not affected by size.

Grain Farms. The cost–size data for two grain farms, one located in the Corn Belt and the other in the Pacific Northwest, are from a U.S. Department of Agriculture study.[1] Costs were estimated for four hypothetical farm sizes in each region. The farm sizes, measured as the number of acres of cropland farmed, were chosen to approximate roughly the midpoints of the four quartiles of the farm size distributions. In other words, the four study farms in each region were developed to correspond to the smallest 25 percent of farms, then the next 25 percent, and so forth.

The Corn Belt grain farms were assumed to produce corn, soybeans, winter wheat, and oats (Table 10-2). The acres of cropland for the four study farms ranged from 77 to 639 acres, with gross income ranging from $16,000 to $145,000 per farm.

The Pacific Northwest grain farms were assumed to produce winter wheat and barley. The size of the four study farms in this region ranged from 120 to 1887 acres of cropland (Table 10-3). Gross income for these farms ranged from $14,421 to $155,617 per farm.

The results for both of these study farms indicate the same general economies of size relationships. As farm size increases, per-unit costs decline at first and then are relatively constant over a wide range of sizes.

The most efficient farm size for the Corn Belt grain farm is farm D, with $145,000 in gross income. The average cost is $0.51 per dollar of gross income. This is only $0.08 lower than the average cost for farm C but is $0.15 less than the average cost per dollar of gross income for the smallest farm.

[1] T. A. Miller, G. E. Rodewald, and R. G. McElroy, *Economies of Size in U.S. Field Crop Farming*, U.S. Department of Agriculture Economics and Statistics Service Agricultural Economic Report No. 472, 1981.

TABLE 10-2. Average Costs Per Dollar of Gross Income for Four
Corn Belt Grain Farms

	Farm A	Farm B	Farm C	Farm D
Cropland (acres)	77	141	272	639
Gross income ($)	16,000	32,000	53,000	145,000
Total costs ($)[a]	10,492	19,936	31,160	73,808
Average costs ($)	0.656	0.623	0.588	0.509
Difference from lowest cost ($)	0.147	0.114	0.079	0

SOURCE: T. A. Miller, G. E. Rodewald, and R. G. McElroy, *Economies of Size in U.S. Field Crop Farming*, U.S. Department of Agriculture Economics and Statistics Service Agricultural Economic Report No. 472, 1981.
[a] Includes all relevant costs except for the operator's net worth and management.

The results from the Pacific Northwest grain farms are similar. Again, the farm with the lowest average cost per dollar of gross income ($0.60) is the largest farm. This average cost is only $0.01 less than that of the next largest farm and is $0.17 less than that of the smallest farm. These relationships are illustrated in Figure 10-2.

These results do not provide a conclusive indication of the least-cost farm size, because the cost for the largest farm (farm D) is lowest in both regions. However, because of the small cost decrease from farm C to farm D, farm D appears to be large enough to include almost all available efficiencies.

This study indicates that mid-size farms with gross incomes of $50,000 to $90,000 per year achieve most technical cost efficiencies. However, a farm of this size may not provide net income sufficient to support the farm family. Farmers in this mid-size category may have to expand, not to improve efficiency but to increase net income through a greater volume of output.

TABLE 10-3. Average Costs Per Dollar of Gross Income for Four
Pacific Northwest Grain Farms

	Farm A	Farm B	Farm C	Farm D
Cropland (acres)	120	290	630	1,887
Gross income ($)	14,421	34,851	75,712	155,617
Total costs ($)[a]	11,029	24,473	46,438	93,438
Average costs ($)	0.765	0.702	0.613	0.600
Difference from lowest cost ($)	0.165	0.102	0.013	0

SOURCE: T. A. Miller, G. E. Rodewald, and R. G. McElroy, *Economies of Size in U.S. Field Crop Farming*, U.S. Department of Agriculture Economics and Statistics Service Agricultural Economic Report No. 472, 1981.
[a] Includes all relevant costs except for the operator's net worth and management.

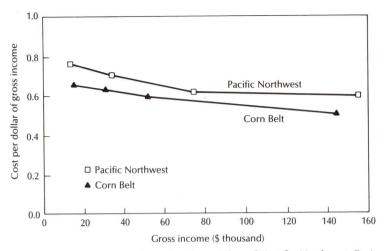

FIGURE 10-2.   Average Costs for Corn Belt and Pacific Northwest Grain
Farms.

SOURCE: T. A. Miller, G. E. Rodewald, and R. G. McElroy, *Economies of
Size in U. S. Field Crop Farming*, U.S. Department of Agriculture Economics and Statistics Service Agricultural Economic Report No. 472, 1981.

Dairy Farms.   The relationship between cost and size in animal production is illustrated by
a study of dairy farms located in the Chino Basin of California.[2] The dairies in
this study were specialized, single-enterprise units producing fluid milk. Alternative
combinations of milking parlor and housing configurations and equipment and
labor complements were considered, with capacities ranging from 375 to 1200
cows. Costs were estimated for the various functions involved in dairy production:
milking, housing, feeding, waste disposal, management, and record keeping. Then
the average costs per cow were calculated for each dairy (Table 10-4).

Figure 10-3 indicates the costs per cow for various sizes of the dairy cow herd.
Dairy farms with a double 5 Herringbone parlor are estimated to have a cost of
$1132 per cow with a herd size of 275 cows and a cost of $1065 per cow with a
herd of 375 cows. The general relationship holds for the other milking parlor
designs and housing units.

The estimated average costs (Figure 10-3) reveal significant economies of size
in the range of 375 to 750 cows. The estimated average total cost per cow for a
375-cow dairy is $1065; for a 750-cow dairy it is $1001; and for a 1200-cow dairy
it is $999. Thus, there is no significant cost advantage for herd sizes of more than
750 cows.

Although not shown in Figure 10-3, the analysis was extended to a herd size of
3600 cows. The average cost at that size is $994 per cow, indicating only slight
cost reductions for dairies with more than 1200 cows. The advantage of a larger

[2] S. C. Matulich, H.F. Carman, and H.O. Carter, *Cost–Size Relationships for Large-Scale Dairies
with Emphasis on Waste Management*, University of California Giannini Foundation of Agricultural
Economics Research Report No. 324, October 1977.

TABLE 10-4. Average Costs Per Cow for Drylot Dairies with Waste
Incineration, by Herd Size, Southern California

Herd Size (no. of cows)	Parlor Design[a]	Housing Units[b]	Cost Per Cow ($)
375	H5S	4/100	1065
450	S03-2	4/120	1024
500	S03-2	5/100	1041
600	S03-2	5/120	1019
625	S03-2	7/100	1019
750	S04-2	8/100	1001
900	H12A	8/120	1002
1000	H10A	10/100	1010
1050	H12A	9/120	1009
1200	H16A	10/120	999

SOURCE: S. C. Matulich, H. F. Carman, and H. O. Carter, *Cost–Size Relationships for Large-Scale Dairies with Emphasis on Waste Management*, University of California Giannini Foundation of Agricultural Economics Research Report No. 324, October 1977, p. 85.

[a] The abbreviations for milking parlor designs are as follows:

H5S = Double 5 Herringbone parlor with swinging machines
H10A = Double 10 Herringbone parlor with automated machines
H12A = Double 12 Herringbone parlor with automated machines
H16A = Double 16 Herringbone parlor with automated machines
S03-2 = Double 3 side-opening parlor with automated machines and a wash stall
S04-2 = Double 4 side-opening parlor with automated machines and a wash stall

[b] Number of units / capacity (number of cows) of each unit.

herd is in having more cows contributing to net income, rather than in having more net income per cow.

The average costs in Figure 10-3 indicate the level of costs that are possible. However, whether these costs are achieved in practice depends on the management capabilities of the dairy farmer.

Price Economies of Size

The technical economies of size illustrated in these two studies of grain farms and dairy farms is not the only cost effect of increasing farm size. Economies associated with lower prices for purchased inputs, such as fertilizer, due to quantity discounts are also important. They are referred to as price economies of size.

A survey of farm operators in Ohio provides data indicating the magnitude of these price economies of size.[3] The survey was designed to measure the purchasing

[3] C. R. Zulauf and K. F. King, "Input Purchasing Advantages of Large Farms in Ohio," *Ohio Report*, September-October 1983, pp. 69–71.

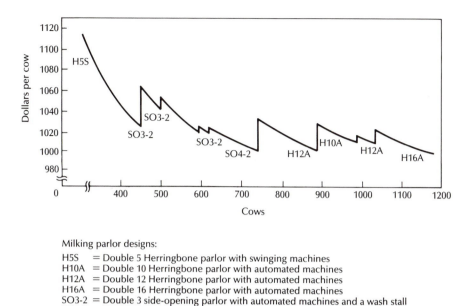

Milking parlor designs:

H5S   = Double 5 Herringbone parlor with swinging machines
H10A = Double 10 Herringbone parlor with automated machines
H12A = Double 12 Herringbone parlor with automated machines
H16A = Double 16 Herringbone parlor with automated machines
SO3-2 = Double 3 side-opening parlor with automated machines and a wash stall
SO4-2 = Double 4 side-opening parlor with automated machines and a wash stall

FIGURE 10-3.   Average Costs for Southern California Dairy Farms.

SOURCE: S. C. Matulich, H. F. Carman, and H. O. Carter, *Cost–Size Relationships for Large-Scale Dairies with Emphasis on Waste Management,* University of California Giannini Foundation of Agricultural Economics Research Report No. 324, October 1977, p. 87.

advantages associated with farm size. The farm operators were asked to indicate the average price discount received on seed, fertilizer, pesticides, and crop machinery.

Table 10-5 reports the relationship between farm size and the average discount received by input category. To indicate the total advantage accruing to farm size in the purchase of seed, fertilizer, pesticides, and crop machinery, the average discounts are applied to the costs per acre of growing corn. The estimated total cost reductions per acre are shown on the bottom row of Table 10-5. The reduction in costs resulting from price economies range from $2.35 per acre for farms with 100 to 179 acres to $9.73 per acre for farms with 1000 acres or more.

Economies of size in product selling are also possible. For example, grain producers may be able to earn a premium on their grain by contracting it in large lots for delivery on a regular schedule. Purchasers will pay such a premium because this arrangement reduces their procurement costs.

Managerial Economies of Size

Another important consideration in determining the relationship between cost and size is the effect of size on managerial requirements. A study of 97 North Dakota

TABLE 10-5. Average Input Price Discounts by Farm Size, Ohio

	Farm Size (acres)				
	100–179	180–259	260–499	500–999	1000+
Discounts by input (%)					
Seed	2.4	3.6	4.3	6.5	10.8
Fertilizer	2.2	2.2	2.8	3.6	6.6
Pesticides	1.4	2.3	2.2	3.0	7.8
Machinery	1.3	2.2	3.8	5.6	7.9
Total discount ($/acre)					
Cost reduction for corn production	$2.35	$3.05	$4.12	$5.77	$9.73

SOURCE: C. R. Zulauf and K. F. King, "Input Purchasing Advantages of Large Farms in Ohio," *Ohio Report*, September–October 1983, 69–71.

grain farmers indicates how managerial requirements are related to farm size.[4] The farms ranged in size from 850 to 5600 acres, and all of the farmers received two thirds or more of their income from grain production. Management activities were defined as purchasing inputs, acquiring land, keeping and using records, gathering information and consulting, marketing products, supervising labor, and planning. Planning and labor supervision were the two most time-consuming management tasks, taking about 40 percent of the farmers' management time.

Analysis of the data for these 97 farmers indicates the economies of size associated with the management function. This result is expected because the time required for many management activities is relatively fixed regardless of the size of the farm. As a result, the time required per unit of output to accomplish these managerial tasks decreases as size increases. For example, the amount of time required to decide whether to market 50,000 bushels of grain is approximately the same as that required to decide about 100,000 bushels of grain. The study results show that the time used per dollar of gross sales to purchase inputs, market products, keep records, plan, and gather information declines with farm size.

On the other hand, the management time devoted to the supervision of labor indicates diseconomies of size. The relationship between the hours required for labor supervision and the size of the labor force is illustrated in Figure 10-4. Per unit of labor hired, the time required to supervise labor decreases as farm size increases up to 15 employee-months. Then the supervision time required per unit increases with the amount of labor hired up to a total of 48 employee-months. However, the diseconomies for labor supervision are outweighed by the economies of size for other management activities.

When farm size is measured by gross sales, management time per dollar of gross sales declines with farm size. Although economies of size may be possible with other managerial activities, labor supervision appears to exhibit diseconomies of size. These relationships between managerial requirements and size may vary with

[4] R. G. Johnson and S. C. Hvinden, "Labor and Management Components in Economies of Farm Size Studies," *Western Journal of Agricultural Economics* 3 (December 1978):205–210.

FIGURE 10-4.   Average Annual Time Spent in Supervision of Labor by Amount of Labor Supervised.

SOURCE: R. G. Johnson and S. C. Hvinden, "Labor and Management Components in Economies of Farm Size Studies," *Western Journal of Agricultural Economics* 3 (December 1978): 205–210.

the type of farm, but they are nevertheless important to the consideration of farm size and growth.

These four studies of technical, price, and managerial economies of size are consistent with many other studies of farm size that have been done in various parts of the United States. Large farms tend to have lower average costs and higher net incomes, but increasing the size of a particular farm does not necessarily mean that its net income will increase. A farmer's management ability, amount of resources, family situation, and attitude toward risk and uncertainty must be taken into account in determining whether or not to increase the size of the farm business.

Determining the Optimal Farm Size

The long-run average cost relationships illustrated in Figures 10-1, 10-2, and 10-3 are useful as a general planning device because they reflect the least costly way to produce different quantities of output and give some idea of the size of farm necessary to achieve economies of size. In addition, the long-run average cost serves as a goal. However, because most decisions are made for the short run, such a goal may not be immediately attainable. Once the decision has been made to farm, size is usually expanded until some limiting factor is encountered and costs begin to rise. A great variety of factors limit farm size, but over time these limitations can often be relaxed or bypassed, allowing progress toward the long-run average cost goal.

Generally, enterprises should not be included in the farm organization unless they are to be carried on at a scale sufficiently large, under existing technology, to permit economies of size to be achieved. Complementary and supplementary enterprises are exceptions, however.

Machinery items that represent a large investment can be analyzed in the same way. Fixed costs per hour of use decline rapidly as use is increased. Variable costs per hour of use remain relatively constant. Unless the machine can be used enough to achieve a low cost per hour, it probably should not be included in the farm plan. Custom hiring or other alternatives should be investigated if the farm is not large enough to make owning the machine efficient.

After a combination of enterprises has been chosen and an efficient combination of machinery, labor, and land has been determined, budgets can be used to vary the volume of production and observe the effect on revenue and expenses. Personal ambitions and income goals should be kept in mind. If a farm manager believes that $20,000, for example, of annual spendable income would produce a standard of living equal to that from nonfarm employment, the feasibility of a $20,000 income can be estimated. The capital and time required to increase the farm business to the necessary size can also be determined. Realistic, level-headed planning at the time the farm is being organized helps to prevent disappointments later.

The Farm Business Growth Process

Growth refers to increases in the size of the farm business as a result of the acquisition of additional resources over time. The rate of growth indicates how fast the size of the business is changing and how long it will take to reach the manager's size goal. The various measures of farm business size presented in Chapter 5 are also used to measure growth. For example, growth can be measured as changes in acres of land, total value of production (revenue), and total invested capital (total asset value).

Before discussing strategies for achieving business growth, it is helpful to discuss why growth occurs. By understanding the various factors influencing growth, the development of strategies to effect growth will become more apparent.

Why Farms and Ranches Grow

The increasing size and decreasing number of farms and ranches in the United States are partly a response to narrower profit margins in agriculture and partly a result of the requirement to keep up with the competition in order to survive. Farm businesses grow to increase the efficiency of the operation and to achieve economies of size. But achieving economies of size is not the only reason that farms continue to grow; they grow for several other reasons, such as attaining a net income that is adequate to support the farm family.

Economies of size aside, profit margins provide an incentive to grow. A farmer producing 20,000 bushels of wheat, with a net farm income of $0.40 per bushel, may not consider the total $8000 sufficient to provide the desired living standard for the family. Increasing the size of the business would not affect the net farm income per bushel if there were no effect on the price per bushel (generally true because individual farm businesses are small compared to the total market) and on the cost per bushel (true if costs are constant). If $16,000 is considered an adequate net farm income, increasing the size of the farm business to produce 40,000 bushels of wheat would solve the problem.

The relatively low net income per unit of agricultural output is a strong incentive for growth. These low net income margins result from an inelastic demand for farm products and from technological change that has the effect of increasing the total supply. Another incentive for growth is the increasing capacity of tractors and other machinery and equipment. In order to utilize this new technology effectively, farmers must usually expand the size of the farm business.

Finally, farm managers seek growth to exercise their management ability. They want greater control over their environments, and expanding the business is one means of exerting this control.

Necessary Conditions for Growth

For farm growth to occur, certain necessary conditions must be met:[5]

1. *Excess managerial ability.*   The operator must be willing and able to take on the duties that a larger farm business requires.
2. *Profitability of the business.*   The current farm operation must be profitable; growth will not turn losses into profits. However, highly profitable operations have less incentive to grow. Growth is also influenced by the manager's goals and the particular financial management strategies employed.
3. *Minimum starting size.*   For growth to take place, the farm business must be large enough to support the farm family and to provide some surplus cash to acquire additional resources.
4. *Unused resources.*   Businesses that have the opportunity to grow are those that have some unused resources, perhaps unused machinery capacity. A farm business already well organized, with no unused resources, has less incentive to grow, unless, of course, there is unused managerial capacity.
5. *Available resources.*   Because growth requires control over additional resources, these resources must be available to, or procurable by, the farmer. The farmer in need of additional land must be able to find land for rent or sale if the business is to grow.

Comparing Strategies for Growth

Both the timing and the methods used to achieve farm growth will influence the profitability and financial stability of the business. The various effects of alternative growth strategies were estimated using a simulation analysis of a farm situation representative of southern Georgia.[6] The principal enterprises were tobacco, peanuts, cotton, corn, soybeans, and hogs. Owned assets initially included 200 acres

---

[5] W. R. Bailey, "Necessary Conditions for Growth of the Farm Business Firm," *Agricultural Economics Research* 19 (January 1967):1–6.

[6] G. S. Smith, W. N. Musser, and F. C. White, *An Analysis of the Effects of Farm Size and Expansion Decisions on the Financial Conditions of Farm Firms in South Georgia,* University of Georgia Agricultural Experiment Station Research Bulletin 321, December 1984.

of farmland, machinery, equipment, and working capital. These assets were assumed to be worth $149,888.

The time period for the study was seven years, from 1974 to 1981. The average prices and yields for the farm enterprises were assumed to conform to trends that were actually observed over this time period. The potential variability of the prices and yields during this time period was also considered. Given the observed average trends and variability, price and yield outcomes were randomly generated. Twenty sets of random outcomes were generated to estimate the results for each growth strategy.

The alternative growth strategies analyzed in this study included (1) continuing the base farm operation of 200 acres, (2) expanding by purchasing 200 acres in 1974, and (3) expanding by purchasing 200 acres in 1978. Each strategy was analyzed with initial debt at 30 and 50 percent of assets, giving a total of six situations.

The final-year results for the alternative growth strategies are summarized in Table 10-6. Continuation of the initial farm size of 200 owned acres results in a substantial erosion of net worth over time. The ending net worth in this case is a little more than one half of the initial net worth, assuming a beginning debt to asset ratio of 30 percent. This decline in net worth is primarily a result of low net farm income. Although land value increases contributed to net worth, the farm size is too small to have a significant impact. The probability that the business would be insolvent, that is, that it would have a debt to asset ratio greater than 100 percent, is less than 1 percent. However, if this farm started out with a 50 percent debt to asset ratio and did not grow, the probability of failure is 17.2 percent.

TABLE 10-6. Final-Year Average Net Worth, Average Debt to Asset Ratio, and Probability of Insolvency for Alternative Growth Strategies, South Georgia Crop Farm, 1974–1982

Farm Size (acres)	Average Ending Net Worth ($)	Average Debt to Asset Ratio (%)	Probability of Insolvency (%)
Initial debt to asset ratio of 30%			
Own 200 acres	55,886	82.2	0.2
Own 200 and buy 200 in 1974	445,574	17.3	0.0
Own 200 and buy 200 in 1978	34,545	93.5	13.2
Initial debt to asset ratio of 50%			
Own 200 acres	20,906	93.3	17.2
Own 200 and buy 200 in 1974	430,858	19.2	0.0
Own 200 and buy 200 in 1978	−24,023	104.5	78.1

SOURCE: G. S. Smith, W. N. Musser, and F. C. White, *An Analysis of the Effects of Farm Size and Expansion Decisions on the Financial Conditions of Farm Firms in South Georgia,* University of Georgia Agricultural Experiment Station Research Bulletin 321, December 1984.

Expanding the farm by purchasing 200 additional acres in 1974 is the most desirable alternative for growth. Expanding the acreage allows the achievement of significant economies of size, and the net worth is also enhanced by favorable land value increases. As a result, the net worth increases to $445,574 when starting with a debt to asset ratio of 30 percent and to $430,858 when starting with a debt to asset ratio of 50 percent. In both cases, the probability that the business would be insolvent is zero.

The growth strategy of purchasing land in 1978 is the least desirable. The early years at 200 acres under this strategy are not economical. Then, when the farm is expanded in 1978, debt levels have increased and land purchase prices are higher, so the business's financial condition continues to deteriorate. With a 30 percent beginning debt to asset ratio, the ending net worth is $34,545, with a 13.2 percent probability of being insolvent. If the farm begins with a 50 percent debt to asset ratio, the probability of being insolvent is 78.1 percent. Although revenue increases with expansion, the large acquired debts resulting from buying land at the higher 1978 prices result in net incomes too low to support the farm.

Another alternative considered in this study is that of increasing the farm size by renting rather than purchasing an additional 200 acres. Although the probability of insolvency is zero for both of these alternatives, the ending net worth is lower for renting the additional 200 acres; renting does not provide the advantage of the increasing land value.

This analysis, based on the economic conditions of the late 1970s, indicates that purchasing additional land early is the most favorable growth strategy, resulting in the steady growth of net worth and no financial stress. The least favorable strategy is no expansion at all, resulting in a decrease in net worth and a high probability of failure, because the initial farm size is too small to realize any economies of size. By expanding to 400 acres, the efficiency of the farm machinery can be increased with only minimal increases in machinery capacity.

This study emphasizes the importance of considering expected economic trends as well as economies of size, cash flow, and debt position when developing plans for growth; for example, land purchases are obviously more desirable if land is expected to continue increasing in value.

Planning for Growth

The growth stage in the farm and family life cycle (Figure 9-1) is the most critical and, in some ways, the most perilous. It is not uncommon for small and mid-size farms that have prospered for many years to go broke while attempting to expand. Why? Some farmers have the ability to operate a small business profitably, but not a larger business. Some expand too much or too quickly. Others expand too little or too slowly. Some choose the wrong time to expand, and still others expand in the wrong ways.

Increasing the size of the business has many pitfalls, but this does not mean that growth should be avoided. However, to be successful, the manager needs a well-planned strategy for achieving growth. Planning can make the difference between growing successfully and "growing broke." The total farm and cash flow budgeting procedures outlined in Chapter 6 can be used here. If expansion means borrowing substantial amounts of money, this long-range plan will be needed when the manager applies for financing. The plan will help both the manager and the lender to estimate how much debt can be handled comfortably and profitably.

The following sections outline some of the considerations involved in planning for farm business growth.

Improving Managerial Capacity.   What happens to the profitability of the farm business as its size increases? Is net farm income expected to increase, remain constant, or decrease? The answer depends, in part, on the realization of economies or diseconomies of size. It also depends, in part, on how quickly management capacity adjusts to the increase in farm size.

The capacity of the business's management is reflected in its profitability and depends on the manager's ability to set realistic goals, define relevant problems, obtain information, consider alternatives, make decisions, take action, accept responsibility, and evaluate performance. Managerial capacity must increase with the growth of the farm business if profitability is to be maintained or improved. In fact, extra managerial capacity may be needed to plan the growth process, including resource acquisition.

Matching managerial capacity to new managerial requirements is a special problem when growth involves new and unfamiliar enterprises. Any lags in management's adjustment to the new requirements may severely jeopardize the business.

Acquiring additional managerial capacity is part of the growth-planning process. Managerial capacity can be increased by education or by hiring additional managerial services. Hiring can include getting advice from a private consultant, hiring a new employee to share management responsibilities, and retaining the services of an accountant. Management education can include attending classes, participating in extension workshops, self-education activities, and experience. (However, experience can be an expensive teacher.) Educational activities are also available through suppliers and product buyers, including commodity brokers and processors.

The important point is that improving the profitability of a growing business involves increasing the capacity of the business's management before, or coincident with, the growth of the business and management requirements.

Obtaining Capital to Grow.   The growth plans are likely to require additional capital investment. There are four basic sources of (or methods of obtaining) additional capital resources; each has advantages and disadvantages.

1. The owner's net worth is the manager's own capital, available for investment in the growth of the business. It includes net farm income after withdrawals plus gifts, inheritances, and nonfarm income.
2. Debt capital is borrowed from a lender who requires interest payments and repayment of the loan.
3. Outside equity is money that others invest in exchange for a share in the ownership of the expanded business.
4. Leasing is another way of acquiring the use of capital.

Acquiring capital is discussed in more detail in the next chapter, and leasing as a method of acquiring land is discussed in Chapter 12.

Family living requirements and income taxes use net farm income and cash flow. As such, they influence reinvestment in the business and therefore farm business growth. The way to maximize growth is to minimize current expenditures for family living. But current consumption and life-style are also important to the manager and the manager's family. Thus, growth decisions must include the allocation of income between current living expenditures and investment, investment that will provide for greater consumption in the future. Family living expenditures should also be examined to be sure that they contribute to the family's goals and that the maximum benefit is obtained from each dollar spent.

Tax management can also play a significant role in achieving farm growth. The manager seeking to increase growth rates should examine all opportunities for reducing the income tax rate. It should be stressed, however, that the objective is not to minimize taxes but to maximize net income after taxes.

Managing Risks.   The farm business continually faces risks, but the middle of an expansion plan is one of the worst times to have something go wrong. A growing business that has extended itself with additional debt may not yet have realized the projected increase in income expected from the expansion. Increasing the rate of growth often exposes the farm business to more risk. Risk can also limit the amount of debt capital available to finance growth. (Various methods for controlling or counteracting risk were discussed in Chapter 8.)

Other Growth Planning Considerations.   There are several other considerations when deciding whether to grow, how much to grow, how to grow, and how fast to grow:

1. What are the manager's goals? How big does the manager really want the business to be? How soon? Some managers are better off and happier owning and operating a small business rather than a larger one.
2. What sacrifices would have to be made? Is the manager willing to make them? Managing a larger business may mean doing less desirable jobs; it may mean working longer days. It may involve financial risks that the manager is reluctant to take. It may mean more decision-making responsibility.
3. What would happen if the business were not expanded? Would it be able to continue as it is? Would it be able to provide an adequate net income to meet family living expenses? Is there a chance that the land currently being leased could be lost to another tenant? In other words, is expansion something the manager wants to do or has to do?
4. What are the neighbors doing or likely to do? How would the expansion of their businesses affect the manager's expansion plans? Would they be competing for the same land and other resources?
5. What are the chances of failure? Too often when planning growth, managers consider only the prospects for success, neglecting the possibility of failure. Consider the sources of risk, and list those events that might hamper successful expansion of the business.

6. Is now a good time to expand? What are the advantages and disadvantages of initiating the growth plan now, as opposed to waiting?
7. What are the best sources of help? A decision to enlarge the business is important. Managers should consult their lender, attorney, accountant, and other advisers.

Answering these questions and carefully preparing a long-range plan is essential for successful growth.

Summary

Economies and diseconomies of size describe the relationship between the cost per unit of output and the total size (or output) of the farm business. Economies of size refers to the reduction in average costs associated with increasing size and explains why farms grow bigger. But economies of size are not the only reason for increasing farm size. When economies are constant, farms may also grow to increase net income through greater volume, to take advantage of new technology, and to use excess management capacity.

Studies of farm size reveal the effects of technical, price, and managerial economies in agriculture. These economies are significant for mid-size farms compared to small ones, but they are less so for large farms, indicating smaller cost advantages as farm size is expanded. In other words, the typical long-run average costs exhibit decreasing costs for smaller farms sizes and constant costs for larger ones.

The growth process refers to how farm businesses increase in size. Increasing the size of the business does not guarantee reduced average costs or greater profitability. The strategy and timing of growth are important in influencing the success of the expansion over time. Growth is risky, but this risk can be reduced through careful long-range planning.

Different factors of production can become limiting at various times in the growth process. In the short run, land may be limited, but in the long run this situation can be changed. Capital limitations can be overcome. Managerial limitations may also be important, but a desire to succeed plus the willingness to improve managerial capacity will enhance farm profitability and growth potential.

Recommended Reading

CARTER, H. O., W. E. JOHNSTON, ET AL. *Farm-Size Relationships, with an Emphasis on California*. University of California Giannini Foundation of Agricultural Economics Project Report, 1980.

CHAPTER 11

Acquiring Capital and Credit

To generate a net income, a farm or any other business must control resources, or capital. Farming generally requires a larger capital investment than other types of businesses to produce equivalent income. For this reason, the process of acquiring capital in farming is of great importance.

The traditional classification of factors of production as land, labor, capital, and management is sometimes convenient. In a broader sense, however, land is merely a form of capital. In this chapter, *capital* is defined as the total economic value of the resources available for use in the farm business. In succeeding chapters, land and labor are treated separately, but before land can be acquired or labor hired, the required capital must be available.

The Role of Capital

The management of capital includes two related considerations: the proper amount of capital to use and the correct allocation of a given amount of capital. The principle of equimarginal returns (discussed in Chapter 2) is the guiding principle in the allocation of capital, and budgeting (discussed in Chapter 6) is the tool used to apply this principle. The factors affecting the amount of capital used and how farm capital is acquired are discussed next.

Factors Affecting Capital Use

In general, the amount of capital to use is determined by comparing the marginal rate of return to capital to the interest rate. The *marginal rate of return to capital* is the return earned on the incremental, or additional, input. It is comparable to the marginal value of the product (MVP), as discussed in Chapter 2, except that it is calculated as a percentage, or rate, like the rate of return to total farm assets (as discussed in Chapter 4). The difference between the marginal rate of return to capital and the return to total farm assets is that the latter is the average return to all capital invested. In comparing the marginal return to capital to the rate of interest, both farm and nonfarm opportunities should be considered, as well as problems of risk and uncertainty.

Economic principles suggest that capital invested in commercial agriculture should earn a return at the margin that is about equal to the current rate of interest. This means that additional investment in commercial farms normally earns its cost, or perhaps just a little more. Because of uncertainty, however, managers may quite naturally decide to use less capital than would be the most profitable. That is, they may decide to be more conservative, passing up potential profits because they are not willing to risk a loss.

Some inefficient use of capital can be traced to more fundamental maladjustments. Farmers are sometimes unable to acquire the most profitable amounts of capital because of limits on the availability of credit. Some farmers may underinvest because their goals conflict with an increased debt load and expanded farm business. On the other hand, rapid inflation and income tax advantages may encourage farmers to overinvest. Another reason for the inefficient use of capital is related to the life cycle of the family farm business. Unless provision is made for transferring farm assets from one generation to the next, farm managers will find it difficult to increase the size of the business and maintain an efficient size through succeeding generations.

How Capital Is Acquired

How do farmers acquire the capital to finance the assets of the farm business? Farmers obtain control of capital in the following ways:

1. The owner's net worth—the manager's own capital invested in the business.
2. Debt capital—money borrowed from a lender who extends credit in exchange for interest payments.
3. Outside equity—money that others invest in exchange for a share in the ownership and profits of the business.
4. Leased assets. (Leasing land is discussed in Chapter 12 and leasing machinery in Chapter 16.)

The farm manager's own net worth or equity in the business is an important source of capital for investment in agriculture. There are several ways in which this equity capital accumulates:

1. Original and additional capital contributions. These contributions may come from personal savings derived from nonfarm sources of income or from gifts and inheritances, or they may result from marriage. Another important source is parents who may be willing to transfer a portion of their net worth to help their son or daughter get into farming.
2. Reinvestment of a portion of farm business net income. The portion of net income available for reinvestment depends on the amount of withdrawals made for income taxes and family living expenses.
3. Inflation in the value of assets. These gains in asset values resulting from inflation increase the market value of a farmer's net worth (cost net worth is

not affected) and are available for reinvestment by selling the assets or by using the higher net worth to borrow money.

Managers use their own capital to prevent an erosion of their control of the business; using credit, outside equity, and leasing dilutes a farm manager's control. A second advantage of using one's own capital is that there is no repayment requirement, as there is for debt financing.

Unfortunately, farmers usually do not generate enough capital from farm and nonfarm income to finance the farm business adequately, and savings, gifts, and inheritances may not be available to contribute to net worth. After paying income taxes and providing for family living expenses, the net farm income remaining for reinvestment may not be enough to provide for any noticeable growth. And gains resulting from increased asset values can be fickle—here one year and gone the next. Because of these problems, farm managers must rely on other sources of capital. The primary source is agricultural credit suppliers, but outside equity capital is becoming a more important alternative.

The following sections concern the use of credit, the sources of agricultural credit, and the acquisition of credit. The final section of this chapter discusses the potential for using outside equity capital in agriculture.

The Amount of Credit to Use

Unless a farmer's net worth is sizable, many farm assets are likely to be financed with credit. *Credit* refers to the ability to control the capital of another in return for a promise to repay it by a specific future date. The use of credit is constrained by the business's borrowing capacity. The *borrowing capacity*, determined by the lender, is the maximum amount of credit the business can obtain. However, the amount of credit to use, that is, money to borrow, within this constraint is a decision made by the farm manager.

The difference between the borrowing capacity and the amount borrowed is the *credit reserve*, a backup source of liquidity available for unforeseen events and investment opportunities. The credit reserve is the manager's unused borrowing capacity. For example, if a farm business has a borrowing capacity of $250,000 but borrows only $200,000, a credit reserve, or financial cushion, of $50,000 is left. The larger the reserve, the more able the farm manager is to assume risk. However, there is an opportunity cost associated with this credit reserve; this cost is the income forgone by not investing this capital in the business.

Any increase in borrowing, of course, reduces the credit reserve. However, by increasing the business's debt-carrying capacity, it is possible to increase the credit used from borrowing without necessarily reducing credit reserves. To increase the farm business's borrowing capacity, the manager must be familiar with alternative sources of agricultural credit, understand loan terms and their implications, and approach the lender with a well-prepared presentation. (These topics are discussed later in this chapter.)

TABLE 11-1. An Example of Three Balance Sheets with Increasing
Financial Leverage

	A	B	C
Total assets	$200,000	$400,000	$600,000
Debt capital	0	200,000	400,000
Net worth	200,000	200,000	200,000
Leverage ratio	0	1 to 1	2 to 1

To determine the proper amount of credit to use, the farm manager should consider its profitability and riskiness as well as the repayment capacity of the farm business.

Leverage and Profitability

Leverage refers to the amount of credit (borrowed capital) relative to equity capital used in financing the farm business. Leverage is expressed as the ratio of debt to net worth, which is obtained from the farm business's balance sheet. For example, the three situations in Table 11-1 illustrate three balance sheets with the same net worth, but with increasing leverage.

As Table 11-1 shows, a net worth of $200,000 can be combined with varying levels of debt capital to finance farm assets up to a total of $600,000. The leverage ratio for this $600,000 farm is 2 to 1, indicating $2 of debt for every $1 of net worth.

By using the leverage provided by borrowed capital, farmers can stretch their own net worth to finance a larger operation. If the return to total farm assets exceeds the cost of the borrowed capital, leverage is an advantage; that is, the farmer's share of the return, or the return to net worth, will be increased.

Part A of Table 11-2 illustrates the advantages of leverage. Assuming a rate of return to total assets of 16 percent, if the farmer financed the $600,000 investment in farm assets solely with equity capital, the rate of return to net worth would be 16 percent. However, by using a combination of $200,000 debt capital and $400,000 equity capital, the return to net worth would increase to 18 percent. When the leverage ratio is 2 to 1 (when two thirds of the assets are debt financed), the return to net worth increases to 24 percent. In general, when the rate of return to total assets exceeds the cost of debt capital, the debt capital acts as a lever that increases the rate of return to the farmer's net worth.

Leverage and Risk

Just as leverage works for farmers in good years, it unfortunately also works against them in bad years. The use of additional debt capital relative to equity capital exaggerates both high and low profit situations.

A leverage ratio of 2 to 1 yields a 24 percent rate of return to net worth when the return to total assets is 16 percent. However, when the rate of return to total

TABLE 11-2. An Illustration of Leverage and Increasing Risk

	Leverage Ratio			
	0.0	*0.5*	*1.0*	*2.0*
Net worth (NW)	$600,000	$400,000	$300,000	$200,000
Debt capital	0	200,000	300,000	400,000
Total assets	$600,000	$600,000	$600,000	$600,000
Part A: rate of return to total assets is 16%				
Return to total assets	$ 96,000	$ 96,000	$ 96,000	$ 96,000
Cost of debt (12%)	0	24,000	36,000	48,000
Return to NW	$ 96,000	$ 72,000	$ 60,000	$ 48,000
Rate of return to NW	16%	18%	20%	24%
Part B: rate of return to total assets is 8%				
Return to total assets	$ 48,000	$ 48,000	$ 48,000	$ 48,000
Cost of debt (12%)	0	24,000	36,000	48,000
Return to NW	$ 48,000	$ 24,000	$ 12,000	$ 0
Rate of return to NW	8%	6%	4%	0

assets is 8 percent, the rate of return to net worth is reduced to zero. By comparing the four situations in part B of Table 11-2, it can be seen that the higher the leverage, the greater the effect on the return to net worth. This relationship is called the *principle of increasing risk*; that is, increases in leverage will cause unfavorable events to have a greater adverse impact on business profitability. In general, when the rate of return to total assets falls below the cost of debt capital, as it can when yields are down or market prices are depressed, leverage works against the farm manager; the debt capital decreases the rate of return to net worth.

Repayment Capacity

Another factor affecting credit use is repayment capacity. Every business must generate sufficient cash income to pay farm expenses, income taxes, and family living expenses and to repay debts. The amount of cash required for debt repayments depends on the length of the repayment period and the interest rate. The longer the repayment period and the lower the rate of interest, the greater the amount of debt that can be carried by a business. Table 11-3 illustrates this relationship.

Given the repayment period and the interest rate, the amount of debt that can be repaid depends on the cash flow available after all other needs have been met. This repayment capacity can be estimated easily by using the information from the financial statements of the business (see Chapter 4).

TABLE 11-3. Maximum Debt That Can Be Repaid Per $1000 of
Available Net Cash Income

Years for Repayment	Interest Rate			
	10%	12%	14%	16%
1	909	893	877	862
5	3791	3605	3433	3274
10	6145	5650	5216	4833
20	8514	7469	6623	5929
40	9779	8244	7105	6233

Table 11-4 illustrates the procedures for computing the term debt repayment capacity, using data from the Frank Pharmer example in Chapter 4. *Term debt* refers to loans repaid with periodic payments made over more than one year. Computing the term debt repayment capacity begins with the net farm income: Depreciation and interest on existing term debt are added and withdrawals for

TABLE 11-4. Estimated Term Debt Repayment Capacity for
Frank Pharmer

Farm sources of repayment capacity[a]		
Net farm income		$ 46,000
Depreciation (+)	$20,000	
Interest on term debt (+)	14,600	
Family living (−)	8,000	
Income taxes (−)	9,000	
		17,600
Available for principal, interest, and capital purchases		$ 63,600
Uses of repayment capacity		
Current term principal payments	$26,500	
Current term interest expense	14,600	
Down payments on capital purchases	5,000	
		−46,100
Net available from farm business		$ 17,500
Other sources of repayment capacity		
Net nonfarm income	$ 3,000	
Gifts, additions, other	0	
		$ 3,000
TERM DEBT REPAYMENT CAPACITY		$ 20,500

[a] See Chapter 4 for the Frank Pharmer example.

family living and income taxes are subtracted to determine the total cash income available for principal, interest, and capital purchases. This is the total repayment capacity of the farm business ($63,600).

However, the farm already has term debt outstanding. Subtracting the current principal and interest payments, as well as an allowance for making down payments on capital purchases, leaves a net amount available for repaying new term debt ($17,500).

There are other sources of repayment capacity besides the farm business; these sources provide $3000 for Mr. Pharmer. Adding this amount to the net available from the business gives a total term debt repayment capacity of $20,500. This is the amount available annually to make principal and interest payments on additional term debt. Whether the manager will actually use all of this repayment capacity depends on the profitability and perceived riskiness of the increased leverage.

Determining Credit Use

Determining the appropriate amount of credit to use can be difficult. Leverage, the use of borrowed capital to augment the farm manager's net worth, can increase the rate of return to net worth, but it also creates cash flow commitments to meet the loan payments. Furthermore, it increases the variability of returns to net worth while using up the credit reserve, which could otherwise be used as a cushion to absorb risk.

The appropriate level of debt depends on the anticipated variability in yields, costs, and income and on the willingness of the farm manager and the lender to accept the risk inherent in these variations. It is unusual to find farm businesses with leverage ratios that exceed 2 to 1. Many managers restrict their use of credit in response to risk considerations.

An important consideration in using credit is the lender's willingness to continue to extend credit after unfavorable outcomes. If prices or yields are unfavorable, not because of anything borrowers did or failed to do but because of forces over which they had no control, lenders should be willing to continue providing financing through succeeding production periods if the longer-term prospects are favorable. Borrowers are justified in asking what their lenders' course of action will be: Under what circumstances will they continue to provide credit and under what circumstances will they not? Successful business managers often take chances, but only after considering all probable outcomes, including the effects on their future borrowing capacity and credit reserve.

Sources of Agricultural Credit

There are many sources of credit available to farmers and ranchers. The "best" source depends upon the type of credit desired and each manager's circumstances. Farmers and ranchers should know and understand the alternative sources of credit available, who the lenders are, and how they operate.

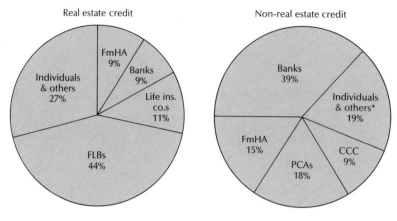

Real estate credit

Non-real estate credit

*Includes merchants and dealers

FIGURE 11-1.   Sources of Agricultural Credit, United States, December 31, 1984.

SOURCE: U.S. Department of Agriculture, *Economic Indicators of the Farm Sector: National Financial Summary, 1984*, Economic Research Service ECIFS 4-3, January 1986, pp. 68–71.

Agricultural credit is typically classed in one of two categories: real estate and non-real estate. Real estate credit is usually associated with land purchases, construction of major buildings, or the refinancing of non-real estate debt. Loan repayments can take from 10 to 40 years. Non-real estate credit includes short-term operating loans to be repaid in 1 year or less and intermediate-term loans for capital purchases to be repaid in less than 10 years. As of January 1, 1984, real estate credit accounted for 52 percent of total agricultural credit, and the remaining 48 percent represented non-real estate credit.

There are seven major categories of lenders serving farmers and ranchers: (1) the Farm Credit System, which includes the production credit associations (PCAs) and the federal land bank associations (FLBs); (2) commercial banks; (3) life insurance companies; (4) the Farmers Home Administration (FmHA); (5) the Commodity Credit Corporation (CCC); (6) merchants and dealers; and (7) individuals and other organizations. Some of these lenders offer both real estate and non-real estate credit. Figure 11-1 indicates the relative importance of the different types of lenders and the percentages of real estate and non-real estate credit that each supplies.

Farm Credit System

The Farm Credit System, a system of farmer-owned cooperatives supervised by the Farm Credit Administration, a federal agency, was developed during the period 1916–1933 by the federal government to meet the need for agricultural credit. Loans are made by local cooperatives, which are owned and controlled by the farmers who borrow from them. Although the government's capital that helped create the cooperative lending agencies has been repaid, the government is still involved in regulating and supervising the System.

Loans to farmers and ranchers are handled through either the federal land bank associations (FLBs) or the production credit associations (PCAs). The local FLBs extend long-term real estate credit for land purchases, real estate improvements, and refinancing. PCAs extend short- and intermediate-term non-real estate credit primarily for financing production expenses and machinery purchases.

The PCAs and FLBs obtain funds for extending loans through their district federal intermediate credit bank and federal land bank, respectively. The Farm Credit System is organized into 12 districts. These district banks acquire loanable funds by selling securities in the national financial markets. These bonds and notes are treated as government agency securities even though they are not guaranteed by the federal government against default. As a result, the Farm Credit System has easier access to financial markets and can acquire low-cost funds for lending to farmers.

The FLBs and PCAs serve only agriculture, and their officers know and understand the agriculture industry well. The System's interest rates are very competitive with its cost of money, averaging only a few basis points above the rates paid by the United States Treasury. Because the interest rates charged to farm borrowers are based on the average cost of all outstanding bonds and notes, including some with quite long terms, they are more stable than commercial bank rates based on the prime rate. They tend to be lower than the prime rate when the prime rate is rising and higher when the prime rate is falling.

One complicating factor that borrowers should consider when comparing the rates charged by other lenders with those charged by PCAs and FLBs is the stock purchase requirement, which makes the effective cost of the PCA or FLB loan higher than the contract rate of interest. (See p. 232, "Comparing Credit Costs.")

Commercial Banks

Almost 90 percent of the commercial banks in the United States extend credit to agriculture, primarily for non–real estate purposes. Some commercial banks also offer real estate credit.

As a source of farm credit, banks have some advantages. For example, they offer a complete line of financial services for the business and the family, and they employ knowledgeable loan officers with expertise in agriculture. However, the commercial banks' desire to keep their funds liquid may lead some to require shorter loan repayment schedules.

The future role of commercial banks in agricultural lending will be affected by the deregulation of financial markets. Deregulation permits greater competition among financial institutions, but it also eliminates the limits on interest that banks and other institutions can pay on deposits, which will possibly increase the cost of agricultural loans.

Compared to other loans, agricultural loans can be low in cost and risk for banks.[1] Because agricultural loans represent a profitable part of a bank's loan portfolio, commercial banks will continue to be a major source of credit in agriculture.

---

[1] E. L. LaDue, "Competing in the Ag Credit Market," *Agri Finance*, October 1981, pp. 12–13, 16, 18.

Life Insurance Companies

Life insurance companies were a leading source of farm real estate credit as recently as 1970, but their involvement has declined appreciably since then. Agriculture is viewed as an investment outlet for the funds of these companies, but many have found alternative investments with higher returns and comparable risks. Thus, most life insurance companies have limited funds to lend to agriculture at competitive rates.

Insurance companies employ trained personnel to appraise and service their loans. The farm mortgage programs offered by these companies are changing from long-term, fixed-rate loans to shorter-term, variable-rate loans.

Farmers Home Administration

The Farmers Home Administration (FmHA), an agency of the U.S. Department of Agriculture, offers real estate and non-real estate credit to farmers, ranchers, and rural residents who cannot obtain adequate financing from other lenders. The FmHA has its roots in the Resettlement Administration and Farm Security Administration of the 1930s, which provided supervised loans to farmers seeking to re-establish their businesses during the Depression. In 1946, the Farm Security Administration was reorganized, renamed the Farmers Home Administration, and given new authority to insure loans made by other lenders, as well as to provide direct government loans.

The FmHA is authorized to make several types of loans. Farm ownership loans may be made to purchase, improve, or enlarge farms, to develop water resources, and to refinance existing debt. Operating loans provide short- and intermediate-term credit. Emergency loans are made to farmers in officially declared disaster areas and to farmers facing scarce credit due to economic stress. The FmHA also guarantees agricultural loans made by commercial lenders.

Farmers considering the FmHA as a source of credit should be aware of two principal characteristics: First, to be eligible, the borrower must be unable to obtain credit at reasonable rates elsewhere; thus, the FmHA is viewed as the agricultural lender of last resort. Second, the FmHA expects the loan to be repaid, which means its personnel can only make loans they believe to be economically sound. Consequently, they require relatively complete records and financial statements. They also provide close supervision when extending credit.

Commodity Credit Corporation

The Commodity Credit Corporation (CCC), another U.S. Department of Agriculture agency, was created to implement government price support operations for agricultural commodities. Although it is not primarily a credit agency, its effect has been to increase the availability of capital to farmers. The total levels of credit provided by the CCC depend on the provisions of the federal farm programs and farmers' participation in these programs.

The CCC makes loans to farmers for several agricultural commodities at pre-determined loan rates per unit. These loan rates establish minimum price levels for farmers who participate in the programs. If an agricultural commodity is eligible for a price support loan, it is taken as collateral and the farmer receives a loan based on the announced loan rate. The farmer then has the option of either letting the CCC keep the commodity or paying off the loan and reclaiming it. If the price rises above the loan level, the commodity can be reclaimed, the loan paid off, and the commodity sold on the open market. If the price remains below the loan rate, the farm manager will not reclaim the collateral, and the debt is canceled. Such loans are called *nonrecourse loans*. In the event that the farmer does not reclaim the commodity, the CCC merely acts as a government purchasing agent.

Merchants and Dealers

Merchants and dealers selling fertilizer, feed, supplies, and machinery also extend a considerable amount of credit to farmers for purchases. Lending is not the primary business of these merchants; they lend to attract buyers or sellers. As a result, they may be unable to extend credit for a lengthy period of time, and their interest rates may be high. In any case, interest costs should be carefully checked and compared with those of other credit sources.

The credit offered by merchants and dealers can take several forms. For machinery and other capital purchases, sales contracts or promissory notes extending over several years may be used. Farm machinery dealers often offer special financing packages to encourage purchases in the off season. For feed, fertilizer, and other noncapital production inputs, 30-day open accounts are common; they can be extended, in some cases, for an interest charge.

It is important that purchasers find out about their suppliers' policies regarding discounts and finance charges and then use this credit accordingly. For example, if a bill is paid within 10 rather than 30 days, there may be a discount of 1 or 2 percent. Whether to pay early and take this discount or to pay the total amount at the end of 30 days depends on the cost of borrowing money to make this earlier payment. If the discount is 2 percent for payment within 10 days, farmers who can borrow money for less than 37 percent interest would be wise to borrow the money and make the payment.[2] A 1 percent discount would be attractive for farmers who are able to borrow money for less than 18 percent.

Individuals and Others

Individuals are an important source of credit for farmers. This is particularly true for real estate credit. Often the seller is willing to provide financing for the purchase of farmland. Some individuals may have a good understanding of agricultural problems and may extend credit in accordance with the demands of farming; others, who do not, may insist on unrealistic credit terms. It should be recognized that

[2] The formula for calculating this break-even interest rate (I) is as follows:

$$I = [D/(100-D)] \times [360/(T-N)]$$

where D is the percentage discount, T is the total number of days until the bill is due, and N is number of days within which the bill must be paid to receive the discount. For example:

$$I = [2/(100-2)] \times [360/(30-10)] = 37\%$$

most individuals have not had much experience with lending. Also, some individuals may depend on the interest and principal repayments for their livelihood and, therefore, may insist on shorter repayment periods.

An installment sales contract is commonly used when real estate credit is provided by the seller. This contract provides that the seller retain title to the property and the buyer make a down payment and periodic principal and interest payments. Even though the seller retains title until the total obligation is paid, the buyer receives immediate possession of the property.

The installment sales contract may have tax advantages for the seller because the income from the sale is spread over a number of tax years. Also, the interest rate on the contract may be higher than on alternative investments available to the seller. A disadvantage to the seller is the risk of default, which is avoided when the seller does not provide credit.

The advantages of an installment sales contract to buyers are a smaller down payment and perhaps an opportunity to negotiate the interest rate and payment terms. A significant disadvantage is that if buyers default, they can lose any and all equity in the property.

Acquiring Agricultural Credit

The first important factor in acquiring credit is a positive attitude toward borrowing. The manager of a farm business must accept the role of credit in financing the business. In acquiring credit, the manager should carefully consider all possible credit sources, prepare the loan application, compare credit costs and terms, and develop credit management skills.

Preparing the Loan Application

A farm manager's borrowing capacity is an asset, and like any other asset it must be developed and maintained. The loan application process is a crucial step in acquiring credit. Farmers should view themselves as salespeople in this application process; that is, they are selling their credit worthiness to the lender.

The manager's first consideration in preparing the loan application should be to understand what is important to the lender. The prospective borrower should be prepared to explain clearly how much credit is needed, how it will be used, how it will affect the business, and how and when it will be repaid (with interest). And, to be most effective, the borrower's proposal should be written. Lenders will normally consider five factors when evaluating loan applications:

1. *The purpose of the loan.* Lenders want to know what the loan will be used for. They consider the purpose of the loan in terms of its effects on net income and cash flow. A loan that increases net income and cash flow enough to repay the loan, a self-liquidating loan, is preferred. The purpose of the loan may influence the length of the repayment period, and there may be special risks associated with the loan's purpose that must be considered.

2. *Reputation and managerial ability.* Prospective borrowers must demonstrate their character, honesty, reliability, and management skills. They should begin by presenting specific goals and a well-prepared plan for using the credit. A cooperative attitude, straight-forward communication, and full and accurate disclosure of financial information helps to establish the necessary trust between lender and borrower.

3. *Financial position.* Lenders are interested in the business's current and past financial position in terms of its profitability, solvency, and liquidity. The most recent financial statements, along with those for the past three to five years, communicate this information most effectively.

   a. The *balance sheet* shows the net worth and solvency of the farm business. A comparison of several years' balance sheets shows the growth and progress of the business.

   b. *Income statements* demonstrate how well the farmer has managed capital.

   c. The *cash flow statement* for the past year shows the sources and uses of cash, that is, the cash available from nonfarm sources, the cash required to repay current debt, and the cash required to pay family living expenses and so forth.

   In addition to these financial statements, lenders may request copies of income tax returns. They give particular attention to the completeness and accuracy of the financial statements. Are all debts fully disclosed? Are assets realistically valued? Does the borrower understand the statements and their implications?

4. *Repayment capacity.* The ability to repay the loan is determined primarily by examining the *cash flow budget.* These projected cash flows over the next year, or over a longer period of time for larger loans, indicate when money should be available to repay loans and are therefore helpful in scheduling the loan repayment to coincide with cash availability. The cash flow budget also shows the lender how the borrowed funds will be used, as well as the other sources and uses of cash.

5. *Collateral.* Lenders need security to cover the loan in case things do not go as planned. They are responsible for protecting the owners of the loaned money from default. Therefore, lenders usually insist that farm assets be pledged as security.[3] The balance sheet indicates the assets that might be used for security.

---

[3] Non-real estate loans are usually secured with personal property. Security interests in personal property are regulated by the Uniform Commercial Code (UCC). Under the UCC, a security interest is the claim or lien the lender has on the personal property of the borrower. This security interest is created by a security agreement, which must be in writing, containing a description of the personal property serving as collateral and signed by the borrower. For a security interest to be enforceable, either the collateral must be in the lender's possession or the lender's security interest must be reported by filing a public notice at the designated county or state office. Real estate loans are often secured with a mortgage. The mortgage creates a lien on the property as security that the loan will be repaid. The title to the property is held by the owner-borrower. When the debt is repaid, the mortgage becomes void. In some states, a trust deed is used instead of a mortgage. The trust deed conveys the property's title to a trustee, who holds it as security for the payment of the debt. When the debt is paid, the trust deed becomes void, and the owner receives the title. However, if the borrower defaults, the trustee is authorized to sell the property to pay the debt.

With a legitimate use for credit, borrowers should not be apologetic about asking for a loan. The better the case they can make, the better their bargaining position for desirable loan terms. A complete application and a well-thought-out farm plan cannot fail to impress a lender. They provide relevant data that explain the circumstances surrounding a loan request. One of the more valuable uses of farm accounts and records is to support the loan application.

The process of selling one's credit worthiness does not end with the approval of the loan application. After the loan has been made, it is essential that the borrower abide by the agreed-upon terms. The loan agreement specifies what the lender agrees to do and what is expected of the borrower. The agreement may spell out when and from what sources repayment is to be made. If the borrower diverts funds that were to be used to repay the loan, future credit extensions can be seriously affected. If circumstances prevent a borrower from meeting the terms of the agreement, the borrower should notify the lender immediately (lenders do not like surprises) and cooperate in developing a new plan.

Lenders like to see that borrowers are monitoring their plans by comparing actual and budgeted cash flows each month and the year-to-date cash flows. The comparisons allow potential problems to be identified so that adjustments can be made before the problems become serious.

Borrowers should keep lenders informed about their long- range and short-range plans. The borrower should not make major commitments to buy machinery or breeding stock without first consulting the lender and checking for any consequences to credit availability. Keeping the lender regularly informed about the performance of the business can also help to build a farmer's borrowing capacity. Generally, farmers who communicate their plans effectively and document their financial progress have more credit made available to them and at more favorable terms.

Finally, risk management can improve the availability of credit. Practices that reduce risk and the lender's uncertainty about repayment can expand the availability of credit.

Comparing Credit Costs

Interest is the price paid for the use of someone else's money. Farmers should shop for credit with the same care that they use in acquiring other inputs. Despite truth-in-lending regulations, which require lenders to disclose the annual percentage rate of interest, borrowers should be familiar with the procedures and terminology used in calculating interest and other loan charges.

Simple Interest.   The simple interest method is used when the loan is to be repaid in a lump sum. Simple interest is expressed as an annual percentage of the amount borrowed, or the principal. The total interest cost depends on the principal, the interest rate, and the length of the loan period. If $10,000 is borrowed for a six-month period and the interest rate is 12 percent per year, the total interest cost will be $600. The basic formula for determining the dollar cost of such a loan is as follows:

$$\text{Interest cost} = \text{principal} \times \text{interest rate} \times \text{years}$$

The example gives the following:

$$\$600 = \$10,000 \times 0.12 \times 6/12$$

Interest on Unpaid Balances.   Simple interest is charged on the unpaid balance. When the loan is not repaid in one lump sum, interest is paid only on the amount owed for the length of time that amount was borrowed. For example, if $10,000 is borrowed for one year at a 12 percent annual interest rate on the unpaid balance and if $5000 is paid on the principal at the end of six months, with the remainder of the principal being paid at the end of the year, the interest cost is calculated as follows:

$$\$10,000 \times 0.12 \times (6/12) = \$600$$
$$\$\ 5,000 \times 0.12 \times (6/12) = \underline{\$300}$$
$$\text{Total interest} = \$900$$

Discount Method.   Another way to compute interest is the discount method. The discount rate is the percentage of principal charged for the loan period and deducted from the principal when the loan is made. In this case, the actual, or effective, interest rate is higher than the discount rate because the borrower does not have use of the entire amount of the loan.

   If the borrowed amount is $10,000 to be repaid in six months and if the lender discounts the loan in advance at a rate of 12 percent per year, the interest cost will be $600. However, the borrower will have the use of only $9400. By transposing the interest cost formula, the effective interest rate can be calculated as follows:

$$\text{Interest rate} = \text{interest cost}/(\text{principal} \times \text{years})$$

For this example:

$$12.8\% = \$600/(\$9,400 \times 6/12)$$

Add-On Interest.   A common way to finance machinery, equipment, and automobiles is with an equal-payment installment loan. Add-on interest is based on the original principal for the entire period of the loan. If a borrower obtains a $10,000 loan for three years at 8 percent add-on interest, the interest cost is calculated as follows:

$$\$10,000 \times 0.08 \times 3 \text{ years} = \$2400$$

The monthly payment is calculated as follows:

$$(\$10,000 + \$2400)/36 = \$344.44$$

The following formula can be used to compute the effective interest rate:

$$\text{Interest rate} = \frac{2 \times \text{interest cost} \times \text{payments per year}}{\text{principal} \times (\text{total payments} + 1)}$$

For the example, the effective interest rate is calculated as follows:

$$\frac{2 \times \$2400 \times 12}{\$10,000 \times (36 + 1)} = 15.6\%$$

Generally, if interest is charged on the original principal and it is repaid in equal monthly installments over the life of the loan, the effective rate of interest is approximately twice the quoted rate.

Stock Purchase Requirements. Farmers borrowing from PCAs and FLBs are required to participate in the ownership of the local cooperative by purchasing stock in amounts proportional to the amount borrowed. For example, to borrow from a PCA, a stock purchase equal to 10 percent of the amount of the loan may be required.

The 10 percent stock purchase requirement means that the borrower will have available only 90 percent of the amount borrowed. If a borrower wants to have $10,000 available for use, the amount that must be borrowed is calculated as follows:

$$\$10,000/0.90 = \$11,111$$

So, to have the use of $10,000, interest must be paid on $11,111. If the contract interest rate is 12 percent and the loan is repaid in one lump sum after nine months, the effective interest rate is calculated as follows:

$$\$1000/(\$10,000 \times 9/12) = 13.3\%$$

Table 11-5 indicates the effective interest rates for various combinations of contract rates and stock purchase requirements. These requirements increase the effective interest rate from 0.5 to almost 2 percentage points, and the effect is greater at higher interest rates.

The effect of the stock purchase requirement on a loan repaid over several years is more difficult to compute. Long-term loans are more common for FLB real estate loans. Table 11-6 indicates the equivalent interest rates for various contract rates and repayment periods. The stock purchase requirement is assumed to be 5 percent. Interest is charged on the unpaid balance, payments are made annually, and the total principal plus interest paid each year is constant over the life of the loan (a level-payment plan).

TABLE 11-5. Effective Interest Rates for Various Contract Rates and Stock Purchase Requirements

Contract Rate (%)	Stock Requirement	
	5%	10%
10	10.5	11.1
12	12.6	13.3
14	14.7	15.6
16	16.8	17.8

TABLE 11-6. Effective Interest Rates for Various Contract
Rates and Repayment Periods with a 5 Percent Stock
Requirement and Annual Payments

Contract Rate (%)	Repayment Period (Years)		
	5	15	30
10	10.8	10.7	10.6
12	12.9	12.8	12.7
14	15.1	14.9	14.8
16	17.2	17.0	16.9

SOURCE: E. L. LaDue, "Influence of the Farm Credit System Stock
Requirement on Actual Interest Rates," *Agricultural Finance
Review* 43 (1983): 50–60.

Variable Interest Rates.   Fluctuating interest rates have caused many lenders to adopt
variable-rate loan programs. These programs allow lenders to adjust the interest
rate on the outstanding principal balance to keep up with changes in the prime
interest rate or the cost of money in national markets. Depending on the program
and the type of loan, the interest rate may be adjusted monthly, quarterly, or
yearly.

Variable interest rates pass some of the risk of money market fluctuations from
the lender to the borrower. Lenders can maintain a constant profit margin on their
loans. Borrowers can also benefit if interest rates go down. But if interest rates go
up, variable-rate loans are a disadvantage to borrowers.

Before agreeing to a variable rate program, borrowers should evaluate the possibilities. How high can the interest rate go? Are there limits on the amounts by
which the interest rate can increase or decrease at one time or during the entire
length of the loan? If the interest rate increases, will the total payment for principal
and interest also increase, or will the loan be extended with the same periodic
payment?

If the total payment increases, will there be adequate cash to meet the payment?
For example, if the interest rate on a $100,000 loan increases from 12 to 14 percent
next year, will the borrower be able to afford the additional $2000 in interest and
continue to meet the repayment schedule?

On the other hand, if the total payment does not increase with an increase in
the interest rate, negative amortization may result. *Negative amortization* occurs
when the payment is not large enough to cover the interest on the loan, let alone
to reduce the principal. The interest cost that is not covered is added to the unpaid
balance, and even after making several payments, the borrower can owe more than
when the loan was first acquired.

Borrowers should consider the terms and conditions of the loan and the alternatives available. If interest rates are near historical highs and are expected to
decline, a variable-rate loan may be preferable. However, if rates are expected to
increase, borrowers will want to lock in interest costs with a fixed-rate loan, that
is, if the fixed-rate interest is not much higher than the initial variable rate.

Other Loan Costs.   In addition to interest, some lenders include loan fees and other charges. Such fees may be charged to offset the administrative costs of initiating the loan and investigating the borrower's credit history, the government fees for recording loan documents, and premiums for special insurance required on certain loans.[4]

Sometimes commercial banks require that borrowers maintain minimum deposits, called *compensating balances* because they compensate the lender for what would otherwise be a higher interest rate on the loan. If this required deposit is larger than the borrower's usual balance, the opportunity cost of the difference is an additional cost of the loan.

Prepayment penalties are costs associated with paying off the loan ahead of schedule. They are most frequently imposed on long-term real estate loans, but their use varies considerably among lenders. These other costs, as well as interest, should be considered when comparing the cost of credit from alternative sources.

Comparing Repayment Plans

The objective in comparing loan repayment plans is to develop a repayment schedule that is consistent with the availability of cash flow to meet those payments. In other words, the dates for making principal and interest payments should coincide with the sale of farm products. Payment dates of short-term loans can be determined by studying the cash receipts from previous years and by budgeting cash flow for the coming year (Chapter 6).

Determining repayment schedules for long-term loans is more difficult because of the uncertainty of cash flow over the period of the loan. The most commonly used method for repaying long-term loans is called *amortization*, whereby a series of installment payments is established. The payments are usually made annually or semiannually, although quarterly and monthly payments are also used. Each payment consists of principal and interest: The interest payment just covers the interest due on the outstanding balance of the loan; the remaining portion of each payment is the amount by which the principal is reduced, or the amortization.

There are two types of amortization plans: the level-payment plan and the decreasing-payment plan. The *level-payment plan* of amortization requires equal payments over the period of the loan. With each succeeding payment, a larger portion of the total represents principal and a smaller portion represents interest. The first payment is mostly interest, and the last payment is mostly principal. To determine the amortization payment, tables of amortization factors like Appendix Table 4 (Chapter 7) can be used. To find the yearly payment, the amortization factor is multiplied by the amount of the loan. For example, if the loan is $20,000 at 8 percent interest and the loan is to be repaid in 15 years:

$$\$20,000 \times 0.1168 = \$2336 \text{ (yearly payment)}$$

The amortization factor, 0.1168, is found in Appendix table 4 at the intersection of 8 percent and 15 years. To find the total interest paid, first calculate the total payments (principal and interest):

---

[4] By including these charges with the interest cost in the formula used to compute the effective interest rate for add-on interest loans, the effect of loan fees and other charges on the effective interest rate can be compared.

$$\$2336 \times 15 = \$35,040$$

Then, to find the interest, subtract the $20,000 principal. This leaves $15,040, which is the total interest cost for this loan.

The *decreasing-payment plan* of amortization provides for constant principal payments and declining interest payments based on the unpaid balance. Under this plan, the principal payment is the same each period, but the amount of interest, and hence the total payment, declines over time.

Table 11-7 compares the two types of amortization plans for a five-year loan. Although the total interest cost is higher for the level-payment plan, it is usually preferred because the initial payments are lower.

Managing Credit

Credit management involves identifying and evaluating practices that expand or restructure the business's credit, and proper credit management allows for additional borrowing and maintenance of liquidity.

The lender's influence on the amount, terms, and use of borrowed capital is an important element of credit management. In general, the lender would prefer higher rates of interest over a relatively short repayment period, because loans represent earning assets to the lender. The borrower, on the other hand, prefers lower rates of interest for relatively longer repayment periods, which minimize cash flow problems. Lower annual payments are easier to make in poor years. However, borrowers must discipline themselves to make larger payments in the good years to reduce interest charges. Prepayments also rebuild the credit reserve, which can then be used to cover shortfalls in low-income years.

A common problem in credit management is that short-term loans are often used to finance long-lived farm assets. As a result, borrowers find it difficult to generate adequate cash flow to meet the repayment schedule. However, not all of the fault is the lenders'. Many borrowers do not aggressively seek longer loan repayment schedules. The length of the repayment period should correspond to the expected useful life of the asset being financed. Current liabilities should be used to finance current assets, intermediate liabilities to finance intermediate assets, and long-term liabilities to finance fixed assets.

TABLE 11-7. Comparison of Level-Payment and Decreasing-Payment Plans for a $10,000 Loan Amortized over Five Years at 10 Percent Interest

Year	Level-Payment Plan			Decreasing-Payment Plan		
	Principal	*Interest*	*Total*	*Principal*	*Interest*	*Total*
1	$ 1,638	$1,000	$ 2,638	$ 2,000	$1,000	$ 3,000
2	1,802	836	2,638	2,000	800	2,800
3	1,982	656	2,638	2,000	600	2,600
4	2,180	458	2,638	2,000	400	2,400
5	2,398	240	2,638	2,000	200	2,200
Total	$10,000	$3,190	$13,190	$10,000	$3,000	$13,000

Many agricultural lenders make non–real estate loans for one year, knowing that the loan will not be repaid in full when it falls due. At the end of the year, they review the loan with the borrower and decide whether to extend it. This loan may be for any number of purposes, such as purchasing machinery or financing production expenses. It is unrealistic to expect that loans to purchase machinery can be repaid in one year. Thus, at least a portion of the loan is likely to be extended. However, when borrowers sign loan agreements not expecting to repay them when due, there should be a clear understanding regarding the amount of the loan that the lender expects to be repaid by the end of the year. The borrower should be assured that the lender is willing to make an extension if the borrower needs more time.

Outside Equity: A Potential Source of Capital

Although reinvested net income is an important source of capital in agriculture, it is often not sufficient to meet the capital needs of a growing farm business. Debt financing can provide a portion of the capital required to make a farm business viable. It has the advantages of leveraging net income, but its use is limited by rigid repayment schedules and uncertain business income. Thus, for new operators trying to get started and for operators seeking major expansion, outside equity capital should be considered.

Inflation and technology have increased the cost of getting a start in farming. Unless beginning farmers have significant financial resources of their own or get family help, they will need equity supplied by retiring farmers or outside investors. If loans were the only source of capital, it would not be possible to borrow enough (and the risk would be too high even if it were possible to borrow enough) to start a viable farm business.

Outside equity financing has all the advantages of investing the manager's own capital, with one major exception: The manager has to give up a portion of ownership and possibly some managerial control of the business. Even so, it is worth considering because outside equity capital, like the farmer's own capital, does not create an interest expense or a demand for repayment. Income that would otherwise be expended for these purposes can be kept and reinvested in the business. Equity capital investments from outside the business, combined with the added debt financing they will support, allow for faster growth and economies of size while shifting some of the risk to outside investors.

In spite of the potential advantages, outside equity has not been considered a viable alternative for financing farm assets, mainly because of the lack of organized markets and institutions to facilitate the investment of outside capital. Farmers who are interested in attracting off-farm equity capital may have trouble locating investors. Developing convincing net income and profitability projections takes time, and from the viewpoint of the investors who may be reluctant to assume an interest in a closely held farm business, such investments involve limited management control and are difficult to sell. Another reason many farmers and ranchers have resisted the use of outside money is their desire for independence. They are reluctant to share ownership and management with outside interests.

Although farmers have not made extensive use of business arrangements to generate outside equity, this situation is changing. With farm size and capital requirements increasing, traditional sources of capital, such as reinvested net income, savings, gifts, inheritances, and debt, may not be adequate. Consequently, it is anticipated that outside equity capital will receive increased attention as farm and ranch managers seek to acquire capital in the future.

Outside Equity Capital Alternatives

There are several alternatives for introducing outside equity into the farm or ranch business. A merger or joint venture between two or more farm businesses with complementary assets is one possibility. Outside equity can also be obtained by forming a corporation and selling shares of stock. Another possibility is to take in a partner. Additionally, outside equity capital is introduced when investors buy land, acquire other farm assets, and hire labor and management.

The opportunities for bringing additional equity capital into the business depend on the arrangement under which the business operates. The farm business arrangements discussed in Chapter 9 can be used to bring outside investors into the farm business to share its ownership, profits, and risks.

Partnerships.   General partnerships between farmers and outside investors provide a way to transfer equity capital to the farm business. Because partnerships do not pay income taxes, any tax preferences or losses available to the partnership are passed through to the individual partners, including outside investors. In a partnership, the owners share managerial responsibilities, which means that when outside capital is brought into the business the farmer must give up some control.

Limited Partnerships.   To access equity capital, limited partnership interests can be sold to outside investors. In order to sell these interests, the partnership must first be registered in its home state. Moreover, if interests in the limited partnership are to be sold in more than one state, approval must be received from the Securities and Exchange Commission.

Because of their indefinite duration, limited partnerships may be less stable than general partnerships or corporations as a way of obtaining long-term commitments of outside equity capital. Limited partnerships attract investors with short-term goals and a preference for relatively high liquidity. Limited partnerships, like general partnerships, allow a pass-through of tax deductions.

Regular Corporations.   Regular corporations may attract outside equity capital by selling common or preferred stock. They are taxed at state and federal corporate tax rates, so outside investors may be taxed twice on returns to investments—corporate taxes and individual taxes on the dividends (Chapter 9).

The public sale of corporate stock is a complex and costly procedure, requiring a review of the prospectus by state and federal security agencies. Consequently, public offerings are economically justified only when large amounts of outside equity capital are sought. The private sale of stock is hampered by the limited marketability of the stock. Investors are reluctant to invest in small corporations with few stockholders because of the risks involved, the lack of control if they hold minority interests, and the difficulty of selling the stock if the investment proves unprofitable or if they need the cash. However, the last disadvantage can be alleviated by including provisions in the bylaws or articles of incorporation that establish procedures for pricing and purchasing the stock of investors who choose to withdraw.

Subchapter S Corporations.   A subchapter S corporation can issue stock, but by law it is restricted to only one class of stock. It has the same tax advantages as a partnership and the limited liability of a corporation. This type of corporation has limitations, however, in attracting external equity capital to agriculture because shares are not easily marketed. However, subchapter S corporations are an important tool for transferring farm assets from one generation to the next. Shares of stock given to or inherited by off-farm family members is an example of outside equity investment in agriculture.

How to Acquire Outside Equity

The following steps outline the general process involved in attracting outside equity capital to an agricultural venture:

1. Identify potential investors. Direct personal contact is the most effective way to determine the interest of local investors—for example, local attorneys, physicians, or others who might find a farm investment attractive. Friends and relatives should not be overlooked. The place to start might be the local lender, who is undoubtedly aware of interested capital sources. Attorneys, accountants, and investment brokers are also aware of individuals and organizations interested in investing in new ventures.
2. Find out what investors are looking for. Are they seeking short- or long-term investments? Tax shelters or profits? Capital gains or current income? To attract equity capital to agricultural ventures, several considerations are important:
   a. Tax advantages. Investors in high income tax brackets are looking for investments with tax shelter potential.
   b. Glamour. People are attracted to cattle ranching and vineyard investments but are not as enthusiastic about hog production.
   c. Quick, speculative profits. Cattle feeding is an example of an agricultural venture that has attracted speculative interest.
   d. Long-term growth and capital gains. Land investments are appropriate for these investors.
3. Prepare and present an effective proposal. The proposal should be documented with projected balance sheets, income statements, and cash flow budgets. Anyone interested in putting money into a business wants answers to at least the following questions:
   a. How much money do you want?
   b. How long do you want it?

c. What will it be used for?
d. What return can be expected?
e. What can go wrong (risk)?

Competent legal and accounting advice will greatly improve the chances of securing capital. Managers seeking outside equity capital must sell themselves. Their managerial abilities are the key to the success of the venture, and potential investors must be convinced of these managerial abilities.

Many farm operators avoid turning to outside investors to help finance their businesses. Regardless of the reason for it, this attitude can lead to problems if a farmer attempts to build a farm business with inadequate capital. Once financial trouble occurs, it is too late to attract outside equity investors.

Summary

Capital is the total economic value of a farm business's resources. Farm managers generally seek to increase the amount of capital used as long as its marginal rate of return exceeds the interest rate. To obtain capital, they use borrowed money, outside equity, leasing, and their own net worth. Although the net worth of the farmer is a major source of capital in U.S. agriculture, beginning farmers and those who are expanding their businesses must borrow or attract outside equity to accomplish their goals.

The use of credit for debt financing is advantageous if it leverages net income, which it does when the rate of return to total assets exceeds the cost of the debt. However, debt financing as a source of capital is limited by rigid repayment schedules, the increasing cost of debt as borrowing increases, and the principle of increasing risks. Simply stated, when things go wrong, large debts will make them worse.

There are several sources of real estate and non-real estate credit available to agricultural borrowers. They vary in the purposes for which they extend credit, the terms of their loans, how they operate, and the cost of credit. Comparison of credit costs and repayment plans is complicated by the different methods used for calculating interest, so it is important to understand how they work and their implications to the borrower.

The loan application is the first step in acquiring credit. Prospective borrowers should anticipate the factors to be considered by the lender and communicate the purpose of the loan, their managerial abilities, the business's past and current financial positions, their capacity to repay the loan, and the collateral available to secure the loan.

Using outside equity to finance farm capital has the advantages of no interest or repayment requirements. However, farm managers may be reluctant to pursue this alternative because they are concerned about giving up control of the business and managerial independence. Also, attracting outside equity requires expertise in developing business arrangements that accommodate outside investors, as well as in locating and persuading investors to commit their money to an agricultural venture.

Recommended Reading

BARRY, P. J., J. A. HOPKIN, and C. B. BAKER. *Financial Management in Agriculture*, 3rd ed., Danville, Ill.: Interstate Printers & Publishers, 1983. Chapters 14, 16, and 19.

KLINEFELTER, D. A., and B. HOLTEL. *Farmers' and Ranchers' Guide to Borrowing Money*. Texas Agricultural Experiment Station MP-1494, October 1981.

CHAPTER 12

Buying and Leasing Land

One of the most important decisions made by beginning farmers is to acquire land. An error made here can be fatal to the new business. The characteristics of the land obtained and the way in which it is acquired will shape many subsequent farming decisions. If the farmer decides to lease the land, the kind of leasing arrangement developed will affect the enterprises selected and the way labor, capital, and management will be used. If the farmer decides to buy land, the price paid and the size of the mortgage will influence the business's capital position and, therefore, the type of farming organization that can be developed. The importance of the decision does not end with the initial purchase because established farm managers may add land to the farm business several times during their careers. The considerations involved in decisions about land acquisition are outlined in this chapter.

From a farm management standpoint, land is a factor of production. It serves as the site for buildings and facilities, but in addition to this rather inert function, it contributes directly to the production of crops by providing an environment that supplies water, air, and nutrients for plant growth.

Land can be thought of in a number of ways, but the following definition is widely accepted: *Land* is "that portion of the earth's surface over which ownership rights might be exercised."[1] Real estate is generally defined as land and the improvements that are usually transferred with it. Thus, real estate represents a complex of resources: land, buildings, fences, water management structures, mineral deposits, timber, soil improvements, and so forth.

Land has, however, certain characteristics that distinguish it from other factors of production. In the first place, it is durable; that is, land is not used up in the production process, although it may be depleted by use. It can usually be used for more than one purpose. It responds in different ways to various applications of labor and capital. In addition, existing institutional arrangements influence how land is used.

---

[1] R. Barlow, *Land Resource Economics*, Englewood Cliffs, N.J.: Prentice-Hall, Inc., 1958.

To Buy or to Lease

Other things being equal, most people prefer to own rather than to lease. A certain amount of prestige is associated with owning land, and many people experience a feeling of satisfaction when they make an improvement on something that belongs to them. Budgeting analysis can determine if this satisfaction comes at a cost, and if so, the amount of that cost.

Land is a preferred form of security for loans. It may be easier to borrow money with land as collateral than with other farm assets as security. Because a portion of the cost of a farm can be borrowed, farmers with limited capital are sometimes tempted to buy land. This decision may leave them short of capital to use for other assets or to reinvest in the business in a way that will allow them to develop the long-run potential of their management abilities. Tying up capital in this way represents one of the principal disadvantages of owning land.

However, owning rather than leasing provides a greater opportunity to enjoy windfall gains. If farmland *appreciates* (increases in value), the owner may eventually benefit, but the tenant may have to pay more for the privilege of farming land that has become more valuable. By the same token, the owner may suffer losses as a result of decreases in land values.

Land ownership is usually considered to be a less flexible means of operation than leasing. Farmers generally find it easier to expand and reach an efficient farm size by leasing rather than by purchasing land. The data in Table 12-1 indicate how ownership increases and tenancy decreases over the life cycle. For beginning farmers, leasing is a way to avoid the large capital investments required to own land. Moreover, leasing provides flexibility for contracting the operation when necessary.

Some lease arrangements provide that the landowner and tenant share the risk associated with the uncertainty of crop production and prices. The extent of the risk sharing depends on the type of lease. Also, lease agreements can be renegotiated

TABLE 12-1. Tenure Characteristics by Age of Operators Whose Principal Occupation Is Farming (1982)

Age of Operators (years)	Full Owners (%)	Part Owners (%)	Tenants (%)
Under 25	22.8	21.9	55.3
25–34	28.0	40.2	31.8
35–44	37.5	48.8	13.8
45–54	43.9	46.8	9.4
55–64	53.2	38.8	8.0
Over 65	73.3	20.9	5.9

SOURCE: U.S. Department of Commerce, Bureau of the Census, *1982 Census of Agriculture*, Vol. 1, Part 51, Washington, D.C., October 1984, p. 44.

in response to changes in the agricultural economy; this is usually not possible with land purchases involving long-term mortgages. On the other hand, tenants have less managerial freedom than owner-operators.

Insecurity of tenure is often mentioned as a disadvantage of leasing. It appears that both landowners and tenants want short-term leases and are unwilling to sacrifice flexibility to achieve security. Writing the lease for a three- to five-year period, with a one-year notice for cancellation by either party, can give considerable stability to both owners and tenants without sacrificing a great deal of flexibility. Most farm leases are one-year oral agreements.

Family living conditions may be less satisfactory on tenant farms. The landowner receiving a share of the production or a fixed cash payment may have little incentive to maintain and improve the house and buildings. Separating the lease payments for land from those for buildings may be one way to increase the landowner's incentive. Also, better housing and buildings might be viewed by the landowner as a way to attract better tenants. However, if leasing a farm yields a higher net income for the farmer than owning it, the farm family might be compensated for the sacrifice of living in a rented house on rented land.

Whether it is better to own than to lease depends on the financial position of the farm family, their preferences, the availability of land for lease and purchase, and lease rates. From a farm management point of view, the problem should be viewed as objectively as possible. That is, the way in which land is held should be viewed as a means to an end rather than as an end in itself. Only by adopting this attitude is it possible to separate profit considerations from the other goals of the farm family.

Purchasing Land

The critical question when deciding to purchase land is what price to offer. Answering this question requires careful analysis because the decision can have a major impact on the profitability and net cash flow of the farm business. Offering a price that is too low may result in losing an opportunity to purchase a particular tract of land. Offering a price that is too high may jeopardize the future of the farm business if debt commitments cannot be supported by the business's cash flow. The land purchase decision, therefore, requires a thorough analysis, which should include the following four considerations:[2]

1. *The market price of the land.* The price of the land is not determined until the land is actually sold. However, its market value can be appraised; the appraisal can be based on comparable sales of farmland in the same geographic area. Appraisal of market value is a highly specialized process and

---

[2] This approach for analyzing land purchase decisions is adapted from G. S. Willett and M. E. Wirth, *How to Analyze an Investment in Farmland,* Washington State University Western Regional Extension Publication 34, 1980; and R. M. Klemme and R. A. Schoney, "Calculating Bid Prices Based on Profitability and Cash Flows in Times of Constant or Declining Land Values," *Journal of the American Society of Farm Managers and Rural Appraisers* 48(April 1984):18–25.

will not be dealt with in great detail here.[3] Market value appraisals are valuable in formulating a bargaining strategy; they can be obtained by researching comparable sales or by hiring a professional appraiser.

2. *The maximum bid price considering profitability.*   Regardless of the land's market value, a farmer considering the purchase of land should determine the price above which the purchase would not be profitable. This break-even value of the land can be estimated on the basis of the effects of this decision on the farm's net income and the return to total farm assets. Although estimating the precise contribution of land to potential income is difficult, a useful approximation can be made. Land, because it is long-lived, will yield a return for many years. This future contribution must be taken into account and translated into terms that can be used for current decisions. The process of converting future returns to present terms, net present value analysis, was discussed in Chapter 7. (A worksheet for estimating this bid price, representing break-even profitability, is presented later.)

3. *The maximum bid price considering financial feasibility.*   Financial feasibility refers to the ability of the business to finance the land purchase with equity or debt capital, or both, and to repay the debt. To be financially feasible, the bid price cannot exceed the sum of the added real estate debt that can be supported by the farm's projected cash flow plus the equity capital available for the down payment. If cash flow projections indicate difficulty in servicing the added debt, the investment is not feasible. (A worksheet is provided later for analyzing financial feasibility.)

4. *The risk of financial loss.*   Typically, large amounts of debt capital are used to finance a land purchase. Thus, additional strain is imposed on the farm's cash flow as a result of making principal and interest payments on the land loan. Such fixed financial commitments, coupled with the uncertainty of future prices, yields, and land values, may increase the risk of financial loss beyond what the farmer is willing to assume.

Once profitability and financial feasibility have been estimated, the farmer is in a good position to make a decision. If the analysis indicates that the maximum bid price considering profitability and the maximum bid price considering financial feasibility both exceed the expected market price, the land purchase will be attractive. However, if either the maximum bid price considering profitability or the maximum bid price considering financial feasibility is below the expected market value, the land purchase will not return the desired profit or will pose cash flow problems.

Estimating the Maximum Bid Price Considering Profitability

The worksheet presented in Table 12-2 is designed to simplify the process of estimating the maximum bid price that could be offered when considering the profitability of the purchase. If the land were bought at this maximum bid price, the

---

[3] Some references on the appraisal of farm real-estate values are W. G. Murray et al., *Farm Appraisal and Valuation*, 6th ed., Ames: Iowa State University Press, 1983; and R. C. Suter, *The Appraisal of Farm Real Estate*, 2nd ed., Danville, Ill: The Interstate Press, 1980.

TABLE 12-2. Worksheet to Estimate the Maximum Bid Price to Offer
for Farmland Considering Profitability

1. Average annual before-tax return to land per acre (do not include interest on land debt)	$ 80.00
2. Enter 1.000 minus marginal income tax rate	0.560
3. After-tax annual return to land (multiply line 2 by line 1)	$ 44.80
4. After-tax discount rate based on weighted average cost of capital	0.08
5. Average annual rate of change in the land return appearing on line 3	0.04
6. Subtract line 5 from line 4 and divide the difference by the total of 1.000 plus line 5	0.038
7. Enter factor from Appendix Table 3 (Chapter 7) for interest rate on line 6 and number of years in planning period[a]	15.958
8. Present value of after-tax returns to land (multiply line 7 by line 3)	$ 714.92
9. Enter the factor from Appendix Table 2 (Chapter 7) for interest rate on line 4 and number of years in planning period	0.146
10. Enter factor from Appendix Table 1 (Chapter 7) for interest rate equaling annual rate of change in land market price and number of years in planning period	4.292
11. Estimated current market price of land	$ 1,600.00
12. Multiply line 10 by line 11	$ 6,867.20
13. Multiply line 12 by capital gains tax rate[b]	$ 1,208.63
14. Subtract line 13 from line 12 and multiply the total by line 9	$ 826.15
15. Multiply line 9 by capital gains tax rate	0.026
16. Subtract line 15 from 1.000	0.974
17. Add lines 8 and 14	$ 1,541.07
18. *Maximum bid price* (line 17 divided by line 16)	$ 1,582.21

[a] Because the rate on line 6 may not be an even number, finding the factor for line 7 may require interpolation or calculation using the formula in Appendix Table 3 (Chapter 7) and a pocket calculator.

[b] The capital gains tax rate is assumed to be 40 percent of the marginal ordinary income tax rate.

purchase would be profitable. If bought at a higher price, it would not. The process entails net present value analysis procedures (presented in Chapter 7) tailored specifically to the land purchase decision. The worksheet includes the following variables that affect the profitability of a land purchase decision:

1. Annual return to land and projected annual rate of change in the return.
2. The farmer's desired after-tax rate of return on the land investment.
3. Number of years in the farmer's planning period.
4. Market value of land at the end of the planning period.
5. Income taxes on annual land returns and on capital gains when the land is sold.

Using the worksheet to estimate the maximum bid price considering profitability is illustrated by the following example: A farmer is considering the purchase of 160 acres of land that will annually return $80 per acre above variable expenses and before payment of income taxes (line 1).

The return to land is obtained by subtracting all production expenses, except interest on the land investment, from gross revenue. Because land returns should be based on current yields, prices, and costs adjusted to normal circumstances, it is necessary to prepare detailed budgets for the crops to be produced.

The marginal tax rate of 44 percent (line 2) is based on the prospective land-owner's taxable income and tax status (see Chapter 7). Multiplying line 1 by line 2 gives the estimated after-tax annual return to land (line 3).

The after-tax discount rate for the land investment (line 4) is based on the weighted average cost of capital (also discussed in Chapter 7) adjusted for income taxes. The before-tax weighted average cost of capital of 14.3 percent is multiplied by 1 minus the marginal tax rate (1–0.44) to give the after-tax discount rate of 8 percent.

The average annual rate of change in land returns (line 5) recognizes the fact that returns are likely to change over the planning period and may not increase at the same rate as the general price inflation. From the information on lines 4 and 5, the rate used to discount future land returns is computed and entered on line 6.

To find the factor (line 7) from Appendix Table 3 (Chapter 7), both the discount rate (line 6) and the length of the planning period are needed. The length of the planning period is an important determinant of profitability. The planning period used should be based on the farmer's age and business objectives; for this example, the planning period is assumed to be 25 years. The factor (line 7) is used to compute the present value of the future land returns after taxes (line 8).

Lines 9 through 14 are used to compute the after-tax market value of the land at the end of the planning period. The land's market value at the end of the planning period is based on its current market value (line 11) and the anticipated annual rate of change in land value (line 10). Such value changes can be expected as a result of nonfarm demand for land use, technology changes, optimistic commodity price projections, and low interest rates. The annual rate of change in land value should be estimated conservatively because of future uncertainty and normally should not exceed the rate of general price inflation or the rate of change in land returns (line 5). In fact, the number entered on line 10 might appropriately be 1 if significant changes in the market for land are not expected.

The factor on line 10 assumes, for this example, that the land's market price will increase 6 percent annually over the 25-year planning period—a rate equal to the rate of general price inflation. An estimate of the land's current market price (line 11) should be based on recent sale prices of comparable land nearby. Then line 10 is multiplied by line 11 to calculate the land's market value at the end of the planning period (line 12). Lines 13 and 14 adjust the land's future market value for capital gains tax and discount it to a present value.

After completing lines 15 through 17, the final value of the land is computed and entered on line 18. As indicated on line 18, the land is worth $1582 per acre to the farmer in this example. If the farmer pays this amount, an after-tax rate of return of 8 percent will be realized on the investment. If more than $1582 is paid, a lower rate of return can be expected. Conversely, a higher return is implied if the land is acquired for less than $1582. Because the market price was estimated to be $1600, the farmer may have difficulty purchasing the land at a price that will provide the desired profit.

Completing the worksheet for various assumed levels of the required rate of return, return to land, rate of increase in land value, and level of general inflation will give an indication of the risk associated with this land purchase.

Estimating the Maximum Bid Price Considering Financial Feasibility

If the land purchase appears to be profitable, the next step is to consider its financial feasibility. Financial feasibility involves looking at the cash flow and calculating a bid price based on the business's repayment capacity. The question is whether the cash on hand together with the unused borrowing capacity, or the credit reserve, is sufficient to finance the purchase at the expected market price. If a large amount of new debt is involved, the ability of the business to repay that debt should be estimated as well.

The worksheet in Table 12-3 provides a procedure for estimating the maximum price that the farm business could offer for land when considering the financial feasibility of the purchase. It is assumed that this price is equal to the cash on hand that can be used as a down payment plus the largest loan that can be repaid by the net cash flow of the farm business after the land purchase. The debt repayment requirement depends on (1) the rate of interest on the new real estate loan and (2) the number of years over which the loan is amortized.

This worksheet starts with the term debt repayment capacity of the farm business on line 1 (this is described in Table 11-4). This figure is an estimate for the

TABLE 12-3. Worksheet to Estimate the Maximum Bid Price to Offer for Farmland Considering Financial Feasibility

1. Average annual term debt repayment capacity for farm business after land purchase (do not include interest on land loan)	$ _18,000_
2. Enter factor from Appendix Table 3 (Chapter 7) for interest rate equaling after-tax contractual rate of interest on loan and number of years in loan repayment period[a]	$ _10.675_
3. Maximum loan size (multiply line 2 by line 1)	$ _192,150_
4. Cash available for down payment on land purchase	$ _50,000_
5. *Maximum price that can be paid for total acreage* (line 3 plus line 4)	$ _242,150_
6. *Maximum per-acre bid price* (line 5 divided by number of acres in land purchase)	$ _1513.44_

[a] The after-tax interest rate equals the before-tax rate multiplied by 1 minus the average income tax rate.

expanded farm business, assuming the land is purchased. Therefore, this analysis of cash flow feasibility takes into account the cash from all sources and all the cash requirements in determining what will be available to repay the debt to finance the land purchase. Line 1, then, represents the amount of cash available to make principal and interest payments on additional real estate debt.

The factor (line 2) indicating the amount of debt a $1 annual loan payment will support can be found in Appendix Table 3 (Chapter 7). The factor selected should correspond to the after-tax interest rate on the land loan and the number of years in the loan repayment period. For the example, this would be 8 percent [12 × (1−0.33 average tax rate)] and 25 years. The factor is multiplied by line 1 to give the maximum size of loan the expanded business can support (line 3). Assuming that $50,000 in cash (equity capital) is available for a down payment (line 4), a maximum of $242,150 can be paid for the 160 acres of land (line 5). The maximum per-acre price of $1513 (line 6) is found by dividing line 5 by 160 acres.

If the maximum per-acre price the business can pay for the land (line 6) exceeds the expected market price, the purchase is financially feasible. However, if this maximum bid price is less than the market price, it is likely that the business will experience cash flow difficulties if the land is purchased.

By completing the worksheet for different projections of future crop prices and yields, cash operating expenses, and real estate loan terms, the farmer can better evaluate the risk of not meeting repayment requirements in poor years.

Using the Results

After the worksheets in Tables 12-2 and 12-3 have been completed, the farmer is better prepared to make the land purchase decision. If the analysis indicates that the land's maximum bid price considering profitability and the maximum bid price considering financial feasibility both exceed the expected market price, the purchase is attractive. However, if either or both of these maximum bid prices fall below the expected market price, the land investment will not yield the desired profit or will result in cash flow problems.

The analysis presented in Tables 12-2 and 12-3 is an example of an unattractive land investment. Both the maximum bid price considering financial feasibility ($1513) and the maximum bid price considering profitability ($1582) are below the expected market price ($1600). The analysis indicates that the farmer is not in a strong position to purchase the land at its appraised market value. However, this farmer might make an offer in the range of $1450 to $1500, recognizing that even if the bid is successful, the risk of not meeting repayment commitments is significant. If the land is purchased, careful attention should be given to risk management (see Chapter 8).

In the case of a land purchase, the consequences of an unfortunate decision may be disastrous. As a result, the potential buyer may wish to be somewhat conservative and not base the decision on what is believed to be the average or most likely outcome. The assumptions made in estimating the maximum bid prices would be made accordingly. Just how conservative these assumptions should be depends on the individual, because each person reacts differently to risk.

Leasing Land

A farm lease is a contract by which a landowner gives a tenant the right to use a specific tract of farm real estate for a specific time period in exchange for a lease payment or a share of production. The owner retains all other rights to the real estate, such as the right to transfer ownership. A lease may be a written document, or it may be simply an oral agreement between the parties.

Custom and tradition, which have had great influence on the leasing of agricultural land, have provided stability and have undoubtedly prevented gross unfairness, but they can at times become an obstacle to better farming methods. Lease provisions established by custom usually reflect long-run economic influences. These provisions are slow to change and reflect changed economic conditions only in a sluggish fashion. It is probably good that they do not reflect every short-run variation in the economic system, although when fundamental changes do occur, leasing rates should reflect these changes. Otherwise, lags will result in the inefficient use of agricultural resources.

Prospective tenants should be aware of the characteristics landowners want in a tenant. It is not unusual for a number of tenants to be interested in leasing desirable farmland. Although there are no precise qualifications, experience indicates that good tenants are likely to have the following traits: (1) honesty, (2) a cooperative attitude toward planning with the landowner, (3) knowledge of farm enterprises and practices on the leased farm, (4) ability and energy to accomplish tasks in a timely manner, (5) sufficient machinery and capital to operate efficiently, (6) willingness to adopt new practices and methods after their effectiveness has been proven, and (7) diligence in keeping records and reporting to the landowner.

Tenants are also interested in landowners who are honest, cooperative, knowledgeable about farming, and open-minded about new practices. Landowners who have inherited farm real estate with no knowledge of agriculture often hire agents or professional farm managers to represent them.

A satisfactory farm lease should provide a profitable and fair system for all parties involved. Initial lease negotiations usually begin with the terms prevalent in the neighborhood and with what is fair to both parties.

The custom in a local area influences the terms of a lease. But because farms vary greatly in productivity, a fair lease arrangement is more likely to result from an honest evaluation of what each party contributes to the total farm business. A fair lease is one that compensates the landowner and the tenant according to their contribution of resources to the farm business; each party shares the income in the same proportion as each contributes to the operation. If, for example, the landowner contributes one third of the inputs and the tenant two thirds, then fairness suggests that income should be shared in the same proportion. A fair lease discourages dishonesty; a poor one invites its own destruction. A lease biased against the tenant will make it difficult for the tenant to justify making improvements that might be profitable to both parties.

Even though fairness is an important consideration, in the final analysis lease rates are determined by the market, subject to supply and demand. In areas where farmland for lease is scarce or in short supply and prospective tenants are many, lease rates will be bid up. When land is plentiful and prospective tenants are few, the rates will be lower. Lease rates are subject to negotiation, and the final price paid for the lease will be determined by the relative bargaining positions of the two parties.

The types of land lease agreements used vary widely from one geographic area to another and even among farms within one area. What is desirable or fair for one farm, tenant, or landowner may not be acceptable to another. The most appropriate lease for any particular property will be determined by the characteristics of the property and by the experience, capital position, and goals of each party.

Farm leases are classified according to the method of payment, that is, a cash payment or a share of production. These two general types of leases are discussed next, including brief descriptions, their advantages and disadvantages, and how to determine the lease payment.

Cash Lease Agreements

Under the terms of a traditional cash lease, tenants pay an agreed-upon cash amount for the use of specified farm real estate. As a result, tenants receive all of the income and pay all of the expenses except taxes, insurance, and major building repairs. Because of its simplicity, the cash lease is usually easier to negotiate than the share lease.

A cash lease has certain advantages and disadvantages to both landowners and tenants. Paying cash for land represents a fixed expense to the tenant, which must be met regardless of yields or crop prices. Tenants therefore bear the risk of these fluctuations. Because of their guaranteed income, owners may be willing to accept lower payments in exchange for the reduced risk. Tenants who pay cash usually have more management freedom than crop-share tenants. Good tenants therefore receive the full benefit of their management ability. Compared to a share lease agreement, however, more operating capital is required because all or part of the lease payment is paid before harvest.

Landowners, in addition to receiving a stable income, are freed from most management responsibilities and from the risk of low prices and poor crops. On the other hand, a fixed cash lease payment removes the possibility of sharing in high prices or good crops.

Tenants with cash leases may exploit the land and property, particularly if there is no assurance that the lease will be extended. To encourage good husbandry, landowners may want to include a provision in the lease to reimburse the tenant for unused fertilizer applied during the past year. Such a provision gives the tenant incentive to maintain the land's fertility.

A cash lease can be ideal for owners with limited farming experience, for absentee landowners, and for family members who wish to lease land to a family corporation. Multiple-owner tenants might use cash leases to have more managerial freedom.

The cash lease payment does not automatically adjust to changes in prices and production unless otherwise stated in the lease. Cash lease rates become outdated and must be either renegotiated periodically or adjusted according to some procedure specified in the lease. Appropriate procedures can help mitigate these problems.

Establishing Cash Lease Rates.   There are at least three methods that can be used to establish the cash lease payment for a particular tract of real estate: (1) the market rate, (2) the landowner's cost, and (3) the tenant's ability to pay.

The market rate is based on the local area's going rate, which reflects, in a general way, the supply and demand for leased farmland. Using the market approach requires knowledge of the local cash lease market, which may not be readily available. Often there is no consistent mechanism for collecting and reporting cash lease rate data. However, even if these data are available, they must be adjusted for specific conditions such as the productivity of the land, the buildings, the timing of lease payments, and the duration of the lease.

The landowner's cost approach involves an estimation of the minimum lease payment required to cover the landowner's annual costs. The costs are those associated with owning the real estate, including the opportunity cost of the capital investment. The landowner in the example in Table 12-4 would break even at a lease rate of $104.50 per acre. This amount should be the least the landowner is willing to accept for the lease.

The most important consideration in calculating the landownership costs is the opportunity cost of the real estate investment. The annual cost is based on the market value of the land and the expected rate of return. The desired rate of return is not necessarily the interest that could be earned by selling the land and investing elsewhere; such interest rates do not allow for the appreciation in land values, which benefits the landowner. If 9 percent could be earned from another investment, and if the long-run increase in land values is expected to average 4 percent annually, the adjusted rate of return would be 5 percent (9 percent minus 4 percent). The landowner should not expect to receive the entire 9 percent return from the lease payment. The increase in the value of the land, which is expected to be 4 percent, should be subtracted when estimating the lease payment required to cover the landowner's costs.

The cash lease rate that the tenant can afford to pay is the residual net income from expected yields and prices after all expenses not related to land have been paid. These expenses include the tenant's operating expenses, depreciation on machinery, interest on machinery investments, and opportunity costs of labor and

TABLE 12-4. Landowner's Per-Acre Costs for Establishing Cash
  Lease Rates

Interest on investment ($1800 per acre at 5%)	$ 90.00
Depreciation (fences, drainage systems)	1.50
Repairs	2.00
Property taxes	10.00
Insurance	1.00
Landowner's cost per acre	$104.50

management. The estimated expenses are subtracted from expected crop receipts to determine the residual net income available for making lease payments. This amount should be the most the tenant is willing to pay for the lease.

Table 12-5 compares the results that might be expected from these three approaches. It is unlikely that all three will indicate the same lease rate. Arriving at a cash lease rate agreeable to both parties involves a bargaining process and some compromises. To effect an agreement for the example in Table 12-5, either the landowner must accept a lower return on the land investment or the tenant must accept a lower return for labor and management, or both must accept lower returns. Whether the final lease rate is closer to the landowner's cost or to the tenant's residual net income will depend on their relative bargaining positions. If there are other prospective tenants, the landowner will have the advantage; if there are other farms to lease, the advantage will go to the tenant. Of course, it is also possible that no agreement will be reached. The landowner may decide to sell the land if leasing will not provide the desired return. Likewise, the tenant may quit farming because of the high cost of leasing land.

Flexible Cash Leases.   A flexible cash lease agreement, an adaptation of the traditional cash lease, adjusts the lease rate to price or yield levels or both. This adjustment allows some sharing of risk between the landowner and tenant, and overcomes some of the problems associated with cash and crop share leases. A common arrangement is to determine the cash payment by multiplying the prevailing price of the crop by a previously agreed-upon number of units of production per acre.

For example, suppose the lease payment has been $60 per acre for the last three years. The main crop is wheat, and the average price received is $4 per bushel. The cash lease rate under the flexible agreement, then, would be equal to 15 bushels of wheat multiplied by the average price paid by the local elevator during harvest. If the average price paid is $4.50 per bushel, the cash lease payment would be $67.50 ($4.50 × 15 bushels). With a higher wheat price, a higher lease rate would be paid; with a lower price, a lower rate would be paid.

This type of arrangement is relatively simple to understand and use, but the landowner and the tenant must have previously agreed on the number of bushels or units per acre upon which to base the lease payment and how the price for the current year is to be determined. Price might be determined by using the average price offered by the local elevator on specific dates or the average price on specific dates and locations as published by an independent agency such as the U.S. Department of Agriculture. This arrangement places the yield risk on the tenant and part of the price risk on the landowner. A flexible cash lease agreement also enables the landowner to share in the additional income resulting from unexpected increases in crop prices.

TABLE 12-5. Comparison of Results from Three Approaches to Establishing Cash Lease Rates

Adjusted per-acre market rate for local area	$103.25
Landowner's cost per acre	104.50
Tenant's residual net income per acre	98.00

Prospects for higher or lower crop prices make it difficult for landowners and tenants to agree on fixed cash lease rates, particularly if the lease is for more than one year. If the lease rate is set on the basis of high prices, tenants face the possibility that prices may be substantially lower and their production costs higher in ensuing years. If the lease payment is based on low prices received in prior years, the landowner's income from a fixed cash lease may be less than that from a crop-share lease. Therefore, making the lease flexible may be beneficial to both land-owners and tenants.

Cash lease payments are often made in advance, but some adjustments have to be made with flexible lease rates because the final payment cannot be computed until after the harvest. Partial payment could be made at the beginning of the year and final settlement made after the harvest, thereby reducing the tenant's credit needs.

There are several methods for making the cash lease flexible:

1. Adjust the base rate when prices are outside a specified range. With this method, the landowner and the lessee agree to a base lease rate that applies as long as the crop price is within the specified range. For example, the base rate might be $60 per acre for prices between $100 and $110 per ton; for every $1 change in the crop price above or below this range, the lease rate would increase or decrease by $0.75 per acre.

2. Adjust the base rate for both price and yield changes. Again, a base rate must be agreed upon. Then a formula like the following can be used to make adjustments:

$$\text{Adjusted rate for current year} = \text{base lease rate} \times \frac{\text{current crop yield}}{\text{base crop yield}} \times \frac{\text{current crop price}}{\text{base crop price}}$$

This formula adjusts the lease rate according to changes in crop yields on the leased farm compared to the average or established base yield for that farm. The price is adjusted similarly. The lease agreement should specify how the yield is to be measured and how the current year's price is to be determined.

When several crops are grown, the base rate can be adjusted by using a published price index or a weighted average of the prices for the combination of crops typically grown. The latter procedure is more complicated, but it reflects changes in prices received over time and corresponds to the actual mix of crops grown. Likewise, a crop yield index (see Chapter 5) can be used to adjust the base rate for yield variations for a combination of crops.

Share Lease Agreements

The predominant type of lease in many parts of the country is the share lease. Under this arrangement, landowners receive their lease payments in the form of a share of the production. From the tenants' standpoint, its major advantage is that they do not bear the entire risk of price and production fluctuations. Tenants may have a disadvantage if they want to lease additional land to expand their operations because landowners may prefer that tenants concentrate their efforts on fewer acres and produce higher yields or more livestock.

Farmers producing both crops and livestock can choose between livestock-share or crop-share leases. A livestock-share lease operates like a partnership between the landowner and the tenant. Each owns one half of the livestock and feed inventory; they share livestock expenses, feed purchases, and seed, fertilizer, and chemical costs equally; and each receives one half of the livestock and crop income. If a crop-share lease is used, the tenant owns all of the livestock, and the landowner does not participate in the livestock enterprise. The landowner receives a share of the crops produced, and the tenant also makes an additional cash payment for the use of buildings, livestock equipment, and land used for pasture or forage production. Farmers who want to produce livestock face the challenge of finding a farm for lease that has adequate buildings and facilities and an owner interested in participating in the livestock enterprise. As a result, the use of livestock-share leases is declining. The crop-share lease is the predominant share lease agreement for farmers producing crops as well as for those producing both crops and livestock.

A share lease permits landowners to benefit from the good years, but in exchange for bearing some of the production and price risk. Overall, landowners receive a higher income under this type of lease. Because they share in the production, owners usually participate in management decisions such as what crops to plant, the number of acres of each, and how much fertilizer to apply. The usual arrangement is that owners furnish the land and buildings and pay the ownership costs associated with these facilities. The tenant is usually expected to pay most of the other expenses and furnish labor and machinery. However, it is not uncommon for the landowner to share some of these expenses, such as fertilizer, with the tenant.

Sharing Expenses.   The division of production and expenses between landowner and tenant can lead to inefficient resource use. This is illustrated by using the fertilizer example in Table 2-2. As demonstrated by that example, an owner-operator would apply fertilizer up to the point where the marginal value of product (MVP) is equal to the marginal input cost (MIC), which is 40 pounds of nitrogen per acre to achieve a yield of 36.4 bushels of wheat. But how much would be applied by a share lease tenant who receives only two thirds of the yield?

The tenant's most profitable application of fertilizer depends on whether the landowner shares the fertilizer expense (Table 12-6). Without expense sharing, the tenant's most profitable fertilizer application is 30 pounds per acre. At this level, the tenant's MVP is $2.67, compared to an MIC of $2.50. Increasing the application another 10 pounds would not be profitable because the tenant's MVP would be less than the MIC.

However, if the landowner shares expenses in the same proportion as production, the tenant's most profitable application is 40 pounds, which is the same as that of the owner-operator. At 40 pounds, the tenant's MVP is $2.13 and the MIC is $1.67. A general principle in designing share lease agreements is that expenses for variable inputs that influence yields should be shared in the same proportion as production is shared. Examples of such variable inputs for producing crops are fertilizer, seed, herbicides, and insecticides.

TABLE 12-6. The Crop Share Tenant's Fertilizer Decision with and without the
Landowner's Sharing One Third of the Expense

Nitrogen Fertilizer (lb)	Yield of Wheat (bu)	Without Expense Sharing		With Expense Sharing	
		2/3 MVP[a] ($)	MIC[b] ($)	2/3 MVP[a] ($)	2/3 MIC[b] ($)
0	32.0				
		3.73	2.50	3.73	1.67
10	33.4				
		3.20	2.50	3.20	1.67
20	34.6				
		2.67	2.50	2.67	1.67
30	35.6				
		2.13	2.50	2.13	1.67
40	36.4				
		1.60	2.50	1.60	1.67
50	37.0				
		1.07	2.50	1.07	1.67
60	37.4				
		.53	2.50	.53	1.67
70	37.6				

SOURCE: Data developed by Leroy D. Luft (see Table 2-2).
[a] Marginal value of product with wheat price at $4.00 per bushel.
[b] Marginal input cost with nitrogen cost at $0.25 per pound.

The problem illustrated by the preceding fertilizer example also arises when a new farming practice is being considered. When fertilizer first came into use, many tenants were reluctant to apply it in the quantity desired by landowners if the tenants had to pay the total cost. When landowners recognize this reluctance and the reasons for it, they will tend to share in the variable costs associated with new techniques. In this way, the terms of the lease will reflect changes in economic conditions.

Establishing Production Shares.   As previously stated, a fair lease compensates the landowner and the tenant according to their contributions of resources to the farm business. Thus, the share of production going to the landowner in exchange for the lease should be based on the relative contributions of the two parties.

Consistent with the principle of sharing yield-increasing variable expenses, the first step in establishing production shares is to determine what expenses will be shared. It is important that both parties agree on the inputs for which costs are to be shared. The next step is to estimate the contributions of each party exclusive of these shared expenses.

In determining the fairness of the lease, it is difficult to measure and value the contributions of the landowner and the tenant. It would be easy if all contributions from both parties were in cash. However, the annual contributions of land, machinery, labor, and management do not have explicit values and therefore must be estimated.

Table 12-7 shows how the contributions of the landowner and the tenant might be determined in a hypothetical situation. In this example, the landowner contributes 40 percent of the resources to the operation and should therefore receive 40 percent of the production. But what if the customary share for a landowner in this neighborhood is 45 percent and the two parties do not want to violate custom?

TABLE 12-7. Simplified Example of Determining the Contributions of the Landowner and Tenant with a Share Lease Agreement

	Contributions ($/acre)		
Resource Contributed	Total	Landowner	Tenant
Seed	(10)	Shared	
Fertilizer and chemicals	(33)	Shared	
Fuel and repairs	13	0	13
Hired labor	2	0	2
Operating capital interest	5	0	5
Machinery ownership	34	0	34
Land taxes	2	2	0
Land investment	46	46	0
Tenant labor and management	18	0	18
Total Unshared Contributions	$120	$48	$72
Percentage of Contributions	100%	40%	60%

In this case, contributions would be adjusted to a 45 to 55 ratio. For example, the landowner could pay 45 percent of the fuel and repair expense of $13; the total contributions, then, would be $54 from the landowner and $66 from the tenant, or 45 and 55 percent, respectively.

Designing fair leases is more complicated for tenants who lease land from more than one owner. Multiple-owner tenants are responding to the principle of economies of size, which encourages tenants to farm more land than is contained in a typical ownership unit. Owners of small tracts of land must recognize the necessity of their tenant's farming additional land. But multiple-owner tenants have a major problem: how to develop leasing agreements that are fair to all owners when the productivity levels, buildings, and other resources differ from farm to farm.

The bargaining process is important when negotiating any kind of lease, and it is most effective when both parties know what their own contributions are worth, what the other party's contributions are worth, and local leasing customs. An unfair lease agreement will not last long, and it tends to discourage honesty and cooperation.

Improvement of Farm Leases

Tenancy plays an important role in U.S. agriculture. It allows many farmers an opportunity to control more resources than they could if they had to own the land. However, improvements in lease agreements are required to overcome some undesirable practices and attitudes. Most difficulties can be traced to five factors:[4]

1. Misunderstandings by either or both parties concerning their rights and obligations under a lease.

[4] Barlow, *Land Resource Economics*, p. 436.

2. Less income than expected by either party.
3. Unfair lease terms.
4. Inadequate provisions for desirable property improvements.
5. Tenants' lack of security.

Many of these difficulties can be overcome if the problem is approached properly by both parties. With respect to misunderstandings, a written lease developed by both parties will do much to solve the problem. Farm lease contracts are often oral but should be in writing. Some of the advantages of a written lease are as follows: (1) it protects not only the original parties but also their heirs or assigns in case either party dies; (2) it provides a record of the terms of the agreement and helps to prevent and settle disputes; and (3) it serves as a checklist to ensure that both parties consider all the provisions of the lease before signing it.

The extension offices in many states have sample lease forms available, but whatever format is used, it is important that certain items be included. Table 12-8 contains a list of suggested items that are of particular importance.

The best time to resolve disagreements and conflicts is when the lease is being negotiated, not after it is in effect. And the best way to identify and resolve them is by preparing a written lease. Because a lease is a legal document, an attorney should be employed to help develop and review it before it is signed.

The problems of less income than expected and unfair leases are more difficult to resolve. Unforeseen economic conditions can cause dissatisfaction on either side. But planning and budgeting will show each party what can be expected under probable yield and price conditions. This planning can do much to prevent disappointment later. The same procedure will help determine a fair leasing

TABLE 12-8. Suggested Items to Include in Lease Agreements

Name and address of the landowner and tenant

Description of the real estate to be leased

Date the lease will start

Number of years the lease will be in effect

Provisions for automatic renewal of the lease unless notice is given to cancel it

Lease cancellation procedures, including the advance notice required (six months or one year, for example) before the lease expires

Lease rates, share and/or cash

When and where cash payment is to be paid or the landowner's share of production is to be delivered

Expenses to be paid by each party

Farming practices to be followed by the tenant

Provisions for compensating the tenant for unused real estate improvements made by the tenant

Landowner's lien, right of entry, and liability

Settlement in case of death

Arbitration of disagreements

Procedure for revising the agreement

arrangement, although it is sometimes difficult to obtain agreement on just what constitutes fairness. Fairness is subjective and depends on the perceptions of the two parties.

Landowners and tenants should review the terms of their farm lease agreement annually and negotiate any changes needed as a result of changes in technology, higher machinery costs, and fluctuating commodity prices. Changing from a share lease to a cash lease, or vice versa, might also be considered.

Compensation for unexhausted improvements can increase the tenant's incentive to improve farm's productivity. One criticism of tenancy is that soil depletion and property deterioration may result. If tenants were assured that they would be paid at the end of their tenancy for any unused portion of improvements they had made, they might be less reluctant to make improvements, but landowners should have the right to approve them before they are made. To illustrate how compensation might be calculated, assume that a tenant remodels a farm building at a cost of $1000 with the prior approval of the landowner. They agree that the improvement should last for 10 years and that the annual rate of depreciation is $100. If after five years the tenant decides to leave the farm, the landowner will compensate the tenant for the unused portion of the improvement, in this case $500. A similar approach can be used for soil-building practices.

One of the reasons for short-term leases and the resulting insecurity of tenure (the risk of losing the lease) has been tenants' desire for flexibility in order to lease a better farm or to buy land. The lack of security would undoubtedly be less of a problem if the foregoing provisions were implemented. Leases that are thoroughly understood and approved by both parties prevent many ruptures in owner–tenant relationships. A complete understanding of the contributions being made by the other party to the business helps to establish more permanent relationships.

Finally, good communication between landowners and tenants is critical. Problems and disagreements will be encountered in the process of negotiating lease agreements, but these are not insurmountable if they are approached openly and frankly by reasonable people seeking solutions. Regular progress reports from tenants during the growing season will keep the landowners informed and their expectations realistic. Improving the communication between landowner and tenant will result in better working relationships and a more profitable farm business—the goal of both parties.

Summary

Land is a unique agricultural resource that also poses some unique management problems for the farmer. The first major decision in acquiring land is whether to buy or lease it. The advantages and disadvantages depend on the farmer's financial position, goals, and preferences and on the availability of land for lease and purchase. Because purchasing land usually requires a larger capital commitment, farmers with limited capital generally lease land and use their capital for machinery and other resources. By doing this, tenants forgo the gains (and losses) resulting from changes in land values, and with short-term leases they also have insecurity of tenure.

A decision to purchase land should involve determining the price to offer considering the market price, profitability, financial feasibility, and risk. The market price can be estimated by hiring a professional appraiser or by comparing the sale prices of land in the area. Computing the maximum bid price considering profitability involves a net present analysis of the expected annual returns to the land and the market value of the land at the end of the planning period. The maximum bid price considering financial feasibility is determined by the cash available for the down payment plus the maximum debt that the farm's cash flow can repay.

Farm leasing is influenced by local custom and tradition, but a satisfactory lease must provide a profitable and fair arrangement for both the landowner and the tenant. A fair lease compensates both parties in proportion to their contributions to the farm business. Although fairness should guide lease negotiations, the final provisions depend on the relative bargaining positions of the two parties.

Cash lease agreements are riskier for tenants and generally result in lower income to landowners. Crop-share leases allow tenants and landowners to share price and production risks. Tenants typically have more managerial freedom with cash leases, and landowners seeking management participation prefer share leases.

Negotiating cash leases is usually simpler; the market rate, the landowner's costs, and the tenant's ability to pay are used to set the lease rate. Negotiating share leases is more complicated: First, the sharing of yield-increasing variable expenses must be considered; then the contributions of each party must be estimated to determine how production is to be shared between landowner and tenant. Flexible cash lease agreements are receiving increased attention because they overcome some of the problems associated with more traditional cash and share leases.

The success of a farm lease depends on the relationship between the landowner and the tenant. Good communication is the key to maintaining trust, understanding, and cooperation, and an important communication tool for handling problems is a carefully prepared, written farm lease.

Recommended Readings

ERICKSON, D. E., P. ROBBINS, ET AL. *Farm Real Estate.* University of Illinois North Central Regional Extension Publication No. 51, 1977.

REISS, F. J. *Farm Leases for Illinois.* University of Illinois Cooperative Extension Service Circular 1199, 1982.

CHAPTER 13

Employing Labor Resources

Efficient labor management is important to the successful operation of modern commercial farms. Many farm managers hire little or no labor, doing the work themselves or with family members. Others hire large numbers of employees for relatively short periods of time to complete critical cultural or harvest operations. Individual crop and animal enterprises vary as to the amount and type of labor required, and various enterprises require vastly different proportions of labor relative to other resources. Because labor must be combined with the other factors of production, it cannot be managed in isolation. The purpose of labor management is not just to maximize returns to labor but to use all the factors of production to achieve the farm manager's goals most effectively.

The objectives of labor management depend upon the particular farming situation. Is labor a fixed input limited to the operator's family, or is regular hired labor available? Can the efficiency of labor be enhanced by increasing the capital input? Is part-time labor available that can be hired by the day or week to augment the fixed or permanent labor input? In general, the objectives of labor management are as follows:

1. To obtain the most efficient combination of labor and capital; this objective may involve substituting capital for labor, thereby reducing costs and increasing profits.
2. To increase output by increasing the input of either capital or labor, or both, when such an increase in output will result in greater profit.
3. To release the operator or family for increased leisure or off-farm work.

These objectives are the focus of the concepts and principles of labor management presented in this chapter.

Supply and Use of Agricultural Labor

Labor, a significant factor in agricultural production, is supplied by farm operators, their families, and regular and seasonal employees.

Operator and Family Labor

A relatively stable component of the labor force, farm operators and their unpaid family members provide almost two thirds of the total labor employed in agricultural production. So, in addition to their management function, operators also contribute a substantial labor input. However, the number of operators and the proportion of total labor provided by them and their families have declined with the increase in the size of farms and the corresponding decrease in the number of farms.

Although most farm work has been done by male family members, over the past decade there has been a significant increase in the amount done by women. Automation of food preparation, laundry, and other housework has allowed women to become more involved in farm work. Often the increased demand for record keeping has been met by the farm manager's wife, and more women are operating farm machinery and caring for livestock.

The future role of operators and their families in supplying labor will depend on the profitability of farming and ranching, as well as on technological advances that reduce the labor required per unit of output.

Hired Labor

According to the 1978 Census of Agriculture, about 47 percent of all U.S. farms employed labor. These farms provided over 5 million wage and salary jobs, which were held by about 2.5 million workers.[1] Thus, many workers held more than one farm job, and many jobs were of short duration.

Although farm managers and their families provide the largest proportion of agricultural labor, as farms have become fewer and larger, the percentage of total farm employment represented by hired workers has increased: In 1970, hired workers constituted about 26 percent of the average annual employment of farm workers; in 1980, they constituted 35 percent.[2]

Regular Employees.   Regular farm employees, those who work 150 days or more at one farm job, made up only about one fourth of the hired labor force in agriculture but accounted for almost 75 percent of the total days worked by agricultural employees in 1981.[3] Although the number of all hired farm workers has remained relatively stable over the past 10 years, they now work more days per year at farm work than they did 10 years ago, and the number working 150 days or more per year increased almost 50 percent during the same period.

The importance of regular agricultural employees is expected to increase in the future. Increased skills will be required to operate complex machinery and to perform precise cultural practices. Specialized training programs, wages competi-

[1] S. L. Pollack and W. R. Jackson, Jr, *The Hired Farm Working Force of 1981*, U.S. Department of Agriculture Economic Research Service Agricultural Economics Report No. 507, November 1983.

[2] L. W. Smith and R. Coltrane, *Hired Farmworkers: Background and Trends for the Eighties*, U.S. Department of Agriculture Economic Research Service RDRR-32, September 1981.

[3] Pollack and Jackson, *The Hired Farm Working Force*.

tive with nonfarm opportunities, and favorable employer–employee relationships are necessary to attract and retain employees with these skills.

Seasonal Employees.   Seasonal employees, who are hired for short time periods (less than 150 days) to meet large labor requirements, are primarily local workers who work close to home. They include young people, housewives, and people on vacation from regular jobs. The others are migrants who travel across county or state boundaries and stay overnight to work. Nearly three fourths of all hired farm workers work fewer than 150 days on farm jobs. In 1981, only about 5 percent of them were migrant laborers.

Employers should be aware of the number of public policy issues related to the living and working conditions of migrant farm workers, who are employed in large numbers during critical harvest periods. These issues include housing, government regulations, working conditions, compensation, and employment of undocumented aliens.

Planning Labor Needs

A calendar of operations as illustrated in Figures 13-1 and 13-2 shows what labor is required, when it will be needed, and when labor requirements cannot be met by the operator's available labor. The purpose of the calendar is to allow more efficient management of both labor and capital inputs. By showing the manager what should be done and when, the calendar facilitates the coordination of work

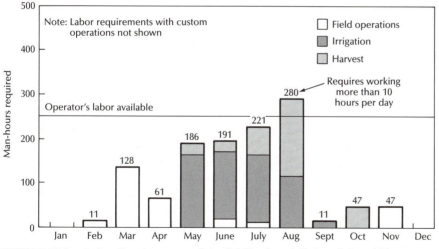

FIGURE 13-1.   Labor Requirements and Availability for a 140-Acre Irrigated Farm.

SOURCE: Frank Conklin, "Factors Contributing to the Success and Failure of Farms in the North Unit Deschutes Irrigation District, Jefferson County, Oregon," master's thesis, Oregon State University, 1959

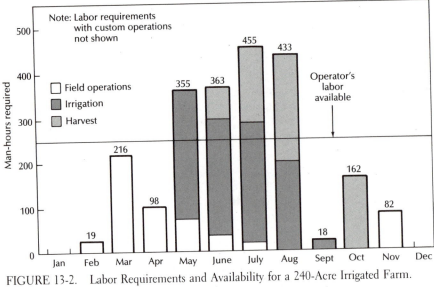

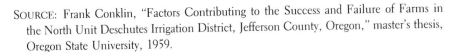

FIGURE 13-2.   Labor Requirements and Availability for a 240-Acre Irrigated Farm.

SOURCE: Frank Conklin, "Factors Contributing to the Success and Failure of Farms in the North Unit Deschutes Irrigation District, Jefferson County, Oregon," master's thesis, Oregon State University, 1959.

and improves communication and performance of labor tasks. It can also serve as a planning tool. By developing a calendar of operations for various combinations of enterprises, the manager can explore the possibilities of reducing labor requirements or making them more uniform throughout the year.

The line in Figure 13-1 labeled "Operator's labor available" assumes that the operator is available for work 250 hours per month, perhaps 25 days at 10 hours per day. The figure depicts the situation for a 140-acre irrigated farm that provides nearly full-time work for the operator for four or five months of the year. Unless the manager wants to work more than 10 hours per day, more labor must be hired in August. If neither of these alternatives is desirable, the combination of enterprises will have to be changed. This chart indicates that an enterprise can be added that would provide winter employment, such as a cattle production enterprise (purchasing cattle in October and selling them in March or April). Because the farm produces both grain and forage, this enterprise would be truly supplementary; it would not reduce the output of any other enterprise on the farm.

The labor requirements of a 240-acre irrigated crop farm are shown in Figure 13-2. For this size farm, labor must be hired. If 250 hours of operator labor are subtracted from the total labor requirements of May, June, July, and August, hired labor is required as follows:

	May	June	July	August
Hired labor needed (hours)	105	113	205	183

This farm manager may wish to hire a full-time employee during these months if dependable help is not available on a part-time basis. The figure shows that the manager does not have full-time work during the winter months; however, obviously there would be some maintenance and repair work on the farm that an analysis of the cropping operation does not show. Even so, it is possible that a supplementary enterprise could be added profitably to this farm business. If the manager is considering even further expansion, the crop acreage could be increased by adding a winter livestock enterprise and hiring someone for the entire year. Further planning would be necessary, of course, to determine if this undertaking would be profitable.

In preparing a calendar of operations, the influence of weather should be considered. The farm area referred to in Figures 13-1 and 13-2 is an arid region in Oregon with annual rainfall of approximately 10 inches, most of it during the winter months. So, the assumption that 25 days of each of the summer months would be suitable for field work is quite reasonable. However, this would not be a realistic assumption for many other areas of the United States. Consequently, a calendar of operations would have to be adjusted accordingly. Although this calendar pertains to irrigated farms, the procedure would be comparable for other types of farms.

Recruiting Farm Employees

In spite of an adequate general labor supply, as indicated by the percentage of workers unemployed, farm managers have trouble recruiting and keeping qualified farm labor. The two major problems are competition with nonfarm employment and the low status of farm labor.

U.S. farm wage rates have risen in recent years but, in general, are below those prevailing in nonagricultural employment. Low pay is frequently the reason workers leave farm jobs. Although agricultural employers have traditionally provided housing and meal benefits, they have lagged behind other employers in providing health insurance and retirement benefits. Farm wage rates vary widely for seasonal and regular labor and for different types of farms throughout the United States. There are also wide variations in nonfarm wages, and some farm jobs pay more than some nonfarm jobs. Although the gap between farm and nonfarm wage rates may be closing, the general perception of low wages is a problem to be resolved in recruiting agricultural labor.

Society has ascribed a relatively low status to agricultural employees; the image of the work and the workers is less than desirable. The term associated with most agricultural employees, *hired man*, is not apt to attract many workers. Unless some of these undesirable aspects of agricultural employment are changed, it will be difficult to attract skilled labor.

Farm managers recruiting farm employees must communicate two main points: (1) farm jobs are competitive with nonfarm jobs in salary, fringe benefits, and working conditions and (2) good labor management can make farm employment desirable. To attract and keep good, productive employees, these two conditions must be met.

This recruiting process is one of the most important aspects of labor management. Mistakes made in hiring are expensive not only to the employer but also to the employee who is not suited to the job.

Job Description

The first step in the hiring process is to decide what will be done by the employee; a description of the job is important for advertising the opening and for determining the qualifications of the employee. It should include the duties, responsibilities, and authority involved.

Next, the qualifications needed by the prospective employee to perform the job satisfactorily should be listed. The higher the qualifications, the fewer the number of applicants. Also, higher qualifications may require higher wages and more benefits. It may be useful to divide the qualifications into required and desired characteristics. Those applicants not meeting the required qualifications would be eliminated; the remaining applicants would then be ranked according to the desired characteristics. The farm manager should write and rewrite the job description until it expresses exactly what is desired.

Advertising the Employment Opportunity

There are several ways to find prospective employees:

1. Inquiring of neighbors who may know someone looking for employment locally.
2. Asking local vocational agriculture teachers, extension agents, lenders, and others who may know of prospective employees.
3. Advertising in local newspapers. The ad should be specific about the job and employment requirements, or much time will be spent unnecessarily screening applicants.
4. Referrals by current employees who know the job to be done and the type of person needed.
5. Checking with the state unemployment or job service agency. This agency can provide a list of qualified people who have been screened to meet specific qualifications.

The local market area and the job description will determine where to look for what type of employee. These two factors will influence the advertising strategy, the sources that are checked, where ads are placed, and the extensiveness of the advertising area.

The advertisement should give a brief description of the size and type of farming operation and the duties and responsibilities to be performed. Qualifications should be listed, including any special experience or skill requirements. Because the wage or salary is a negotiable item that should be discussed later, only a salary range should be indicated. If housing is available, that should be stated; a description of the rural community may also be a selling point.

Application for Employment

Interested persons should be asked to fill out an application form. The form might request the applicant's address, telephone number, education, and special training. Other important information includes a list of former employers, job duties, dates worked, wage or salary received, and reason for leaving. Applicants should be asked to provide at least three references. Although the application will not provide all the information needed to hire someone, it is a useful screening device for choosing the persons to interview.

Interviewing

The interview has two purposes: (1) to give the applicant a clear picture of the job and what is expected and (2) to allow the farm manager an opportunity to compare each applicant's qualifications and aptitudes to those of the others.

The most promising applicants should be invited to the farm, along with their spouses, for an interview. A checklist might be used to guide the interview. First, information on the application form should be clarified, such as the exact nature of the farm work experiences and skills and the reasons for leaving former jobs. Personal goals, aspirations, and reasons for wanting farm work are especially important when hiring a full- time employee.

Certain questions are illegal. Although state laws vary, in general, employers are not allowed to ask about race, color, religion, age, marital status, and sex; managers should realize that many farm jobs can be performed equally well by both men and women. Employers may ask if the applicant is a citizen of the United States, but not about nationality, ancestry, or national origin. They may ask whether an applicant has any disability that would interfere with the performance of the job, but not about disabilities that are not job related. Employers may not ask if the person has been arrested or convicted of a crime, unless this information is substantially related to the reponsibilities of the job.

Checking References

After the interview, the information provided by the applicant should be verified. Telephone calls to the references listed can provide insights regarding the applicant's background and experience, but previous employers, school teachers, or others who may be familiar with the applicant should also be contacted. A prepared list of questions helps to guide the conversation. Information provided by the applicant should be verified first; then more pointed or leading questions about the applicant's abilities can be asked.

The Hiring Decision

A deadline should be set for making a decision and notifying the applicants. However, if more information is needed, this deadline can be extended. For example, the employer may want to interview one or more of the applicants a second time before making a final decision.

In selecting an employee from among a number of applicants, personality, appearance, intelligence, enthusiasm, experience, background, and, above all, trainability should be considered. Often the inexperienced person can be trained to do a job better and more efficiently than a more experienced one who will not submit to retraining. Other supervisors and the employer's spouse should be involved in the final decision.

Once convinced that the right person has been selected, the employer is ready to make a job offer. Whether or not the person accepts will depend on how well the employer has explained the merits of the job and whether the two parties can agree on the wage or salary, benefits, and other conditions of employment. Points of disagreement should be negotiated and compromises made to ensure a mutually satisfactory relationship.

After the job has been filled, all applicants should be notified of the decision and thanked for their interest. Employers should keep the list of qualified applicants, which can be used when recruiting for future positions.

The Employment Agreement

The decisions to make and accept a job offer should be based on the complete understanding and agreement of both parties that the conditions of employment are acceptable. A written employment agreement is the best way to ensure that this is accomplished. Employer–employee relations in agriculture are usually handled informally with oral explanations. However, if agriculture is to compete with other industries for employees, employers must specify the working conditions more clearly or more formally. A written employment agreement gives the worker an opportunity to ask questions about specifics before problems or disagreements arise.

The important components of an employment agreement vary with the type of farm, the job description, and the location. Employment agreements can range from formal contracts signed by both the employer and the employee to a simple memorandum of understanding that describes the basic conditions of employment. Signing the agreement has legal implications that should be checked by an attorney. Some believe that the employment agreement is a more effective communications device if signatures are not required.

The contents of an employment agreement depend on whether the job is regular (year round) or seasonal, and the contents of agreements for seasonal workers depend on the nature of the work. Table 13-1 contains a detailed list of items to be included in an employment agreement. Although salary or wage rates should obviously be included, other factors such as working conditions, overtime pay, vacation, fringe benefits, and incentives can make farm employment more attractive to employees.

The job description can be the same as the one prepared to advertise the opening. It should include a job title, such as *farm machinery operator, assistant farm manager, animal production supervisor*, or whatever is appropriate; do not use the term *hired man*.

The written agreement might indicate the length of a probationary period, perhaps 6 or 12 months, after which a decision will be made about continuing em-

TABLE 13-1. Suggested Items to Include in Employment Agreements

Name and address of the employee and the employer
Detailed job description (duties, responsibilities, and authority)
Salary or wages and when paid
Regular working days and hours
Overtime pay
Bonus or incentive payments
Holidays and other days off
Regular paid vacation
Sick leave
Housing benefits
Health and life insurance
Retirement plan
Other benefits (telephone, utilities, food, and so forth)
Description of payroll deductions
Training and attending short courses
Termination notice
Transportation and use of farm vehicles
Safety rules
Provision for revising agreement

SOURCE: Adapted from D. L. Armstrong, *Employment Agreements in Hiring Farm Workers*, Michigan State University Rural Manpower Center Special Paper No. 9, April 1969.

ployment. The agreement might specify the minimum notice required to terminate the agreement and any other termination procedures in the event that either party finds the situation unsatisfactory. Provision should be made for updating and modifying the employment agreement.

Pay and Benefits Packages

The value of the pay and benefits package is an important consideration in labor management. Although employees consider other aspects of the employment agreement, they are concerned about receiving total compensation comparable to what could be earned in another farm or nonfarm job. Low earnings are a common source of dissatisfaction. From the employer's point of view, the wage offered will influence the caliber of applicants, their qualifications, and their performance if hired. Employees who are dissatisfied with their compensation will not be motivated to perform to their full capabilities.

Farm managers who cannot offer total compensation competitive with other employers because of their low farm income must rely on nonmonetary advantages to make up the difference. This involves emphasizing the advantages of farm employment, such as a healthy work environment, personal contact with the employer, and working independently.

Determining the Wage or Salary

Determining the wage or salary to offer is not easy. It depends on where the farm is located, the local labor market, the responsibilities of the job, the qualifications and experience of the employee, the alternative employment opportunities, and the fringe benefits offered. To get information to determine appropriate salary or wage levels, employers might contact the state unemployment or job service agency or other employers hiring employees with comparable qualifications. Also, applicants can be asked how much they earned on their last job. Employees who are leaving for other jobs should be asked about their new salary or wage.

Three basic methods are available to pay for labor. The option chosen depends on the type of farm, the nature of the job, and whether the labor is seasonal or regular.

Weekly or Monthly Salary.   When farm employees are paid a weekly or monthly salary, the number of hours worked per week or month is traditionally unspecified. Under these circumstances, employers tend to believe that they are paying for all the hours employees can work; therefore they think that workers might as well be kept busy. If salaried workers have no specific number of hours to work, they are put in the position of having to request time off to take care of personal matters. Most nonfarm employers hire labor for a specified number of hours per day or week, with a bonus paid for overtime work.

When a weekly or monthly salary is paid, the hours of work should be specified and adhered to strictly. Because working more hours than usual cannot be avoided during some seasons, the employment agreement should provide for overtime pay or compensatory time off.

Hourly Rate.   When employees are paid an hourly wage for a maximum number of hours per week and 1.5 times this rate for overtime, employers are more justified in asking employees to work extra hours when needed. Compared to a weekly or monthly salary method, an hourly wage system decreases the likelihood that hours of work will become an issue between employer and employee.

Seasonal employees are typically paid an hourly wage, but regular workers paid on an hourly basis want some guarantee of a minimum number of hours per week or month. This is particularly important during annual slack times when farm labor needs are low.

Piece Rate.   Industrial experience has shown that labor efficiency can be increased by adopting a piece-rate method of payment. In agriculture, this method is used to pay seasonal labor hired for fruit and vegetable harvesting. The disadvantages include the increased risk of poor workmanship and a less than conscientious attitude on the part of employees. Unless properly designed, the piece rate method can be unfair to the employee; for example, piece rates should be higher during the early and late stages of strawberry picking than during the peak of the season.

Fringe Benefits

In recent years, much of the change in nonfarm employment compensation has been in the area of fringe benefits. To make farm employment more competitive and to provide incentives for good employees to stay, more attention should be given to fringe benefits. The major fringe benefits are social security, unemployment insurance, workers' compensation (accident insurance), health and life insurance, retirement plans, paid vacations, training, housing, and transportation. Some of these benefits are mandatory, depending on the type of farm, the number of employees, the size of the payroll, and the type of business arrangement (proprietorship, partnership, or corporation). The cost of a full complement of benefits equivalent to that of nonfarm employment can amount to 30 percent or more of wages and salaries.[4] Thus, determining what fringe benefits to provide is an important decision that should include two key considerations: the tax deductibility of the cost and the employees' perceived value of the benefit.

If employers want to reward employees with higher total compensation, they should consider increasing benefits rather than wages, particularly if the cost of these benefits is tax deductible to the employer and not taxable to the employee. If an employer wants to give the employee a $600 raise, increasing the salary by $600 might leave the employee only $400 after taxes. If, instead, the employer provides a $600 health insurance policy, the cost is tax deductible to the employer (as the salary increase would be), but the benefit is also tax free to the employee. The tax treatment of job benefits depends on the benefit and the business arrangement of the employer.

The other important consideration is the value of the benefit to the employee. Which is worth more to the employee, the $400 after-tax salary increase or the $600 health insurance policy? The answer will depend on the employee. A young, single employee may prefer the $400 after-tax salary increase, whereas an older employee with a family may prefer the health insurance benefit. If farm employees undervalue the fringe benefits, it can cause dissatisfaction. Therefore, employers should be sure that employees understand their fringe benefits. The cost of the benefit to the employer should not exceed its value to the employee. Following is a discussion of the more commonly offered fringe benefits.

Work Schedules. A major complaint of farm employees is that they are often expected to work long hours for little extra compensation. The nature of farm work and the number of employees available affect work schedules and days off. Some operations rotate weekends off between the employer and employees. There are any number of possible schedules, but the important thing for employers to remember is that employees place a high value on having weekends with their family and friends.

Many nonfarm employers are experimenting with flexible work schedules. The general idea is to arrange them to benefit both the employer and the employee. Farm managers have usually provided informally for time off during slack periods but have expected extra work during peak periods, such as planting and harvesting

---

[4] K. R. Krause, *Indirect Farm Labor and Management Costs*, U.S. Department of Agriculture Economic Research Service Agricultural Economic Report No. 496, December 1982.

times. Formalizing the agreement in terms of the days and hours to be worked regularly, with overtime pay or compensating time off for extra hours, gives the employer an incentive to manage the labor force carefully. It also helps employees to understand the work schedule more fully.

Paid Vacation.   Two weeks of paid vacation appears to be common policy for regular employees; long-term employees might be offered three to four weeks off. Employers should recognize that a vacation during inclement winter months may not be viewed as much of a benefit by the employee. Likewise, employees should request vacation time when disruption to the farm operation will be minimal. Employees should know in advance when they will have time off or will be able to take vacation so that they can plan and take full advantage of it. Other paid time off can include holidays, birthdays, and pregnancy leave. Leave for personal reasons, such as banking and medical appointments, should also be considered. Paid time off increases the cost of farm labor, but it can reduce the costs resulting from fatigue, accidents, and labor turnover.

Sick Leave.   Because nonfarm employers pay their employees during illnesses, farm managers should consider providing a similar benefit. To prevent employees from unnecessarily taking advantage of sick leave, they might be given half a day off with pay for each unused sick day. Employers should also consider providing disability insurance for protection against prolonged illnesses, particularly for employees who have been with them for a long time.

Housing.   There are both advantages and disadvantages to furnishing the traditional benefit of housing. A rent-free home near the job, which would save transportation costs, would seem to be desirable from the employee's point of view. However, employees and their families may prefer different housing or housing in a different location. Some may prefer to make their own decisions about home repairs and improvements, thus eliminating the possibility of friction with their employer. Including housing as a fringe benefit in the employment agreement is worthwhile only if the employee and the employee's family believe it is. If housing is provided, the employer should also give the employee an estimate of its monthly value.

Employee Training.   Increasingly, farm employers are encouraging additional education and training for their employees. This training might consist of a four-hour workshop on the proper maintenance of farm machinery or a five-day course on the artificial insemination of livestock. Employers might pay all or part of the employee's expenses for tuition and travel while providing paid time off. The employer gets a more skilled employee in return but faces the risk that the employee may find it easier to get a higher-paying job.

Health and Life Insurance and Retirement Plans.   Obtaining group coverage under health insurance programs, paying whole or partial premiums toward a life insurance policy, and setting up a retirement plan for employees are additional fringe benefits that might be provided. As with other benefits, their merits must be judged by their costs and perceived value.

Farm Produce.   It is common practice to provide employees with farm produce such as milk, meat, eggs, and vegetables at no cost. Again, this practice can create dissatisfaction if employees do not value this fringe benefit as highly as employers do. An alternative is to pay higher wages and let employees buy farm produce at cost.

Bonus and Incentive Plans

Bonus and incentive payments have been used by some farm employers to attract, keep, and reward good employees. There is a wide variety of plans, and some have been more successful than others. Following are brief descriptions of some general plans. Guidelines for designing successful plans are then presented, along with an example.

Bonus Plans.   As commonly used, Christmas bonuses are ineffective labor management tools because employees come to expect them. Merit bonuses based on the employee's performance provide more incentive for extra effort, and paying bonuses after critical farming operations, such as seeding and harvesting, are completed, helps to prevent an employee from leaving during the peak season.

Production Incentive Plans.   Production incentive payments are based on the level of output, the gross income of a given enterprise, or the gross income of the total farm business. These payments are in addition to regular wages. In designing such a plan, there must be a standard measure of production, such as pounds or bushels. One approach that encourages employees to strive for higher production is to pay larger incentives per unit for successively higher levels of production.

Equity-Building Incentive Plans.   Equity-building incentive plans are sometimes used to encourage good employees to stay on. Examples of these plans include giving animals in lieu of salary, providing the use of the employer's machinery so that employees can farm their own land, and allowing employees to raise a limited acreage of a crop and to receive a share of the crop's income in addition to their base wages. Employers should wait until they have worked with employees for a year or two before offering a plan that allows them to buy into the farm business.

Profit-Sharing Plans.   Employees with profit-sharing plans receive a percentage of the net income from an enterprise or from the entire farm business in addition to a base salary. These plans, which are commonly used in father-son joint farming ventures (Chapter 9), have the effect of encouraging employees to assume more management responsibilities. Problems arise if the farmer and the employee disagree on the management of the farm business or on the calculation of net income, which many farm managers are reluctant to disclose to employees.

Tenure Incentive Plans.   One type of tenure incentive plan provides that employees receive their basic salaries and, in addition, an agreed-upon amount that is set aside and paid only if they stay until the end of the year. This incentive payment might be increased with each year of employment.

General Guidelines.   The bonus or incentive paid should be clearly above the normal wage rate for performance that is better than average. It should be just what the term implies—a bonus, not a partial substitute for an adequate salary or wage. The plan should be in writing and defined in such a manner that computation of the payment is simple and easily understood by both parties.

The incentive or bonus must be achievable and large enough to encourage extra effort. It should be based on performance that is largely within the control of the employee but should not encourage or promote practices that are uneconomical for the employer. For example, an incentive payment of additional milk produced per cow without consideration for the amount of grain fed could encourage an employee to feed grain to the point where the cost of the additional grain is greater than the value of the additional milk. This practice would be beneficial to the employee but uneconomical for the employer. (The marginal input cost would be greater than the marginal value of the product.)

The incentive should be paid relatively soon after it has been earned to maximize its effect. The purpose of the bonus is to encourage better than average performance from the employee, not to shift some of the financial risk or uncertainty from the manager to the employee. At the same time, the bonus incentive should not hamper the manager in making sound management decisions.

Incentive plans are sometimes mistrusted by employees and consequently can hinder rather than foster good employer–employee relations. The success of any incentive plan can be affected by the degree to which employees participate in drawing it up. The incentive system must be mutually advantageous. It is not a substitute for good labor relations, but if it is properly developed, it may improve them.

An Example.   Lloyd Duyck produces berries on 1200 irrigated acres in Oregon.[5] He also operates a plant to process and freeze the berries. Mr. Duyck undertook a new enterprise when he decided to use his processing plant and labor force to process watermelons into pulp and juice for a beverage manufacturer.

The watermelons were trucked in and processed using a technique he devised: The watermelons were halved and up-ended onto beaters to extract the pulp, and then the rinds were tossed aside. The process involved hard physical work. The processing line was designed to produce 190 barrels of watermelon pulp and juice per day. The employees were paid an hourly rate.

After three days of operation, output was at 140 barrels per day, but Mr. Duyck needed 190 barrels for a profitable operation. He decided to implement an incentive system: a $1 per barrel bonus for every barrel over 140 produced per day. The bonus payment went into a pool that the employees would divide equally. For the next three weeks, until the end of the season, the processing line consistently produced between 240 and 270 barrels per day. Other advantages were fewer breakdowns, less absenteeism, and more cooperation among the workers.

As this example demonstrates, incentive and bonus plans can be beneficial to both employers and employees. However, the plans must be carefully developed

[5] G. Felix, "Group Incentives in Watermelon Processing," *Productivity Primer*, Oregon State University, Oregon Productivity Center, October 1983.

to accomplish the desired goal, and both parties must thoroughly understand how the plan works.

Human Relations Principles

In addition to offering competitive wages and reasonable hours of employment, a manager must apply human relations principles to attract and keep employees. Many farm managers have had less experience in this area than they will need in the future. Good labor management creates a situation that motivates employees to perform to their potential, but to do this managers must first understand what makes people tick.

The subject of human relations, which includes sociology and psychology, deals with what is important to people and what makes them think and act as they do. To manage farm labor effectively, employers should understand that people have basic needs that they strive to satisfy.

Human Needs

As described in the section on decision making (Chapter 1), activities are more effective when they are directed toward specific goals. However, this does not mean that goals motivate people; people work toward goals because of the needs these goals will satisfy. Individual behavior is guided by basic needs, and the needs that are most influential at any particular time are those that are unsatisfied.

Human needs, divided into five categories, are illustrated in Figure 13-3. They are satisfied in a certain order of priority; that is, there is a hierarchy of human needs. The motivation to satisfy these needs develops according to this order: lower-level needs will have greater influence on behavior than higher-level ones until the lower-level needs have been satisfied. For example, basic physiological survival needs influence behavior until they are satisfied; then the need for security assumes top priority.

FIGURE 13-3.   The Hierarchy of Human Needs.

SOURCE: Adapted from A. H. Maslow, *Motivation and Personality*, 2nd ed., New York: Harper & Row, Publishers, Inc., 1970.

New employees, particularly if unemployed for a while, concentrate on making it through the probationary period just to keep the job, but once more secure in the job, their concern will shift to being accepted by co-workers. The employee who does not feel accepted usually experiences emotional isolation and distress, and is likely to seek employment elsewhere.

After an employee feels accepted, the needs for esteem and status emerge; these are the external signs of being liked and appreciated by others. Examples include a promotion, a salary increase, or a new pickup truck to drive to work.

Self-actualization and fulfillment, the highest level of human needs, may take many years to satisfy. This achievement requires finding worthwhile challenges, applying one's knowledge and effort to meet them, and succeeding. An employer can help employees satisfy these needs by increasing their responsibilities and their latitude to fail or succeed, thus encouraging the development of capabilities that would not have emerged otherwise. Employees can never be fully actualized if they must always follow the employer's instructions.

Initially, when an employee is first hired, lower-level needs must be satisfied, but keeping the employee requires that attention be given to the higher-level needs as well. Employees will perform beyond minimum requirements if they are given the opportunity to satisfy their needs for acceptance, esteem, status, self-actualization, and fulfillment. Employers must empathize with their employees' needs to determine if they are being met.

Approaches to Motivation

The way employers motivate their employees depends on their own basic beliefs or assumptions about people in general. There are two contrasting sets of assumptions that determine a manager's approach to supervising and motivating employees: Theory X and Theory Y.[6]

The Theory X manager assumes that the average person inherently dislikes work and avoids it if possible. Because the work or job itself is not motivating, people work mostly for the money and status achieved as a result of the job. Theory X managers assume that the average person prefers to be directed, avoids responsibility, has relatively little ambition, and wants security above all.

Theory Y, on the other hand, suggests that because work is as natural as play or rest, people do not inherently dislike it. To the contrary, work can be satisfying as a result of the pleasure of association, pride in achievement, and the stimulation of new challenges. If people are encouraged to think for themselves and to act on their own initiative, they will learn to accept and even seek out responsibilities. Theory Y managers have high expectations of employees and believe that they will rise to meet them.

Theory X managers manage labor autocratically, regarding employees as basically lazy, unwilling to assume responsibility, and resistant to change. They supervise employees closely to get adequate performance. Theory Y managers regard employees as responsible and hardworking, and try to make jobs more meaningful so that employees can develop their capabilities and satisfy the human need for self-fulfillment; therefore, employers encourage involvement and innovation.

[6] D. McGregor, *The Human Side of Enterprise*, New York: McGraw-Hill Book Company, 1960.

There are very few exclusively Theory X or Theory Y managers. However, those managers who are most successful in developing satisfactory long-term relations with employees have adopted more Theory Y than Theory X assumptions. These farm managers, who pay higher wages, practice more participative management, and generally show more concern for employees' needs, experience less turnover, and achieve higher levels of labor performance.

Labor Management Functions

To apply this knowledge of human needs and motivation requires communication skills. A manager's actions may communicate far more effectively than words; actions speak louder than words. Managers must set a good example in terms of planning, organization, and work habits.

Turnover is expensive, that is, labor efficiency is reduced due to the time required to hire and train new employees. Therefore, once the employee has been hired and the employment agreement has been finalized, the manager's attention should be directed toward developing and retaining the employee. The first important labor management function is training.

Training

The process of training employees can take a few days or several weeks, depending on the size of the operation and the complexity of the tasks. The employee must understand what, when, where, why, and how the farm manager wants things done. Without this knowledge, the employee will be frustrated, which may result in wasted time and resources and in accidents that risk personal injury and property damage.

To provide orderly and adequate training, employers should break down the job into logical steps as follows:

1. *Orient the employee.* The manager should acquaint the new employee with the farm business, explaining how the employee's duties relate to the overall operation.
2. *Teach the employee the job.* The manager should patiently demonstrate the job to the worker one step at a time, stressing and repeating key points and safety procedures. The reason one method is used instead of another should be explained. Unfamiliar terminology should be defined and questions encouraged.
3. *Supervise the employee.* The manager should provide close supervision when the worker first tries to do the job so that corrections can be made immediately.
4. *Follow up.* Once the worker has gained experience doing the job, the manager should check periodically to correct any mistakes before they become bad habits. With continual encouragement and training, employees will develop self-confidence to the point where they are ready to accept added responsibility.

This approach to training, based on the Theory Y model of labor management, assumes that employees want to learn and that they do not have to be coerced.

Work Planning

Without careful planning, jobs will not be finished on time, some jobs will be overlooked or left to the last minute, and employees and equipment will not be fully employed. Planning the work to be done a day and a week in advance leads to a more efficient use of time. Making a list of jobs to be done enables employees to do some things on their own initiative and gives them a feeling of accomplishment when the jobs are completed.

Schedules and instructions should be prepared at a time when the manager can give them careful consideration. The priorities of the work to be done should determine the work schedule. For instance, fences, buildings, and machinery can be repaired when field operations are slack. Instructions should be given so that employees know exactly what is expected of them and when.

Effective work planning involves not only identifying the best timing and sequence of jobs but also providing the resources needed to accomplish the jobs on schedule. A job list identifies the jobs to be done within the next week; a work schedule identifies the employees and machines to do them. Each job listed should include a deadline and priority, and a job not completed by the deadline should be reconsidered before it is rescheduled.

Employer–Employee Relations

Once an employee is recruited, oriented, and properly trained, the employer must continue to show an interest in the employee's well-being and discuss the goals and plans for the farm business. Showing an interest in the welfare and goals of employees and expressing appreciation for a job well done will improve morale and motivate employees. Likewise, managers should also take corrective action at the first sign of a mistake, an oversight, or failure to meet realistic goals, but in a considerate and constructive manner, quietly, and not in front of others.

A common complaint of employees is that "the boss" does not listen to them. The farm workers' knowledge can be used to advantage by the farm manager, so suggestions should be encouraged. Workers may also be asked to participate in planning, particularly as it pertains to their work. If a suggestion clearly adds to the income of the farm, a bonus may be appropriate.

Performance Evaluation

Managers should tell employees how they are doing, how they have excelled, where they have improved, and offer constructive suggestions for improvement. Evaluation, if carried out correctly, encourages the ongoing development of the employee. The job description, which should define what is expected of the employee, is an important evaluation tool, and the first step in evaluating performance is to review the job description. If it does not reflect what the employee is doing or what is expected, it should be revised by the employer and employee together.

During the evaluation process, the manager should explain which aspects of the employee's performance are satisfactory and which are not. A rating form might be used to summarize the evaluation and to provide a written record for future reference. Plans to improve job performance, to train, and to broaden responsibilities should be discussed. And, finally, employees should be encouraged to respond, to discuss their performance and what might be done to make the job more productive and satisfying.

Employers should meet at least annually with their employees to discuss and evaluate their performance. More frequent evaluations might be good for new employees. These meetings should coincide with decisions regarding a salary increase. By evaluating employees and making salary adjustments within the same month, the cause-and-effect relationship between salary and performance is reinforced.

Terminating Employees

Labor management decisions for good employees are much easier than those for employees not suited to their jobs. An employer who must fire an employee is admitting failure; even so, some employees, no matter how carefully they were hired and trained, are not able to live up to expectations and requirements. When it is clear that the employee's performance will not improve, notice of termination should be given. This is best done in a private meeting, with a calm discussion of the reasons for the firing. Many employers allow the employee time to look for a new job while still on the payroll, that is, after notice has been given until the termination becomes effective.

In the event that an employee quits, the employer should conduct a termination interview to find out why. It is imperative that the manager evaluate these reasons and make any changes that would improve or correct the situation for future employees.

Improving Labor Efficiency

There are several ways to improve the efficiency of labor management. Two of them, work simplification and substituting capital for labor, are discussed in this section. Work planning and incentive payments can also improve labor efficiency, as discussed earlier in this chapter.

Labor efficiency (Chapter 5) is usually measured by dividing total labor into output—for example, pounds of milk produced per hour. The most appropriate measure of labor efficiency depends on the type of farm and the enterprise. Capital (machinery and equipment) per unit of labor indicates the extent to which capital has been invested for labor. Poor labor efficiency can result from a capital investment in machinery and equipment that is too small. To improve labor efficiency, the manager must first know how labor is presently used, by job performed and enterprise.

Time and Labor Records

Time is one resource that cannot be bought, borrowed, or stretched; therefore, its use should be the most carefully planned of all. An accurate record of where and how employees spend their time is essential to any systematic analysis of labor efficiency. In addition to serving as an aid in improving labor efficiency, records are required to prepare and file income tax and social security returns.

Weekly time cards provide such records for systematic analysis. (An example was given in Figure 5-1.) The data from these cards, which can be kept by the employer, a foreman, or the employee, can then be posted to the labor section of the expense record or to the specific enterprise. If a computer is being used, the card becomes the input form. This particular card can also be used for recording machine time.

The data from the time cards can be used to develop a calendar of operations for each enterprise. Then, if expansion of one enterprise and contraction of another are being considered, the effect on labor and machine requirements can be anticipated. Such analysis will determine if labor bottlenecks are likely to develop.

Work Simplification

Studying the tasks to be performed with the objective of finding the most efficient method to do them is called work simplification. For example, by changing the grip or leverage on a hammer, fewer strokes may be required to drive a nail; by developing a rhythm, a fruit picker can pick more fruit.

The process of work simplification involves careful analysis of the task, listing the essential parts of the task, determining the resources available to accomplish the task, and developing the most logical method for doing the work. Farm work simplification can be summarized as a systematic study of work methods in order to accomplish the following:[7]

1. Eliminate all unnecessary work.
2. Simplify hand and body motions in doing the work.
3. Provide a more convenient arrangement of work areas and location of materials for doing the work.
4. Improve on the adequacy, suitability, and use of equipment for doing the work.
5. Organize the work routine for full and effective use of employees and machines.

The efficiency with which farm tasks are performed is highly variable. The performance of livestock chores is an example. The location of buildings and the types of corrals, gates, and equipment all combine to create wide variations in work efficiency.

Continuing with the livestock example, the first step in work simplification is to prepare a map of the farmstead, with lines drawn from one point to another as trips are made in the performance of chores. Then any steps that are obviously redundant can be quickly identified, and the operation can be studied to see if all

---

[7] L. W. Vaughan and L. S. Hardin, *Farm Work Simplification*, New York: John Wiley & Sons, Inc., 1949, p. 59.

of the trips are necessary. The analysis can then be broadened to include improve-ments: Can gates be easily opened and closed? Are the buildings properly arranged in relation to the jobs to be performed?

Fairly minor rearrangements may greatly improve labor efficiency. Major re-modeling or additions should be considered only after other, less costly options have been explored. A careful analysis of alternatives, including their costs and benefits in labor savings, is needed before any major investments are made.

Capital–Labor Substitution

The trend historically has been toward increasing the amount of capital that is combined with labor, resulting in increased output per unit of labor. The proper amount of capital to be combined with labor is determined by the substitution principle (discussed in Chapter 2). Because wage rates have consistently increased more than the cost of capital goods, capital has been increasingly substituted for labor in agricultural production.

Obviously, as the cost per unit of labor increases, the incentive to substitute capital for labor becomes correspondingly greater. Acquiring more or different machinery or other capital items to combine with labor inputs should be considered as a way to increase labor efficiency. However, the profitability of such decisions should be tested using the partial budgeting techniques presented in Chapter 6. There are some other possible benefits from combining more capital with labor. It may be that the improved morale of the operator, the family, or the employees will be a major indirect benefit. If work can be made more pleasant by eliminating

disagreeable tasks and replacing outmoded equipment, efficiency and performance are likely to improve. Obviously, it is possible to overinvest in capital equipment for the sole purpose of improving morale.

Another possible benefit relates to off-farm work. If opportunities exist for off-farm work, farm managers may consider more mechanization. This decision can also be analyzed appropriately by using a partial budget. Still another possible benefit from improved labor efficiency is increased leisure time.

Summary

Hired labor is necessary for the success of many farm operations. Although farmers and unpaid family members still provide the largest amount of agricultural labor, the proportion of hired labor is increasing. Moreover, the skills and qualifications required of agricultural employees are growing as a result of technological advances in farming.

Planning labor needs involves preparing a calendar of operations showing what labor is required and when. Not only does this calendar indicate the amounts of labor that must be hired to meet labor requirements, it also serves as a tool for exploring possibile ways of reducing labor requirements or making them more uniform throughout the year.

Recruiting labor is a challenge because of the low-wage, low-status image of agricultural employment. These negative aspects can be overcome with effective communication in preparing the job description, advertising, requesting applications, conducting interviews, checking references, and negotiating the employment agreement.

Farmers must offer pay and benefit packages comparable to those provided by nonagricultural employers if they expect to hire equally competent labor. Employees are paid a salary, hourly wage, salary plus incentive, or piece rate, depending on the nature of the work and whether the job is regular (year round) or seasonal. Decisions regarding the fringe benefits to offer should be based on their cost to the employer and their value as perceived by the employee. Incentive payments should be in addition to a fair salary, large enough to encourage extra effort, and based on performance within the employee's control.

Wages and benefits are important, but effective labor management requires an understanding of human relations, personal needs, and motivational approaches. Training is one of the most elementary and necessary functions of labor management, because it is the best insurance against costly mistakes and accidents. Planning work, acknowledging accomplishments, correcting mistakes, evaluating overall performance, and terminating employees are also important functions that affect labor turnover and efficiency.

Improving labor efficiency requires records of labor use so that problem areas can be identified. Work simplification, substituting capital for labor, careful work planning, and incentive plans are all devices for increasing the output per unit of labor. Proper attention to labor management, from recruitment to evaluation, will also provide employees an opportunity to achieve their potential, thus benefiting themselves, as well as the farm business.

Recommended Reading

BLANCHARD, K., and S. JOHNSON. *The One Minute Manager.* New York: William Morrow and Company, Inc., 1982.

Management of Farm Enterprises

The commodities produced by the farm business should be selected in light of the resources available and the market to be satisfied. Part IV is concerned with enterprise selection and management.

CHAPTER 14

Enterprise Selection

Two important decisions made by farm managers are the enterprises selected and how they are combined in the farm business. Physical, biological, and economic forces determine the most profitable crop and animal enterprise combinations. However, stability of income and risk must also be considered. These decisions of individual farmers are reflected in the production patterns of geographic areas and regions where the production of certain commodities is concentrated. So it is with nations; their economic well-being is often enhanced if they specialize in certain products and trade with other nations for the other products they need.

In order to clarify the basic principles relevant to enterprise selection and combination, this chapter first considers the broader problems of the location of agricultural production and then the individual farm management problems.

The Location of Agricultural Production

An investigation of the location of agricultural production in the United States reveals that certain geographic areas tend to specialize in certain commodities. The wheat fields of the Great Plains, the corn-hog farms of the Midwest, the specialty crops of the Far West, and the dairy farms of the Northeast are all examples of this regional specialization. A variety of crops and animals are raised within each region, however. Although the Midwest is famous for corn and hogs, commodities such as beef cattle, soybeans, fruits, vegetables, and small grains are also commercially produced. These production patterns are not absolutely stable; new forces are continuously changing the location of production and the trade relationships among regions. What are the forces that affect the location of production? Economic principles can be applied to answer this question and to analyze these complex forces.

The Principle of Comparative Advantage

Perhaps the most important concept to understand in this connection is the principle of comparative advantage. This principle can be illustrated by explaining why Kansas is called the "wheat state" when Illinois has higher wheat yields. Because of soil differences, wheat and feed grain yields are not the same in the two states. The following are the per-acre yields of wheat and feed grain for the two states:[1]

	Illinois	Kansas
Wheat (bu.)	48	32
Feed grain (bu.)	120	64
Ratio	2.5	2.0

Farmers in Illinois obtain greater yields of both wheat and feed grain than Kansas farmers do. If farmers in each state were to put one half of their land in each crop, wheat and feed grain production per acre for the two states would be as follows:

	Illinois	Kansas	Total
Wheat	24	16	40
Feed grain	60	32	92

What would be the outcome if Illinois produced only feed grain and Kansas produced only wheat?

	Illinois	Kansas	Total
Wheat	—	32	32
Feed grain	120	—	120

To get an additional 28 bushels of feed grain per acre, 8 bushels of wheat were forgone. Illinois farmers must decide if they should plant half of their land to wheat or if they should grow only feed grain and trade it to Kansas farmers for wheat. They know that in order to get 1 bushel of wheat by growing it, they must sacrifice 2.5 bushels of feed grain because of the difference in yields. As a result, if they can trade fewer than 2.5 bushels of feed grain for 1 bushel of wheat, they will be ahead.

Kansas farmers, on the other hand, know that for every two bushels of feed grain they produce they must give up one bushel of wheat. Therefore, they will be ahead if they can trade one bushel of wheat and get more than two bushels of feed grain in return.

Under these circumstances, Illinois farmers would be expected to specialize in the production of feed grain and Kansas farmers in the production of wheat, and

[1] The yields for this example approximate 1980–1982 average yields for corn in Illinois, grain sorghum in Kansas, and wheat in both states.

to trade wheat and feed grain at a rate between 2.0 bushels of feed grain to 1 bushel of wheat and 2.5 bushels of feed grain to 1 bushel of wheat.

Although Kansas and Illinois farmers do not specialize entirely in wheat or feed grain production, the example does illustrate an important principle: the tendency of any economic unit—an individual, a region, or a nation—to concentrate on the production of those items that give it the greatest relative advantage or the least relative disadvantage. This is the basic reason for economic units to specialize their production and trade with one another.

Other Factors Affecting Location

In addition to soil, which is a principal factor affecting productivity differences between farms and regions, other physical factors such as climate, topography, and distance to market (which is reflected in transportation costs) influence the location of production. Biological factors such as insects, plant diseases, and weeds are also important determinants of location. These biological factors may prevent a crop from being grown, and they can be controlled only at a cost.

It may be helpful to classify the forces affecting location into categories. One category of factors affects the yield or production from a given land area; that is, physical and biological forces combine to set limits on production. New farming techniques can alter production responses from time to time.

The second category includes relative prices and costs, which reflect economic forces. Over time, populations, incomes, and markets influence the location of production. These economic considerations influence where products are produced and even if they are produced at all. For example, a demand must exist for a product to be sold.

Institutional and social elements may also play a role in the location of production. The production of certain commodities may be concentrated in an area in spite of unfavorable physical, biological, and economic factors. In other words, the production of a commodity may be initially located in an area because of the circumstances of the producers or the institutional environment in which they operate. However, if the business is to be successful, the biological, physical, and economic conditions must be favorable. Originally the location of production may have been determined by institutional or social forces, but a shift in location may be necessary because of unfavorable biological conditions that cause production in a given area to become uneconomical.

These factors explain the production patterns that emerge in the various regions of the country and the world. Even though particular regions may tend to specialize in certain products, individual farmers may deviate from the general pattern. For example, a farmer in a region that produces primarily wheat may not grow any wheat; or if wheat is grown, other commodities may also be produced. Moreover, some regions may produce a variety of crops, whereas considerable specialization may occur on individual farms within these areas. This inconsistency is due to different physical and biological forces within an area, so that individual farmers may find that they have a comparative advantage in one commodity and consequently specialize in producing it.

Specialization and Diversification

Discarding the traditional advantages of diversification, farmers have tended increasingly toward specialization. Many areas concentrate largely on single-commodity production; the wheat and beef cattle regions in various parts of the United States are examples. Even where there is considerable diversity within an area, many individual farms tend toward specialization.

Diversification may be either horizontal or vertical. As the term is commonly used, *diversification* refers to *horizontal diversification*—the production of a number of commodities. *Vertical diversification* refers to the many steps involved in producing a product on the farm. The production and distribution of milk reflects a high degree of vertical diversification, including raising crops and replacement cows; producing the milk; and then pasteurizing, bottling, and delivering it to consumers. A low degree of vertical diversification, on the other hand, is exemplified by a dairy producer who buys all feed and replacement cows and who sells unprocessed milk.

The Trend Toward Specialization

In recent years, most farms have become more specialized both vertically and horizontally. Obviously there must be economic advantages to specialization. When other firms or farms can perform a function more efficiently, farmers will give up this function and concentrate on others they can perform more efficiently.

There was a time when farms were almost completely self-sufficient, and a few still are. However, farmers have discovered that they can increase their standard of living by purchasing more of their production inputs; this practice allows them to produce a few commodities well suited to their resources and in a volume that keeps the cost of the operation low.

There are many reasons for the improved efficiency that results from concentrating on a small number of enterprises. One is that it is difficult for farmers to become specialists in the production of many commodities; if they spread their management resources too thin, they find they are doing nothing well. Specializing in a particular commodity permits farmers to increase the volume of production to achieve economies of size and increase their profit per unit. In other words, as long as costs are decreasing, it is economical for them to increase production. The production of any agricultural commodity requires considerable equipment or investment. Often this investment is also specialized. So, farmers with limited capital may do better to invest adequately in the production of a limited number of commodities rather than inadequately in a larger number.

It may be possible to obtain some of the advantages of both diversification and specialization by producing more than one crop—apples, prunes, and cherries, for example. Although such an operation does not specialize in the production of one fruit, it does specialize in fruit production.

In this case, all of the fruit requires approximately the same type of harvesting, spraying, and cultivating equipment. Because harvesting and spraying of the three fruits are performed at different times of the year, less equipment is required because it is not all needed at the same time. Also, diversification permits operators to utilize their own labor over a longer period of time. And by lengthening the harvest period less housing is required for laborers. Finally, more income stability can be expected because price and yield fluctuations of the different fruits tend to balance out.

Enterprise Relationships

The number and combination of enterprises on a farm are influenced by the relationship that exists between and among enterprises. A farm forestry enterprise that utilizes winter labor has a different relationship to wheat production than a beef cow enterprise does. These relationships can be classified as *competitive*, *complementary*, and *supplementary*.

Two enterprises are competitive when an increase in the output of one results in a decrease in the output of the other. They compete for, or use, the same resources at the same time. An example is barley and wheat. A farmer in a wheat-summer-fallow area may have a 1200 acre farm, one half in crop each year. The entire 600 acres can be planted to wheat, to barley, or to any combination of the two. An increase in barley acreage will result in a decrease in wheat production, because wheat and barley compete for land, for labor at about the same time of the year, and for the same type of equipment. Therefore, any resources devoted to the production of one commodity must be taken away from the other.

A different relationship exists between wheat and the winter feeding of cattle. A cattle-feeding enterprise utilizing winter labor does not compete with the wheat enterprise for labor resources. The cattle may utilize some straw. In this case, an increase in the size and output of one enterprise has very little influence on the output of the other. Such supplementary enterprises provide for better utilization of the farm's resources.

Complementary relationships exist when one enterprise contributes to the production of another; that is, an increase in the production of one enterprise results in an increase in the output of another. A legume that stores nitrogen in the soil and breaks plant disease and insect cycles may result in a greater per-acre production of a grain crop that follows on the same land. Complementarity would exist if greater legume production resulted in more bushels of grain produced than if all of the land produced only grain year after year. Of course, there is a limit to the amount of legume that can be grown before total grain production starts to decline. Whether enterprises will be complementary or supplementary usually depends on the size of the enterprise, and most enterprises eventually become competitive if they become too large.

An argument in favor of diversification is that it permits complementarity and supplementarity among enterprises. In those operations where supplementary and complementary enterprises can be added, of course, they should be. A farmer should diversify until all supplementary and complementary enterprises have been considered.

Diversification and Risk

Risk considerations are also relevant to the proper combination of enterprises. In Chapter 8, diversification was referred to as a method for controlling risk; that is, that by producing a number of commodities there is a smaller chance of having a complete failure. If the price of one commodity falls, that of another commodity may remain constant or increase. Or a poor crop year for one enterprise, say raspberries, may be a good crop year for another, such as cherries.

An example is used to examine the effects of diversification on risk. It is assumed that enterprise A generated the following net income:

Year	Income from A
1	$100
2	200
3	0
4	500
5	− 50
Total	$750

These data indicate that the income from this enterprise is highly variable. One statistical measure of variability is the range from the largest to the smallest figure. In this example, the range is $550. Another indicator of risk is the lowest income, which in this case is the $50 loss in the fifth year.

A second enterprise, B, had the following income:

Year	Income from B
1	$500
2	0
3	250
4	− 50
5	100
Total	$800

This enterprise's income is also variable. Again the range is $550, and the low income is a $50 loss. However, in years when A is high, B tends to be low, and vice versa. If the two enterprises were combined, variability would be expected to decrease. This is exactly what happens. In combining the two enterprises (the resources are assumed to be divided equally between them), half of the income from each is totaled, as follows:

Year	Income from A and B
1	$300
2	100
3	125
4	225
5	25
Total	$775

The range is narrowed to $275, and the lowest income is $25. What would happen if a third enterprise were added? It would be difficult to find an enterprise that would not vary with either A or B. However, consider enterprise C, which also has an income range of $550:

Year	Income from C
1	$ − 50
2	100
3	500
4	200
5	0
Total	$ 750

If all three enterprises were combined by devoting one third of the resources to each, the following is the result:

Year	1/3 A Income	+	1/3 B Income	+	1/3 C Income	=	1/3 (A + B + C) Income
1	$ 33		$167		$ − 17		$183
2	67		0		33		100
3	0		83		167		250
4	167		− 17		67		217
5	− 17		33		0		16
						Total	$766

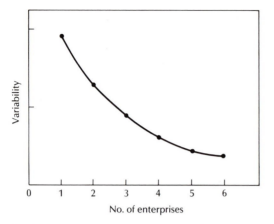

FIGURE 14-1.   Relationship between Income Variability and Number of Enterprises.

In this example, adding a third enterprise reduces variability again. The range is now $234. But adding a third enterprise does not reduce variability as much as adding the second did. Variability was reduced by 50 percent by adding B to A. By adding C to A and B, variability is reduced only about 15 percent.

How this will actually work depends on the income variability and degree of association of the enterprises. However, a generalization is appropriate at this point: The amount that additional enterprises will reduce income variability declines rapidly after the second enterprise has been added, assuming that yields and prices of the various enterprises are neither positively nor negatively associated. Figure 14-1 illustrates this principle.

The enterprises in this graph are assumed to be independent; there is neither positive nor negative correlation among the enterprises. Therefore, by adding a second enterprise to the original one, income variability is reduced by one half. Adding a third enterprise reduces variability to two thirds of that for two enterprises. The fourth reduces variability to three fourths of that of the three enterprises, and so on.

Those familiar with agriculture know that prices and yields for different commodities tend to be positively associated. This means that in most cases variability will be reduced much less than the hypothetical data indicate. Generally, diversification alone does not offer adequate protection against risk in agriculture.

Procedures for Selecting Enterprises

Determining the optimal combination of enterprises requires an understanding of the principle of comparative advantage and the relationships between various enterprises. The first step in the farm planning process, described in Chapter 6, requires that managers define their goals. Once these goals have been examined, the resources available for engaging in these enterprises must be determined.

Identifying Farm Resources

The nature and quantities of the available resources will determine which enterprises can be considered and which cannot. Fixed resources must be distinguished from those that are not fixed. For many farm situations, the total land acreage available represents a fixed resource, because there is little possibility of adding to the acreage either by purchase or by lease. Other resources such as capital, labor, and management may be limiting factors as well. For example, managerial capabilities may limit the size of an enterprise.

Land.   Land quantity and quality are critical factors in enterprise selection. How many acres of land are available? Are they suited for row crops, small grains, forage, livestock, or forest production? Land is a complex resource with many characteristics to be identified: soil types (slope, texture, and depth), current fertility and acidity levels, surface and subsoil drainage, soil conservation structures, existing irrigation facilities, and climatic factors (annual rainfall, growing season, and so forth).

The evaluation of the land resource might start with a good field map of the farm showing field sizes, fences, drainage ways, ditches, tile lines, and other characteristics. (The classification of land capabilities is discussed further in Chapter 15.)

Labor.   The supply of labor will affect the types and sizes of enterprises selected. How many hours of labor are available from the operator, family members, and hired labor, and are they available on a daily, weekly, or seasonal basis? The cost and availability of additional hired labor (regular, seasonal, and part-time) should be determined. The skills, training, and experience of the workers may also be important considerations in selecting enterprises.

Machinery and Equipment.   The number, size, and capacity of available machinery and equipment should be identified. The capacity of machinery that is owned may represent a fixed resource, but consideration should be given to ways of increasing machinery capacity, such as hiring custom work.

The capacity of machinery used for crop production is affected by the distribution of days suitable for fieldwork, as determined by the climate of the area. The days suitable for fieldwork and the size of available machinery can limit the total acreage of a crop when field operations must be performed according to a specific schedule. For example, if the crop must be seeded during a specific two-week period for optimum yields and there is a 90 percent probability that only eight days during these two weeks will be suitable for fieldwork (because there is too much moisture the other six days), then the size of this crop enterprise will be limited to the number of acres that can be seeded in eight days.

Buildings.   Buildings are an important consideration for livestock enterprises and crop storage; some enterprises require costly, specialized structures. Buildings and other facilities, such as feedlots, should be identified by size, capacity, and potential uses. Perhaps some buildings can be renovated so that they can be used for other purposes.

Capital.   Capital is required to purchase animals and feed for livestock enterprises and fertilizer, seed, and chemicals for crop enterprises.

Available capital resources should include capital for additional investment and for meeting the needs of any new enterprises being considered. Capital sources include the farmer's own capital and the estimated amount of credit available; this total represents the upper limit on capital available for the combination of enterprises being considered.

Management.   Although managing some enterprises is relatively easy to learn, managing others is more complex and requires greater knowledge. The farmer's managerial skills, training, experience, strengths, and weaknesses should be considered when selecting enterprises. If the manager lacks training or experience with a particular enterprise or has a dislike for it, the enterprise should not be considered. Evaluating management resources requires an honest self-appraisal of current capabilities and of the potential for improvement through study, training, and educational programs.

Market Availability.   This last factor should probably be considered first. Unless a viable market exists for the enterprise, it should not be considered. Will the product be sold locally or nationally? What affects consumer demand and the price that will be received? Are there many potential buyers or only a few? The potential market for the product is a resource considerably different from those discussed earlier. For some commodities, the market may be relatively fixed and beyond the farmer's control. For others, it may be possible to develop markets by selling directly to wholesalers, retailers, or consumers.

Investigating Enterprise Alternatives

After identifying the resources available, the next step is to list the crop and animal enterprises that the manager is willing to consider and that are feasible, given the resources of the farm and the available markets. The farmer's personal experience, that of other farmers, and the advice of extension personnel will affect this list.

Past and present enterprises should be considered. Have these enterprises been profitable? If not, what might be done to improve their profitability? Should their size be increased or new production methods adopted? Investigating these enterprise alternatives involves obtaining information about resource requirements, preparing enterprise budgets, and estimating the gross margins of the enterprises.

Resource Requirements.   The per-acre or per-animal resource requirements must be estimated. For example, an acre of corn might require one acre of land, five hours of labor, three and one-half hours of tractor time, and $80 of operating capital. A beef cow, on the other hand, might require 10 acres of pasture, six hours of labor, and $90 of operating capital. A calendar of operations (presented in Chapter 13) is useful for summarizing the labor, machinery, and equipment requirements for each enterprise. Combined with the information on available resources, these resource requirements determine what the maximum size and final combination of enterprises will be.

Enterprise Budgets.   In addition to resource requirements, information about the enterprises should include potential revenue and expenses, which are usually summarized in the form of enterprise budgets, the primary management tool for enterprise selection. An enterprise budget provides a detailed list of all estimated revenue and expenses associated with a specific enterprise for the purpose of evaluating its profitability.

This budget provides per-unit estimates of yield or production, revenue, quantity and type of inputs, variable and fixed production expenses, and measures of profitability. Revenue is based on normal yields and expected prices. Variable expenses such as fertilizer, pesticides, seed, custom services, and interest on operating capital vary with the size of the enterprise. Fixed expenses are those associated with resources already committed to the business and therefore do not vary with enterprise size. These include depreciation, interest on building and machinery investments, property taxes, insurance, and interest on land investments. (Enterprise budgets for crop and animal enterprises are discussed in Chapters 15 and 17.)

Gross Margins.   The gross margin, the return over variable costs, is an appropriate measure of profitability to use for comparing enterprises for short-run, annual planning decisions. Year-to-year enterprise selection decisions usually do not affect fixed costs, so the return over variable costs is an appropriate profit indicator. Gross margins can be used to make comparisons among enterprises that place similar demands upon the limiting resources of the business. However, when comparing enterprises with different resource needs, gross margin planning may be misleading.

The gross margin for an enterprise is the difference between the revenue and the variable production costs of a single unit (one acre or one animal). Variable costs are those costs that vary with the size of the enterprise, and the revenue is the total value of the products sold. The gross margin can be calculated by using the following formula:

$$GM = (P \times Y) - VC$$

To obtain the gross margin (GM), subtract variable costs (VC) from the revenue, where revenue is the product of price (P) and yield (Y). For example, if the price of the product is $3 per unit, the yield is 90 units per acre, and variable costs are $200 per acre, then the gross margin is $70.

The objective is to select enterprises with the greatest return over variable costs (gross margin), subject to the appropriate resource constraints. Gross margins should be calculated for each feasible enterprise, and the enterprise with the highest gross margin is, of course, the one that would contribute the most profitability. This approach, appropriate for year-to-year enterprise selection decisions, excludes fixed-cost considerations. Although these fixed costs are not relevant to short-run decision making, the manager must nevertheless meet both variable and fixed costs over the long run to continue the enterprise.

Break-Even Analysis for Short-Run Planning

Translating the gross margins estimated for the alternative enterprises into the annual enterprise combination decision requires the consideration of several other factors. That is, resource availability and risk management warn against planting

TABLE 14-1. Gross Margin Calculations

Enterprise	Price (per unit)	×	Yield (per acre)	=	Revenue	−	Variable Costs	=	Gross Margin
1	$ 3		90		$ 270		$200		$ 70
2	2		120		240		230		10
3	50		25		1250		800		450

all available land to the single most profitable crop, and crop rotations and agronomic concerns must be considered.

Break-even analysis can be used to consider how price and yield changes might affect the selection of alternative enterprises. Table 14-1 shows a comparison of the gross margins of three sample enterprises.[2] The results indicate that enterprise 3 contributes the most to covering fixed costs, such as machinery ownership and the operator's management, and enterprise 2 contributes the least.

Table 14-2 illustrates the comparison of the crops using break-even analysis to determine how sensitive their gross margin ranking is to changes in price and yield. The process involves calculating the price or yield level (the break-even price or yield) for the second crop that is required to generate the same gross margin as the first. This break-even level is found by equating the gross margins for two enterprises and solving for either price or yield.

The gross margin for enterprise 1 is as follows:

$$GM_1 = (P_1 \times Y_1) - VC_1$$

The gross margin for enterprise 2 is as follows:

$$GM_2 = (P_2 \times Y_2) - VC_2$$

Equating GM_1 to GM_2:

$$(P_1 \times Y_1) - VC_1 = (P_2 \times Y_2) - VC_2$$

Solving for P_2 gives the price for enterprise 2 that is necessary to produce the same gross margin as that produced by enterprise 1:

$$P_2 = \frac{(P_1 \times Y_1) - VC_1 + VC_2}{Y_2}$$

[2] This example is from R. Carkner, *Gross Margins Analysis for Evaluating Alternative Crops*, Washington State University Cooperative Extension EM 4617, August 1980.

TABLE 14-2. Break-Even Yields Per Acre

First Crop	Second Crop		
	1	2	3
1	90	150	17.4
2	76	120	16.2
3	516	340	25.0

TABLE 14-3. Break-Even Prices Per Unit

First Crop	Second Crop		
	1	2	3
1	3.00	2.50	34.80
2	2.30	2.00	32.40
3	7.20	5.60	50.00

Using sample values from Table 14-1:

$$\$2.50 = \frac{(\$3 \times 90) - \$200 + \$230}{120}$$

The results indicate that enterprise 2 would generate the same gross margin as enterprise 1 ($70) if the price of enterprise 2 increased from $2 to $2.50.

The break-even yield can be calculated in the same way by solving for yield (Y_2). Examples of break-even yields and break-even prices for the sample data are included in Tables 14-2 and 14-3.

Looking first at the break-even yield table (14-2), if an enterprise is compared to itself, the yield is the same as that in Table 14-1. Comparing enterprise 1 with enterprise 2, the second enterprise must yield 150 units per acre to generate the same gross margin as the first enterprise. Comparing enterprises 1 and 3, the yield of enterprise 3 must be 17.4 to generate the same gross margin as that of enterprise 1. If the first enterprise selected were enterprise 3 and the second, or comparison, enterprise were enterprise 1, the table indicates that the yield for enterprise 1 must be 516 units to generate the same gross margin as enterprise 3.

Break-even price calculations and their interpretation parallel those of break-even yield calculations. The break-even yield and price analyses are used to determine how much the yield or price of a comparison enterprise must change before it generates the same gross margin as the first.

In order to apply break-even calculations to enterprise selection decisions, the following question should be asked: What is the probability of prices or yields changing enough to equalize the gross margins of two enterprises? The answer and judgments that result indicate how sensitive gross margin rank is to price or yield change. If relatively small changes in price or yield result in equal gross margins for two enterprises, then these two enterprises can be considered nearly equal in profitability. On the other hand, if large increases in price or yield are required for an enterprise to generate a gross margin equal to a more profitable enterprise, then, all else being equal, fewer acres should be assigned to the less profitable crop.

The Long-Run Combination of Enterprises

Long-run planning involves a more comprehensive approach to determine the best combination of enterprises. The most profitable enterprise combination will be the one that yields the highest return to limiting resources (the equimarginal returns principle). For most farmers and ranchers, land is the limiting factor of production in the short run. This fact suggests that returns per acre of land should be maximized in most situations. Decisions regarding the kinds and acreages of crops to be grown are often made before animal production enterprises are selected. However, if the farm already has a large fixed investment in animal production facilities (a dairy barn and milking parlor, for example), it may be appropriate to tailor crop production to accommodate the use of these facilities.

The following is a procedure for determining enterprise combinations. It provides a way to consider the relevant concepts and information, and it is designed to guide the budgeting process described in Chapter 6.

1. List all the enterprises to be considered in the farm plan and tentatively determine whether their relationships are competitive, complementary, or supplementary. (An enterprise may be supplementary or complementary to another within a certain production range but may become competitive if production is increased too much.)

2. Using enterprise budgets, select one enterprise that is believed to be the most profitable. All relevant costs should be included in measuring the enterprises' contributions to net farm income. For example, the added costs associated with the new machinery required for an enterprise should be included, along with the opportunity costs of existing machinery and other resources, recognizing that in some cases this opportunity cost will be zero. The most profitable enterprise should be as large as possible in view of the existing resources. If it is grown under contract, the contract may be limiting. Because they enhance the effective use of farm resources and add to profits, supplementary and complementary enterprises should be added until they become competitive. Moreover, these enterprises provide a bonus: reduced income variability (or increased income stability).

3. Select the enterprise that is the next most profitable and determine if adding it will increase income. This enterprise will be competitive with the enterprise originally selected in step 2, so if it is included in the farm plan, a reduction in the output of the original enterprise will result. Whether the new enterprise should be included depends on whether it adds to net income; if so, it may completely replace the original enterprise. The enterprise that originally appeared to be the most profitable may not be when enterprise relationships and available resources are considered. Alternatively, the new enterprise may partially replace the original one. This result may occur if a combination of the two enterprises makes better use of resources and increases income. If the new enterprise completely or partially replaces the original one, the complementary or supplementary enterprises should be recon sidered. This procedure should be repeated for the remaining enterprises that are successively the next most profitable and that are competitive with the main enterprises in the tentative plan. Various combinations of enterprises can be tried until the most profitable combination has been determined.

4. Additional enterprises may be added to reduce income variability until the combination of enterprises gives a level and variability of income satisfactory

to the manager. Because the reduction in variability declines with each additional enterprise, the reduction from an additional enterprise is likely to be small if two or three enterprises are already in the plan.

An Example

To illustrate the use of the preceding procedure, a farm in an area where a number of commodities may be produced is considered. The farm to be planned has 480 acres of cropland, 80 of which are irrigated. After performing a partial budgeting analysis, the farm manager decides that milk production will be the main enterprise. The land can produce feed for a sizable dairy herd, but labor availability limits the number of dairy cows to 75. It would be possible to hire additional labor, but a full-time employee would be needed for the entire year. After some calculation, the manager decides that it would be better to add other enterprises rather than to hire labor to expand the dairy farm beyond 75 cows.

The dairy cows require a substantial amount of land for forage and pasture production. Corn silage can be produced more cheaply than it can be bought; therefore, sufficient acreage is devoted to corn silage to meet the needs of the dairy herd. Some of the land is hilly and best adapted to producing hay, which is also needed for feed. The projected production of hay will exceed the feeding requirements in all except poor crop years, so there will be some hay for sale in most years. Not all the cropland is needed to satisfy the requirements of the dairy cows. Grass seed production appears to be a profitable possibility. This enterprise can be added without reducing the output of the dairy herd.

At this point, considerable care must be taken in adding more enterprises. The farmer already has a sizable number, and labor and land are fairly well utilized. Perhaps a sheep enterprise could be added without adversely affecting the other enterprises. If the size of the dairy herd is limited by the amount of feed that can be grown with family labor, this labor probably would not be fully utilized during the winter months. The sheep enterprise would have peak labor requirements during January and February and would not compete with the dairy herd for feed because the sheep utilize aftermath grazing. If the farmer has sufficient managerial capabilities to handle the sheep enterprise, it would probably be a good supplementary enterprise. Therefore, a farm flock of 100 ewes is considered. It is highly unlikely that additional enterprises would either increase or stabilize net farm income.

In this case, the final farm organization consists of a main enterprise, dairy. Grass seed is supplementary. The sheep enterprise is added because it is complementary to the grass seed enterprise and supplementary to the others. Of course, it would be possible to expand the production of any of these enterprises to the point where it would be competitive with the dairy enterprise. In order to have confidence in the final farm plan, the manager should check and compare the alternatives to this organization. In this case, the principal alternative is to drop the dairy and expand grass seed production substantially. This change would allow an expansion of the sheep enterprise. If the systematic procedure previously outlined is followed, the principal alternatives are considered in the process. The preceding illustration has not included the farm budgets for every alternative. However, these budgets would be prepared according to the procedure explained in Chapter 6.

At each point in planning the farm organization, the reasoning and logic used should be checked with a farm budget. Every time an enterprise is added, the effect that it will have on farm income should be estimated. It may be that several plans will be tried before a final plan is selected. There can never be assurance that the "best" plan has been determined, but if the alternatives are budgeted thoroughly, the manager can be fairly confident of the final result. Farm organization and management may be better described as an art than as an exact science. Therefore, judgment is important. For example, the final acreage of corn silage may not exactly equal the acreage that will meet animal feed requirements, because the size of the fields may make it impractical to limit the amount grown to the amount that can be fed, or a slightly larger acreage may be needed to work out a suitable rotation.

In developing budgets for the future, past records are valuable. If the manager knows how costs and returns will vary if an enterprise is expanded or added, many alternatives can be quickly and easily considered. On the other hand, if past records are not available, it is still possible to construct enterprise budgets using data and estimates from extension advisers and other sources. Decisions must be based on estimates of the future, but a budget provides a systematic method for organizing these estimates.

Obviously, with several potential enterprises and a number of limiting resources, the number of possible combinations is very large. Linear programming is valuable in helping to make decisions of this kind.

Linear Programming

Linear programming is used for a variety of problems, such as finding the least-cost feed mix, as well as for determining the most profitable combination of enterprises. Linear programming is a mathematical technique that finds the combination of alternative enterprises that maximizes the return above variable costs within the resource and other constraints limiting production. A linear programming analysis of enterprise combinations requires four types of information: (1) the alternative enterprises that can be produced, (2) the values to be maximized—the return above variable costs for each enterprise, (3) the resources or other constraints that potentially limit the size of alternative enterprises, and (4) the resources required per unit of each enterprise.

As used to find the most profitable enterprise combination, linear programming is usually viewed as a substitute for, or a supplement to, budgeting. It requires essentially the same data, but it is more systematic than budgeting. Linear programming guarantees that the most profitable enterprise combination is determined, which is not possible using the trial-and-error budgeting approach. Using traditional budgeting techniques, the manager is not likely to spend the necessary time exhausting all the possibilities to find the best plan.

Judgment and skill must be exercised when the linear programming problem is formulated, when the data are developed, when they are arranged in equation form, and when the results are interpreted. However, a farm manager need not be a trained mathematician to do linear programming. It is far more important to understand the nature of the farm management problem and to collect reliable data for use in the analysis.

A detailed description of linear programming is not included here. (Additional references are listed at the end of this chapter.) Most of the university extension services have people available who are trained in linear programming. Also, some professional farm management agencies offer linear programming services. Although some problems involve a large number of calculations, this is not a major obstacle. Computer programs are available to find the optimal solution to linear programming problems.

A Simple Example

A simplified farm management problem is used to illustrate the method. The problem is to determine the optimal combination of two crops, strawberries and pole beans. The optimal combination is the one that maximizes the return above variable costs for the two crops within the constraints of the limiting resources. The potentially limiting resources are as follows:

Resource	Quantity Available
Land	15 acres
June labor	500 hours
September labor	540 hours

June and September labor were selected as potential constraints because these are the two months requiring the most labor exclusive of harvesting labor. Harvesting labor can be hired, but other labor is assumed to be limited to the operator and one employee.

The next step is to establish the requirements of each crop for the resources previously listed. The requirements are expressed on a per-acre basis:

	Strawberries	Pole Beans
Land (acres)	1	1
June labor (hours)	30	40
September labor (hours)	18	48

Strawberries are expected to yield three tons per acre and beans eight tons per acre. For this example, the market price of strawberries is $312.50 per ton and that of pole beans is $125 per ton. Variable costs amount to $750 and $800 per acre, respectively. The return above variable costs, then, is $187.50 for strawberries and $200 for pole beans.

The process of linear programming consists of determining the crop with the highest return above variable costs and expanding it to the limit of the available resources. If strawberries are expanded to 15 acres, the limit of the land available, then 450 hours of June labor and 270 hours of September labor would be required. This means that 50 hours of June labor and 270 hours of September labor would not be utilized. Producing all strawberries would yield a return above variable costs of $2812.50. This amount can be compared with the return above variable costs from producing only pole beans.

September labor would limit production of pole beans to only 11.25 acres, which would yield a return above variable costs of $2250. The question that arises, however, is whether some acreage of pole beans could be profitably substituted for strawberries. In this example, the return above variable costs can be increased—in fact, it is maximized—by producing 10 acres of strawberries and 5 acres of pole beans. The return above variable costs is $2875. The solution is as follows:

Crop	Acreage	Production
Strawberries	10	30 tons
Pole beans	5	40 tons
Return above variable costs:	$2875	

The values of an additional acre of land and an additional hour of June labor are also provided as part of the linear programming analysis. Because September labor was not exhausted, there would be no advantage in acquiring additional units of this input.

Value of an additional acre of land	$150.00
Value of an additional hour of June labor	1.25

These figures represent annual returns. For example, if the farmer could lease another acre of land for less than $150 and farm it with no increase in costs other than the variable costs for strawberry and pole bean production, the lease would be profitable. With this additional acre of land and no increase in the labor available, the maximum-profit combination of enterprises would be 14 acres of strawberries and two acres of pole beans. Also, if labor could be hired in June for less than $1.25 per hour (which is not likely), it would be profitable to do so. The additional labor would allow pole beans to be substituted for strawberries, increasing the total net return.

Obviously, budgeting techniques could be used to solve a simple problem like the one in this example quite satisfactorily. Consider the complications that arise, however, if the number of enterprises were increased from 2 to 40 and the number of restricting resources increased from 3 to perhaps 60. Using budgeting to find the most profitable enterprise combination would be an enormous task. Under such circumstances, computerized linear programming can speedily and systematically determine the plan that yields the greatest net return.

Linear programming analysis imposes certain assumptions on the problem of finding the optimal enterprise combination. Because these assumptions affect the use of linear programming and the interpretation of the results, they should be understood before proceeding to a more complex example. The assumptions are as follows:

1. The relationship between the size of the enterprise and total output is assumed to be linear. For example, as the number of acres of corn is increased, the yield per acre is assumed to be constant and the number of bushels produced on 1000 acres is twice the quantity that would be produced on 500 acres.

2. The resources required by an enterprise are assumed to be used in a fixed proportion; that is, if 1 acre of soybeans requires 4 hours of labor and 100 units of fertilizer, then 100 acres would require 400 hours of labor and 10,000 units of fertilizer.

3. The sizes of the enterprises are assumed to be divisible; that is, the quantity produced can be zero or any positive amount. Linear programming assumes that fractional amounts are feasible, for example, 112.5 cows.

4. The sum of the net returns from each enterprise is assumed to equal the total net return for the farm; the net returns are additive. Linear programming ignores the possibility that two enterprises are complementary. However, these complementary effects can be considered in the linear programming analysis by organizing the data differently. (Essentially this adjustment involves treating the two complementary enterprises as one.)

5. The prices, yields, and other data used in linear programming are assumed to be known with certainty. It is possible, however, to change the price, yield, and other data values and solve the problem again to see how the optimal plan changes for different data values.

An Indiana Crop Farm Example

To illustrate the application of linear programming to enterprise selection decisions, a linear programming model developed at Purdue University is used.[3] It was one of the first linear programming models for crop selection and continues to be one of the most used by farmers. Using the farmer's data, the computer program creates a linear programming model to determine the best farm plan.

The linear programming model represents the production alternatives for corn, soybeans, wheat, and corn silage with the possibility of double-cropping soybeans after wheat. The model allows different techniques and timing to be combined for land preparation, planting, harvesting, and selling these crops.

The model is designed to determine how much of each crop should be produced, taking into consideration the estimates of crop yields, prices, and costs, as well as the availability of labor and machinery for land preparation, planting, and harvesting. The model also considers the expected planting and harvesting rates and the expected number of days that the weather will permit fieldwork. In addition to

---

[3] B. McCarl, W. Candler, et al., "Experiences with Farmer Oriented Linear Programming for Crop Planning," *Canadian Journal of Agricultural Economics* 25 (1977):17–30.

considering the current resources available, the model accommodates leasing additional land, hiring part-time labor, and hiring custom harvest services. The computer is programmed to produce the maximum-profit plan but does not take into direct account weather variability, price variability, or machinery breakdown.

The objective of the model is to formulate a plan for the crop production season. The model is valuable in suggesting both short-run and long-run planning adjustments. It is designed to help farmers answer questions about the optimal crop mix, timeliness of field operations, and changes in farm size. For example:

1. What would happen to profits if more corn and less soybeans were planted?
2. How would profit be affected by a change in the size of the combine, tractor, or planter?
3. Is a different tillage system likely to increase or decrease profits?
4. Should the harvest operation be contracted on a custom basis?
5. What hourly wage can be paid to hire part-time labor at critical periods?
6. What is the most that should be offered to lease additional land?

To illustrate the use of the model, a hypothetical farm is used as an example.[4] Tables 14-4 through 14-7 illustrate the solution for a 750-acre central Indiana crop farm. Important information about this farm entered into the computer includes the number of hours available for fieldwork by time period, the labor available, the number and sizes of tractors, the capacity of other machinery, and the farm size.

Table 14-4 is a summary of the important information determined by the linear programming solution. The "Present Plan" indicates that this farmer is currently receiving a net return to management of $2774 by growing 375 acres of corn and 375 acres of soybeans. The "Optimal Plan" indicates that net returns to management would increase with 381 acres of corn, 319 acres of soybeans, 50 acres of wheat, and 50 acres of double-crop beans (planted after the wheat is harvested).

Table 14-5 summarizes the preparation, planting, and harvesting schedules for corn, soybeans, and wheat. For corn, for example, there are six planting periods

---

[4] The example is from D. Doster, "Purdue B-96 Crop Budget: Explanation of Base Case Farm," Purdue University Department of Agricultural Economics EC-5421, 1981.

TABLE 14-4. Present Plan and Optimal Plan, Indiana Crop Farm
    Linear Programming Example

	Present Plan	Optimal Plan
Net returns to management	$2774	$4121
Acres of corn grain	375	381
Acres of corn silage	0	0
Acres of soybeans	375	319
Acres of wheat	0	50
Acres of double-crop beans	0	50
Acres not used	0	0
Acres leased	0	0

TABLE 14-5. Schedule of Machine Field Operations, Indiana Crop Farm
Linear Programming Example

Time Period	Acres Prepared		Acres Planted			Acres Harvested		
	Corn	Soybeans	Corn	Soybeans	Wheat	Corn	Soybeans	Wheat
3/15–4/4	58	0						
4/5–4/25	83	0						
4/26–5/2	0	0	158	0				
5/3–5/9	0	0	158	0				
5/10–5/16	0	0	65	138				
5/17–5/23	0	0	0	181				
5/24–5/30	0	0	0	0				
5/31–6/6	0	0	0	0				
6/7–6/13								
6/14–7/18				50				50
9/13–9/26							0	
9/27–10/17	0	238			50	99	319	
10/18–11/7	64	81			0	228	50	
11/8–11/28	104	0				54		
11/29–3/14	72	0						
Totals	381	319	381	369	50	381	369	50

and three harvest periods, which means that there are 18 combinations of planting and harvesting times, and for each combination there is a different yield per acre. The model seeks to plant and harvest as many acres as possible during the time periods that will result in the highest yields. However, the acres that can actually be planted and harvested depend on the availability of labor and machinery and on suitable weather. These constraints are all built into the model.

Table 14-6 indicates how profits would be affected if resource constraints were

TABLE 14-6. Value of Additional Units of Limiting Resources, Indiana
Crop Farm Linear Programming Example

Resource	Value in Dollars per Additional Unit
Field hours—planting	
4/16–5/2	$119.27
5/3–5/9	115.29
5/10–5/16	4.65
Value of an extra acre of land	212.63
Value of an extra hour at harvesting time	
9/27–10/17	57.50
10/18–11/7	38.63

TABLE 14-7. Original Optimal Plan and New Optimal Plan with an
Additional 160 Acres, Indiana Crop Farm Linear Programming Example

	Original Optimal Plan	New Optimal Plan
Net returns to management	$4121	$13,229
Acres of corn grain	381	316
Acres of corn silage	0	0
Acres of soybeans	319	544
Acres of wheat	50	50
Acres of double-crop beans	50	50
Acres not used	0	0
Acres leased	0	160

relaxed. The resources listed are the constraints that limit the profitability of the farm plan. The values show the effect on profits if the farmer could somehow add another unit of these limiting resources. The values associated with the hours available for planting indicate that another hour available for planting during the week of April 26 to May 2 would increase profits by $119.27. Weather permitting and with adequate machinery, this figure represents a very high return for an extra hour of labor. Because of the relationship between the date of planting and crop yields, the value of an hour of planting time is much higher from April 26 to May 9 than it is from May 10 to May 16.

An extra acre of land is estimated by the model to be worth $212.63. If the linear programming model were solved for 751 acres, the net return to management would be $212.63 higher. Given the labor, machinery, and other assumptions for this example, the farmer could afford to pay up to $212.63 to rent an additional acre of land.

The value of an extra hour of harvesting time indicates that if the farmer is using harvesting machinery to capacity and could hire custom harvest services for less than $57.50 during the period September 27 to October 17, it would be profitable to do so.

One of the most useful aspects of a linear programming model is the ability to change assumptions and explore the implications. Suppose this farmer had the opportunity to lease 160 acres at a cost of $145.50 per acre. The results are presented in Table 14-7. The model results indicate that the entire 160 acres should be leased. The net return to management would increase from $4121 to $13,229. Compared to the solution in Table 14-4, corn acreage would decrease slightly and all of the added acres would be planted to soybeans.

Summary

The selection and combination of enterprises in the farm business depend on a number of physical and biological forces that limit enterprise possibilities. The principle of comparative advantage, which states that economic units should specialize in the production of those commodities for which they have the greatest relative advantage or the least relative disadvantage, applies not only to individual farmers, but also to regions and nations.

Specialization allows farmers to concentrate their management talents on the production of those commodities they produce best. It also permits farmers to increase the volume of production to achieve economies of size and higher profits per unit.

Through diversification, the complementary and supplementary relationships between enterprises can be exploited. Diversification may also allow better use of available resources, such as labor and machinery. However, too many enterprises can dilute managerial effectiveness.

Risk is an important consideration in diversification, but most of the risk-reducing advantages are realized with two or three enterprises; the reduction in income variability decreases with each successive enterprise added. Enterprise diversification as a method to control risk is hampered by the positive association of the prices and yields of different commodities.

The process of selecting enterprises starts with identifying the quantities and qualities of farm resources. Next, enterprise alternatives are investigated to determine their resource requirements, and enterprise budgets are prepared to estimate their gross margins. The gross margin, the return over variable costs, indicates an enterprise's potential profitability for year-to-year planning decisions. Break-even prices and yields indicate how much they must change for the gross margin of one enterprise to exceed that of another.

The equimarginal returns principle should guide the farm manager in combining enterprises to yield the highest return to limiting resources. The most limiting resource is usually land; however, capital, buildings, or facilities may be limiting for animal production enterprises. The procedure for combining enterprises to maximize profitability while maintaining an acceptable level of risk involves a consideration of enterprise relationships, gross margins, resource constraints, and income variability.

When there are several enterprise possibilities to be considered, linear programming is useful for determining the most profitable combination of enterprises, given constraints on the availability of resources. The data required are the same as those used for budgeting, that is, the prices, costs, yields, and resource requirements of the alternative enterprises and the availability of resources.

An important advantage of linear programming and computerized budgeting models for enterprise selection and combination is that they facilitate the yearly evaluation of these decisions. Farmers should reevaluate these decisions annually and adjust enterprise combinations in order to take advantage of changing economic conditions.

Recommended Readings

BENEKE, R., and R. WINTERBOER. *Linear Programming Applications to Agriculture.* Ames: Iowa State University Press, 1973. Chapters 1 and 2.

BOEHLJE, M. D., and V. R. EIDMAN. *Farm Management.* New York: John Wiley & Sons, Inc., 1984. Chapter 10.

CHAPTER 15

Crop Production Management

Crop production management starts with crop selection. These crops must then be integrated into a cropping system that includes the type of crop rotation selected as well as the decisions made about tillage, fertilization, pest control, irrigation, and other inputs. These decisions are interrelated; for example, the most profitable use of fertilizer is affected by the crops grown as well as by the pest control and irrigation practices followed. Therefore, designing a profitable cropping system involves finding the appropriate combination of inputs to use, rather than concentrating on only one or two inputs.

The primary tool used to analyze the cropping system is the crop enterprise budget, which is described in the first section of this chapter. Next, some of the major aspects of crop production—soil management, crop rotations, fertilization, and irrigation—are discussed. The use of experimental data to evaluate alternative cropping systems is described in the final section.

Crop Enterprise Budgets

An enterprise budget summarizes the enterprise revenue and expenses; this summary helps the manager make decisions such as selecting alternative enterprises, choosing the production practices to be used, and preparing a total farm plan. An enterprise budget indicates the profitability of an enterprise by estimating the annual revenue and expenses for one unit; for crop enterprises, the unit is usually one acre. The revenue and expenses estimated in the crop enterprise budget are those expected if particular production practices are used. The budget typically encompasses the entire crop production cycle, indicating the resources required, the products produced, the costs incurred, and the prices received. It can be an estimate of the enterprise's current or projected revenue and expenses.

Microcomputers and spreadsheet programs can reduce the time required to compute and compare enterprise budgets.[1] However, data requirements are not re-

[1] Examples of spreadsheet programs include VisiCalc, SuperCalc, Multiplan, and Lotus 1-2-3.

duced. Past farm records are a good source of data for estimating revenue and expenses if they are available for the enterprise being budgeted. Also, typical enterprise budgets may be obtained from university extension services; these budgets may be a useful source of data that can be modified as necessary to fit the individual farm.

Preparing Crop Enterprise Budgets

Preparing enterprise budgets requires a working knowledge of the economic principles discussed in Chapter 2. These principles, along with an understanding of the production process, guide managers in selecting the amount and timing of inputs such as fertilizer; and their understanding of the impact of weeds, insects, and diseases on crop yields helps them select the most economical pesticides.

The title of the enterprise budget should describe the particular enterprise, the production system, the year for which the budget is relevant, and the location of the enterprise. Enterprise budgets can be formatted in several different ways. One format is presented in Table 15-1. The six sections in this enterprise budget—revenue, operating expenses, gross margin, ownership expenses, total expenses, and return to management—are explained more fully in the following sections.

Revenue.   The revenue section includes the products that will be produced during the year. The first step in developing a crop enterprise budget is to list all the commodities that will be produced, including the obvious ones such as wheat, strawberries, and potatoes, as well as the not so obvious products such as aftermath for grazing.

Next, total production is estimated. Crop yields are typically estimated on the basis of the history and trend of yields of an individual farm or production area. The yield per acre should be the yield expected under normal weather conditions, given the soil type, seeding rate, fertilizer level, chemical use, tillage practices, and other inputs affecting yield. If the land is share leased, only the tenant's share of the crop is included.

Prices are assigned to each product, and the total revenue is calculated for the crop enterprise. These prices should be realistic estimates based on past trends and current expectations.

Operating Expenses.   Expenses are divided into operating and ownership expenses. Operating expenses are those that normally vary with the size of the enterprise, such as seed, fertilizer, chemicals, fuel and repairs for machinery, and hauling costs. The quantity used per acre is multiplied by the price per unit to calculate the total cost per acre.

These expenses can be estimated from farm records or income tax returns, or they can be based on the actual cost of the purchased inputs or machine hire operations. For share-leased land, only the appropriate share of the expense should be included. The following operating expenses should be considered when building the crop enterprise budget:

1. *Seed.*   The amount paid for purchased seed or the market value of farm-produced seed (including the cost of seed treatment and cleaning) is multiplied by the quantity used per acre.

TABLE 15-1. Sample Enterprise Budget for Nonirrigated Corn
Production, Midwest, 1986

Item	Units	Price per Unit	Quantity	Value or Cost per Acre
1. *Revenue*				
Corn sales	bushels	$ 3.00	120.0	$360
Total Revenue				$360
2. *Operating Expenses*				
Seed	pounds	0.80	25.0	$ 20
Pesticides	acres	21.00	1.0	21
Nitrogen fertilizer	pounds	0.25	140.0	35
Other fertilizer	acres	26.00	1.0	26
Fuel and repairs	acres	34.00	1.0	34
Custom grain drying	bushels	0.13	120.0	16
Hired labor	hours	6.00	1.0	6
Oper. capital interest	dollars	0.12	75.0	9
Total Operating Expenses				$167
3. *Gross Margin (Return Above Operating Expenses)*				$193
4. *Ownership Expenses*				
Machinery depreciation	acres	30.00	1.0	$ 30
Interest on machinery	dollars	0.12	166.7	20
Taxes and insurance	acres	3.00	1.0	3
Operator labor	hours	8.00	2.5	20
Land charge	acres	100.00	1.0	100
General overhead	acres	10.00	1.0	10
Total Ownership Expenses				$183
5. *Total Expenses*				$350
6. *Return to Management*				$ 10

2. *Pesticides.* Pesticide expenses include the cost of herbicides, insecticides, and any other chemicals directly associated with the crop enterprise.

3. *Fertilizer.* The cost per pound is multiplied by the quantity applied per acre. Lime costs per acre should be prorated over its useful life, usually two to six years.

4. *Fuel and repairs.* A portion of the total machinery repair and fuel costs should be allocated to each enterprise according to the percentage that each machine is used for that crop enterprise. Electricity or fuel costs for irrigation should also be included.

5. *Custom hiring.* Costs for custom hiring of field tillage operations, applying fertilizer or chemicals, harvesting, hauling to storage, and drying or processing should be included.

6. *Miscellaneous expenses.* Some expenses, such as crop insurance, are directly attributable to an enterprise. That is, if this crop were not produced, these costs would not be incurred.

7. *Hired labor.*   Hired labor expenses can be either fixed (labor provided by the family or a regular employee) or variable (hired seasonal labor). The wage rate should include social security, insurance, and other benefits.

8. *Operating capital interest.*   Operating capital interest represents the opportunity cost of the capital invested in the operating expenses from the time the capital is invested until harvest.

Gross Margin.   The difference between the enterprise revenue and total operating expenses is the gross margin, which indicates how much the enterprise will contribute to the payment of ownership expenses and to the profitability of the business.

Ownership Expenses.   In the production of crops, ownership expenses normally include machinery, irrigation equipment, buildings, operator labor, and land. The ownership costs of capital assets are depreciation, interest, taxes, and insurance. These expenses are usually fixed if changes in the size of the enterprise are minor. However, for new enterprises requiring specialized equipment or for major expansions requiring additional machinery, they are relevant. Equipment costs should be computed for each machine and allocated according to the percentage of use in each enterprise. The following ownership expenses should be included in the enterprise budget:

1. *Machinery depreciation.*   The loss in machinery value due to age, use, and obsolescence can be taken from accounting records or calculated by dividing the difference between the new cost and the salvage value by the years of the machine's useful life.

2. *Interest.*   Interest on the machinery investment is based on the average value of the machine over its life. If money was borrowed to make the investment, the interest paid is the appropriate cost; otherwise, the rate of return on the best alternative investment should be used.

3. *Property taxes.*   The local personal property tax rate is multiplied by the assessed value of the machinery.

4. *Insurance.*   For insurance, a rate from 0.5 to 1.0 percent should be multiplied by the current value of the machinery and equipment.

5. *Buildings.*   Appropriate depreciation, interest, repair, tax, and insurance charges should be estimated for buildings used to shelter machinery, store harvested crops, and so forth.

6. *Operator labor.*   The hours of operator (and unpaid family) labor used in the enterprise should be included. The cost of unpaid labor should be based on comparable hourly rates for hired labor.

7. *Land charge.*   If land is leased on a crop-share basis, there is no land charge; the charge has been accounted for by the landowner's share of crop production. If the land is leased for cash, the land charge is the cash payment. The land charge for land that is owned should be based on the opportunity cost of its use, that is, what it would return before the payment of real estate taxes if it were leased to another producer.

8. *Overhead.*   Overhead costs should include the farm's share of utilities, telephone, accounting fees, farm organization dues, general insurance, office

expenses, and so forth. These expenses can be allocated among enterprises based on their relative total revenue or total expenses.

Total Expenses.   The total expenses—the sum of the operating and ownership costs—can be used to calculate the average cost of production, which is equivalent to the concept of average total cost discussed in Chapter 2. The estimated average cost of production is found by dividing the total cost per acre by the estimated yield per acre. For example, in Table 15-1 the average cost of corn production would be $350 divided by 120 bushels, or $2.92 per bushel.

The average cost of production is a popular subject for discussion among farm managers, primarily because of its importance in justifying government price support levels. However, the actual costs of producing a given crop are likely to vary significantly among farms, states, and regions.

The Return to Management.   The return to management is the difference between the total revenue and the total expenses per unit of the enterprise. Given the assumptions and estimates made in preparing the budget, a positive return to management indicates a profitable enterprise.

Special Crop Enterprise Budgeting Considerations

Some problems encountered in preparing crop enterprise budgets are as follows.

Storage, Transportation, and Marketing Expenses.   Storage, transportation, and marketing expenses may or may not be included in the budget. It is important to note whether these expenses are included if comparisons are to be made between the price received and the average cost. Both must be on the same basis at the same location for the comparison to be valid. If storage, transportation, and marketing expenses are not included, the average production cost is a "farm gate" cost.

Fallow and Establishment Expenses.   For crops requiring a production period of more than one year, expenses for establishing the crop should be included in the budget. An annual amortized establishment cost is often charged in the producing years (years when the crop is harvested). Sometimes, for orchards, depreciation and interest are charged on the trees. In a grain-fallow rotation, fallow costs (including the land charge) are considered enterprise expenses.

Soil Management

Land capabilities and soil conservation considerations must be accurately assessed if the cropping system is to be profitable.

Land Capability

The development of cropping systems and crop production management require an understanding of soil types and land capabilities. The capabilities of land have been classified by the Soil Conservation Service of the U.S. Department of Agriculture: Soils are grouped into eight land capability classes in accordance with the suitability of each for general farm crops, grazing, forestry, and wildlife. This classification system is based on the needs and limitations of soils, on the risks of damage when they are used, and on their responses to management. These eight land capability classes are as follows:[2]

1. *Class I.*   Very good land that can be cultivated safely with ordinary good farming methods. It is nearly level and easily worked. Some areas need clearing, water management, or fertilization. Usually there is little or no erosion.
2. *Class II.*   Good land that can be cultivated with easily applied practices, including contouring, protective cover crops, and simple water management operations. Common requirements are rotation and fertilization. Moderate erosion is common.
3. *Class III.*   Moderately good land that can be cultivated safely with intensive treatments such as terracing and strip cropping. Water management is often required on flat areas. Common requirements are crop rotation, cover crops, and fertilization. Usually this land is subject to moderate to severe erosion.
4. *Class IV.*   Fairly good land that is best suited to pasture and hay but can be cultivated occasionally, usually not more than one year in six. In some areas, especially those of low rainfall, selected land may be cultivated more often if adequately protected. When it is plowed, careful erosion prevention practices must be used.
5. *Class V.*   Land suited for grazing or forestry, with slight or no limitations. It is nearly level, and usually there is little or no erosion. It is too wet, stony, or otherwise unsuitable for cultivation. This land needs good management.
6. *Class VI.*   Land suited for grazing or forestry, with minor limitations. It is too steep, eroded, shallow, wet, or dry for cultivation. This land needs careful management.
7. *Class VII.*   Land suited for grazing or forestry, with major limitations. It needs extreme care to prevent erosion or other damage. Usually it is too steep, rough, shallow, or dry to be seeded to range or pasture plants.
8. *Class VIII.*   Land suited only for wildlife or recreation. It is usually extremely steep, rough, stony, wet, or severely eroded.

If such a capability classification is available for the farm, it may be helpful in classifying land into homogeneous areas. The classification, however, is based largely on physical factors, and the most economic cropping system will not necessarily be suggested by the class description. It may be necessary on a given farm to develop more than one cropping system in order to take into account the variation in soils. Managers should avoid developing a large number of small fields because of the resulting inefficient use of machinery and of land used for fences and turnrows.

The yield potential of land depends on the soil's properties, location, and management. Soils with similar properties and yield potentials are classified in the same

---

[2] U. S. Department of Agriculture, *Our American Land*, Soil Conservation Service Miscellaneous Publication 596, 1948, pp. 15–16.

soil management group. Soil sampling and testing—the local extension office can provide assistance—should be done to determine the soil management group, yield potential, and fertilizer recommendations for the crops.

Soil Conservation Considerations

A manager with an economic point of view considers soil maintenance and improvement to be a means rather than an end and soil as a resource that can contribute to present and future income.

A distinction should be made between soil depletion and soil erosion. As used here, *soil depletion* refers merely to a reduction in fertility, or nutrient level; *soil erosion* means the physical loss of the soil itself. Soil lost through erosion is virtually impossible to replace economically. Although some soil erosion cannot be stopped entirely, in most cases it can be controlled by changing tillage practices or crops, by strip cropping, or by terracing. Soil depletion, on the other hand, may be easier and economically more feasible to correct if it involves simply supplying the necessary nutrients to depleted soils.

A time-honored tradition persists that soil fertility should be at least maintained, and preferably improved, as a farm passes from one generation to the next. Although this expectation reflects a commendable idealism, it often obscures the economic requirements of operating a farm business. The decision on whether to maintain or to improve soil fertility depends on how the manager compares present income to future income. However, emergencies may require farmers to deplete some of the soil's fertility and defer maintenance or improvement measures until later. In addition, there are many instances in which conservation practices would be profitable but are neglected by farmers who mistakenly believe otherwise; in this case, it is up to extension agents to make the facts known.

In any area where soil erosion is widespread, a conservation hazard exists that may pose a threat to society. Indeed, there are those who feel that soil is a precious national resource, that soil loss will affect future generations, and therefore that current economic conditions should not be the sole criteria for determining the use of this nonrenewable resource. Such an argument must, however, take into account two important points. First, in the absence of economic criteria, what criteria will be used? It is impossible to stop soil erosion completely, and in order to maintain soil fertility, some soils should not be cultivated. The nutrient level of some soils is higher in their virgin state than can be maintained when cultivated. Second, it is doubtful that the majority of farmers or any other business managers would, for any length of time, follow conservation practices that are not in their economic interest. A veteran soil conservationist has said that farmers have three reasons for practicing conservation:

1. They do it because they have been told they should by soil conservation specialists or extension agents, or they read this advice in agricultural publications.
2. They do it to "keep up with the Joneses"—because other farmers are following conservation practices.
3. They do it because it pays.

This soil conservationist's experience indicated that unless it paid to adopt conservation practices, the farmers would discontinue them. Therefore, in determining their farming practices, farmers can be expected—indeed, they owe it to themselves as business managers—to use whatever data are available in order to maintain an economically optimal level of soil fertility. If the optimum for the individual falls substantially short of the optimum for society in general, some kind of public action may be necessary to make further conservation efforts and expenditures profitable for individual farmers. Such public action might include income tax incentives and plans for cost sharing between farmers and the government. In any event, in a society where profit and prices determine the allocation of resources, conservation will tend to be practiced when and to the extent that it is profitable.

Crop Rotations

Crop rotation refers to the sequencing of crops produced in a particular field over a number of years. Growing the same crop continuously, year after year, is a specific type of rotation, referred to here as *continuous-crop rotation*. A rotation involving a sequence of different crops is referred to as a *multiple-crop rotation*.

Multiple-crop rotations have been recommended for many years by agriculturalists. Among their benefits is the maintenance or improvement of organic matter, fertility, and soil structure. In addition they may permit better control of insects, weeds, diseases, and erosion. However, in some areas, single crops have been produced profitably on the same soil year after year.

Deciding whether a particular rotation should be adopted is a complex problem, and it must be examined systematically if it is to be analyzed properly. This particular problem is made more difficult by the fact that not all of the agronomic relationships are known. Following is an examination of the possible benefits of multiple-crop rotations.

Weed, Insect, and Disease Control

Unless weeds, insects, and plant diseases are controlled, it is impossible to farm profitably for any significant period of time. Therefore, the most profitable means of control must be determined.

By growing row crops and small grains in a multiple-crop rotation, it may be possible to control the growth of most weeds. The main cost of controlling weeds by rotation is the opportunity cost of limiting the acreage of a high-value crop. The costs of controlling weeds by weed sprays are out-of-pocket. If a rotation is followed, the sacrifice in acreage of high-value crops must be balanced against the out-of-pocket costs of chemical weed control.

Finding the most profitable methods to control insect pests and plant diseases may be considered in the same way.

Maintenance of Soil Fertility

The contribution of a rotation to the maintenance of soil fertility varies by location. Legumes will fix nitrogen in the soil, which is then available for other crops, but legumes use other soil nutrients that must be replaced. If legumes are being grown because of their nitrogen-fixing ability, the manager must determine whether the nitrogen can be supplied more cheaply by commercial fertilizers.

The improvement and maintenance of organic matter depend upon the return of plant or animal residues to the soil. However, the precise amount of organic matter that should be maintained in the soil for the most profitable level of production is not known. Yields can be maintained for a considerable number of years using a continuous-crop rotation, but they may decline eventually unless offset by improved varieties or better cropping practices. In some cases, growing an occasional "green manure crop" may be more economical than following a multiple-crop rotation for the purpose of maintaining organic matter. Crop residue often returns a considerable quantity of organic matter to the soil even if one crop is grown continuously.

Also related to the maintenance of organic matter are soil tilth and structure, which may affect the water-holding capacity of the soil and consequent crop yields.

Erosion Control

Compared to most cropping systems without a rotation plan, rotations help to control erosion. A sod crop in the rotation may provide soil protection during certain months of the year. However, there may be alternative means for providing erosion control, the effectiveness of which is determined by the severity of the erosion. Some of these techniques are (1) strip cropping, (2) contour cultivation, (3) terraces, (4) diversion ditches, and (5) mechanical structures such as dams, spillways, and flumes. The economics of conservation is discussed in the next section, but crop rotation as a means of erosion control may or may not be the most economical way of achieving this end.

Rotations, then, should be viewed as a means of accomplishing various ends. If physical and biological data are available, the relative profitability of various cropping systems can be compared. Such data are becoming increasingly available as a growing number of research projects are devoted to this subject.

Fertilization Economics

The principles determining the optimum rate of fertilizer to apply were outlined earlier. Farmers who are sure of the response they will get and who have sufficient capital will want to apply fertilizer to the point where the added revenue (marginal value of the product, MVP) is equal to the added expense (marginal input cost, MIC). Farmers with limited capital will want to apply fertilizer to the point where the last dollar invested in fertilizer returns approximately the same amount as the last dollar invested in other inputs (equimarginal returns principle).

Even though some farmers may not have specific information on the incremental response to be expected from fertilization, there are a number of guides that can be used. By using the information and services available, fairly accurate expectations can be formulated. For example, records of past fertilizer applications and yields by fields are useful for indicating trends, problems, and yield responses to additional applications.

If a soil is deficient in a nutrient, the crops grown often reflect this deficiency, either by discoloration of the leaves or by "firing." Information is available to interpret these symptoms. Many state universities provide soil-testing services and make fertilizer recommendations based upon their analysis. Such recommendations should be implemented cautiously because it is difficult to predict yield response from soil tests alone. Extension agents can be another valuable source of information. They are trained agriculturalists who are familiar with local conditions. If they do not have accurate information on fertilization, they may be willing to establish test plots to obtain it.

Marginal Analysis of Fertilizer-Yield Data

Even when information is sparse, the marginal principle may still be helpful. For example, assume that test plots for a particular soil type show the following results:

Nitrogen Application (Pounds Per Acre)	Wheat Yields (Bushels)
0	27
30	33
60	35

If the cost of nitrogen is $0.25 per pound and the price of wheat is $4 per bushel, 30 pounds of nitrogen obviously would be profitable (30 pounds × $0.25 = $7.50 MIC; 6 bushels × $4 = $24.00 MVP). However, the profitability of increasing the application to 60 pounds is questionable. The MVP would be $8.00 and the MIC would be $7.50. With meager data, the farmer may be unwilling to make the additional investment for the indicated return. Unless the tests were conducted for a number of years, the yield responses cannot be predicted accurately. It is also possible that the money spent on fertilizer might return more if used in a different way. The farmer who is willing to gamble a bit might decide to increase the application to 40 or 50 pounds.

Sometimes these kinds of data are analyzed on the basis of averages, which can lead to false conclusions. Using this misleading analysis, the problem would be approached as follows: If 60 pounds of nitrogen were applied, the yield would be 35 bushels per acre and the return would be $140, whereas the cost of fertilizer would be $15 (60 × $0.25 = $15). Therefore, the average return per dollar invested in fertilizer would be $9.33 ($140/15). On this basis, it would appear that the 60-pound application would be very profitable even though the MVP is only $8.00 compared to the MIC cost of $7.50.

To carry this analysis one step further, suppose that 90 pounds of fertilizer would result in a yield of 36 bushels. The cost of fertilizer would be $22.50. The return per dollar invested in fertilizer would be $6.40 ($144/22.50). On the basis of average-return analysis, this undertaking would also appear to be profitable. Yet it is known that the last 30 pounds of fertilizer applied would have been unprofitable; the added cost would be $7.50, but the added return would be only $4. This example illustrates the dangers inherent in average-return analysis. Marginal analysis is, therefore, the appropriate tool in such a situation.

Maximum-Profit Nutrient Combinations

Marginal analysis is also useful for using data such as those presented in Table 15-2. The sample data show the response of corn to applications of nitrogen and phosphate. Assuming that the cost of nitrogen and phosphate (including the cost of application) is $0.30 per pound for each, the MIC of 40-pound increments would then be $12 and $12, respectively. If the price of corn is $2.50 per bushel, a prospective yield of 4.8 bushels of corn would justify the application of 40 pounds of nitrogen, and 4.8 bushels would be needed to justify the application of 40 pounds of phosphate.

The procedure for locating the most profitable combinations of nutrients to apply relies on the marginal principle. It is obvious that either 40 pounds of nitrogen or 40 pounds of phosphate would be profitable. Because 4.8 bushels are needed to pay the MIC of 40 pounds of phosphate, the table is read from 0 pounds of nitrogen to the right until that point is reached. Moving from 40 to 80 pounds of phosphate at a nitrogen level of zero gives a marginal yield (MPP) of 5.6 bushels and a MVP of $14. Increasing phosphate to 120 pounds with 0 pounds of nitrogen, would be unprofitable.

Next, the column representing 80 pounds of phosphate is scanned at variable quantities of nitrogen until the MPP falls below 4.8 bushels, which represents a

TABLE 15-2. Total Yields of Corn for Specified Nutrient Combinations

Pounds of Nitrogen per Acre	Pounds of Phosphate per Acre				
	0	40	80	120	160
			Bushels		
0	0	31.5	37.1	37.5	35.3
40	21.8	72.6	83.9	88.7	90.1
80	25.8	82.3	95.9	102.4	105.4
120	25.9	88.7	102.1	110.1	114.2
160	24.0	88.5	105.4	114.5	119.6
200	20.9	88.6	106.8	116.9	122.9
240	16.8	87.4	106.9	117.9	124.6
280	12.1	85.3	105.9	117.8	126.8
320	6.8	82.5	104.1	116.8	124.9

MVP of $12. This MVP is the same as the MIC for 40 pounds of nitrogen when nitrogen is $0.30 per pound. When 80 pounds of phosphate are applied, the MPP resulting from an increase in nitrogen from 80 to 120 pounds is 6.2 bushels, which is greater than the 4.8 bushels required to cover the MIC. The MPP falls to 3.3 bushels when nitrogen is increased to 160 pounds. Obviously, no more than 120 pounds of nitrogen should be applied when phosphate is applied at a rate of 80 pounds per acre.

The row in Table 15-2 corresponding to the nitrogen application rate of 120 pounds per acre must to scanned to the left to determine whether more phosphate would be profitable. Increasing phosphate from 80 to 120 pounds gives a MPP of eight bushels. However, increasing phosphate to 160 pounds gives a MPP of less than 4.8 bushels, and increasing nitrogen to 160 pounds also gives a MPP of less than 4.8 bushels. Therefore, the most profitable combination of nitrogen and phosphate would be 120 pounds of each at the stated prices.

This example illustrates another way to apply the marginal principle and suggests that when the application level of one nutrient is being considered, the levels of others should be analyzed as well.

Irrigation Economics

A few years ago, irrigation was largely confined to arid sections of the country where it was necessary for intensive agricultural production. In such areas, water management is obviously an important farm management concern. More recently, supplemental irrigation has been widely adopted in more humid areas. As a result, water management is now a subject of interest throughout the United States.

Water as an Input

In an economic sense, water must be viewed as another resource or production input similar to land, labor, and fertilizer. In the simplest case, the principles of marginal analysis would seem to be relevant for determining the application level of water. If a response curve to water can be obtained and if additional increments of water come at a cost, then the marginal principles outlined earlier obviously would apply.

If a response function does exist, cost data can be helpful in determining the appropriate number of irrigations. The additional or incremental cost of irrigation drops considerably after the first irrigation because the fixed equipment cost is allocated to the first irrigation. Once the equipment has been purchased and is in place, the cost of additional irrigations will consist of only labor and energy expenses. The returns necessary to justify an additional application of water can be estimated from such data.

The problem, however, is seldom this simple. Response functions to water are difficult to determine. The time of application is frequently more important than the amount of water applied. The fertility of the soil, as well as the density of the crop stand, are important variables that interact with water in determining yields. As a result, the other variables must be specified or controlled if the results are to be meaningful.

Another complication relates to the cost of water. For many irrigation projects, water is sold on a per-acre basis. A contract is made at the beginning of the irrigation season for the irrigation of a specified number of acres. The amount of water used per acre becomes irrelevant except for the cost of the energy and labor necessary to distribute and apply it. Under such circumstances, the marginal returns to water are not as important to the farm manager as the amount of water necessary for maximum yields. In such a case, the water cost is fixed and the energy and labor costs are variable. This situation, of course, is not true of those farms where water is purchased on an acre-foot basis or pumped from wells. In such cases, the marginal cost of water is relevant.

The proper amount of water to use is closely related to the cost of energy and labor, because these inputs are used to distribute the water. To make the most efficient use of irrigation water may require more labor than would a less careful application. If additional water does not come at a cost but labor is expensive, more water will be used than if the reverse is true. Higher energy costs, on the other hand, encourage the use of less water, because reduced water requirements reduce pumping and distribution requirements. The farmer should seek the optimal combination of water and the other inputs required to use it.

For the preceding reasons, irrigation is affected more by the time of application than by the incremental costs and benefits of different quantities of water applied. Consequently, if a response function for water does not exist, soil moisture measuring devices provide the most promising method for determining the best time and amount of application.

Water Application Systems

Finding the most economic method of water application is important to the managers of irrigated farms. Water application methods can be classified as surface, sprinkler, and drip. Surface methods include basins, borders, corrugations, furrows, and controlled flooding. Surface methods generally rely upon gravity flow to distribute the water, in contrast to sprinkler irrigation systems, which distribute water mechanically. From an agronomic standpoint, all methods that make water available to plants in the proper quantity and at the proper time are equally satisfactory. However, soil and topographic conditions and the type of crop grown may make equally effective jobs impossible under some conditions.

The various surface methods are adaptable to a variety of soil and topographic conditions. Sprinkler systems can be used in most areas except those in hot, windy climates and on very heavy soils where the rate of water absorption is very low, because much of it evaporates or runs off. Drip irrigation systems are receiving more attention because of their energy and water application efficiencies.

An agricultural engineer can be consulted for advice on alternative methods of supplying water on a given site for particular crops. Once the engineering alternatives are isolated, there is the economic problem of determining the least-cost method of applying the water. Again, it is impossible to give general recommendations because the least-cost method varies with the location. There are, however, certain elements common to all problems. In general, land preparation must be more extensive for surface systems than for sprinkler irrigation. Sprinkler systems require a sizable initial outlay for equipment, which may last from 10 to 15 years.

Additionally, a pump may be needed to create the necessary pressure for sprinkler irrigation, in which case power costs must be considered.

There are annual cash costs for each method, and different labor skills are needed for each kind of system. Maintenance and repair costs must also be considered. If reliable estimates can be obtained from qualified engineers, a partial budget can be used to estimate relative costs. The university experiment station may have economic studies of locations similar to the one in question. If the land to be irrigated is not the same every year, sprinklers obviously have greater flexibility and may be preferred for this reason.

Supplemental Irrigation

Supplemental irrigation has been used increasingly in humid areas where lack of moisture limits the choice or yield of crops. It permits a wider selection of alternative crops. Supplemental irrigation provides flexibility in farm planning, permitting farmers to irrigate crops in years when precipitation does not provide sufficient moisture. Farmers often invest in supplemental irrigation to reduce the risk associated with variable rainfall. In fact, some farmers may be willing to incur the higher costs of irrigated production as insurance against crop failures in years with low rainfall.

One of the major problems associated with supplemental irrigation is related to capital requirements: For irrigation, these requirements are frequently so large that the system of farming must be changed to obtain a sufficient volume of business to justify the investment. Therefore, many new management problems are introduced with irrigation. What crops or enterprises should be added? Should any of the existing enterprises be dropped? Where will the additional capital be obtained? Will capital invested in irrigation return more than if it were invested in a different way? In Chapter 6, the use of budgets was illustrated for analyzing a problem of this type. After such budgets have been prepared, farmers can decide if they have sufficient capital to make the necessary adjustments and the management ability, interest, and technical knowledge to change to a new type of farming.

Some Larger Water Problems

Water is becoming an increasingly scarce resource nationally. Population growth, industrial expansion, and recreation interests continually increase the demand for water. Consequently, agricultural irrigation may face competition in its claim on water. This possibility is all the more serious because much irrigation water cannot be reused; this is not true of many other uses of water. Many studies indicate that the value of water used for nonagricultural purposes is greater than that used for agricultural purposes. Additional water for irrigation probably will come largely from the development of additional storage or the utilization of groundwater sources. Farmers must be aware of the laws and regulations affecting the allocation and use of water. They should protect their access to water resources by complying with state regulations. It is their responsibility to know these regulations.

Many states are considering changes in policies related to the allocation and use of water. The arguments about the relative value of water for agricultural and nonagricultural uses should be carefully considered. Overall, comparative figures on water value may not be valid when applied to a particular geographic area. For example, if water were not used for irrigation in certain localities, it might not be used at all. On the other hand, the competition for water in other areas is sufficiently keen that farmers should be alert to the situation if they wish to maintain their present water use level.

Developing the Cropping System

Developing a cropping system involves the development and comparison of alternative systems. This section presents some general guidelines for designing the cropping system and is followed by a comparison of alternative cropping systems based on experimental data.

Selecting the Crops

The following list presents the usual sequence of steps involved in selecting crops to include in the cropping system:

1. Inventory the capabilities of the land resources. Climate, soil conservation, and water resources if irrigation is required should also be considered.
2. Identify the crop alternatives, based on the resources available and the location of the farm with respect to markets.
3. Weight the disadvantages and advantages of multiple-crop rotations. For example, one advantage might be the control of weeds, diseases, and insects.
4. Sequence the crops. For example, if a winter grain is produced, the harvest of the preceding crop must be early enough to allow time for seeding of the grain. Or, if alfalfa is seeded with grain as a nurse crop, the cost is less than that of seeding the alfalfa after the grain has been harvested.
5. Compare the profitability of the cropping systems. Enterprise budgets are used at this time. They are ordered according to their respective importance in the various rotations to assess the overall profitability of the cropping system.
6. Consider labor and capital requirements. Monthly labor requirements can be analyzed from a detailed calendar of operations (Chapter 13). An analysis of the capital requirements of the different rotations can also be made. The per-acre operating expenses from the enterprise budgets can be used to prepare a monthly cash flow budget to assess operating capital needs.

This procedure can help to approximate the combination of crops that is the most advantageous to the farm manager. A more detailed analysis can be made if desired. For example, if the crop yields and prices fluctuate widely from year to year, risk may be an important consideration in evaluating the various cropping systems.

Certain generalizations can be made about cropping systems: The most profitable systems contain a high-value cash crop and are flexible, so it is easy to shift from one crop to another. In other words, a general cropping system could be developed, but the particular crops to be grown and the acreage for each could be varied according to the conditions. Successful farmers have long-term plans and goals that are well thought out, but their plans do not become straitjackets restricting their movements.

The cropping plan should be viewed as a guide rather than as a blueprint. New crops may be introduced to the area, price relationships may change, and machinery for handling certain crops may be improved. In order to take full advantage of new conditions, it is advisable to revise the cropping plan periodically. Developing a detailed 10-year cropping plan may be useful for thinking through the consequences of different actions. However, farming conditions are not likely to be static for 10 years, so rigid adherence to such a plan would probably be unprofitable. Even though it is usually unprofitable to switch crops in response to short-run price changes, adjustments should be made when it appears that the comparative advantage has changed.

In summary, then, the principal crops should be selected as outlined in the previous chapter. The extent to which additional crops are included in the system should be determined by their direct contribution to the net income of the farming business and by their indirect contribution to the main crops through fertility enhancement, prevention of soil erosion, and insect, weed, and disease control. A cropping system can then be developed. Such a system may involve either continuous cropping, a multiple-crop rotation, or a sequence of crops that can be varied, within limits, depending upon conditions at the time.

Comparing Alternative Cropping Systems

The cropping system also includes the total set of techniques and practices related to the production of a specific crop or combination of crops, including the crop rotation, tillage operations, fertilization practices, and pest control measures. Experimental data from Illinois illustrate how these various aspects can be considered in evaluating alternative cropping systems.[3] Using both average net returns and risk criteria, this study compares cropping system alternatives.

Crop production, in most cases, exhibits great variability. Yields are influenced by temperature, quantity and temporal distribution of rainfall, and pests such as weeds, diseases, insects, and so forth. Uncertainty also characterizes the crop prices and costs of production. The income, or net return, generated in the presence of these uncertain conditions is risky.

This Illinois study illustrates how average net returns and risks are affected by three crop rotations, two tillage systems, and three levels of pest control. The three rotations analyzed are continuous soybeans, continuous corn, and a corn-soybean rotation. Tillage systems are designated as either "conventional" or "reduced." Pest control measures are designated as "high," "medium," or "low" to represent various levels of chemical use for corn and soybean production. The high, medium, and low levels of pest control refer to the application of various herbicides, insecticides,

---

[3] L. R. Zavaleta, B. Eleveld, et al., *Income and Risk Associated with Various Pest Management Levels, Tillage Systems, and Crop Rotations: An Analysis of Experimental Data,* University of Illinois Agricultural Economics Research Report 191, April 1984.

and fungicides at various rates and frequencies, or not at all in some cases, so that the relative benefits of these practices could be measured.

The use of herbicides, fungicides, and insecticides for crop pest control should be restrained and limited to those situations where treatment is absolutely necessary. This cautious approach is important to minimize (1) the development of pesticide-resistant strains, (2) the release of toxic chemicals into the environment, and (3) the costs of plant protection. Again, the principles of marginal analysis have application. The use of these pesticides should be restricted to those situations where the marginal returns exceed the marginal costs.

The yield data for these various cropping systems were obtained from test plots in Urbana, Illinois, during the crop years of 1980 to 1982. The variability of crop prices and input costs is not considered in this evaluation. Thus, the only source of variability and outcome is the physical and biological environment.

Table 15-3 compares all the cropping system alternatives. The data used for making these comparisons are the average yields, average net returns, and lowest net returns. The net returns represent the returns to land, assuming prices of $2.68 and $6.80 per bushel of corn and soybeans, respectively. The nonland costs were estimated using enterprise budgets. The lowest net return is a measure of risk; the lower the lowest net return, the greater the risk. The lowest net returns were calculated so that there would be a 97 percent probability that the net returns would be at this level or above.

A comparison of the yields obtained under the alternative crop rotations indicates that, with only one exception, average crop yields were higher in the corn-soybean rotation than when either crop was grown continuously. The exception was a slightly lower corn yield (in the corn-soybean rotation) with conventional tillage and a high level of pest control. Comparing the average net returns shows that the corn-soybean rotation was substantially more profitable for all but one, tillage and pest control combinations. These results appear to demonstrate clearly the economic advantage of growing corn and soybeans in rotation. The highest average net return was produced by the low pest control, conventional tillage alternative, using the corn-soybean rotation. However, managers are concerned not only with average returns but also with the risk associated with these returns.

A comparison of the tillage systems shows no consistent pattern of superiority for either one. Whether the conventional method or the reduced tillage method gives the higher average net returns depends on the tillage method's interaction with the pest control level.

A comparison of the pest control levels shows that lower average yields were associated with lower levels of pest control. This finding was consistent across all tillage method and rotation combinations. However, this pattern did not exist for average net returns. In four of the six situations, the highest average net return was associated with the medium level of pest control. For these two exceptions, the average net returns were highest for the low level of pest control.

The riskiness associated with these alternative cropping systems can be compared using the lowest net returns reported in Table 15-3. The four cropping systems with the lowest risk—the systems with the highest "lowest net returns"—were the low and medium pest control alternatives with either conventional or reduced tillage under the corn-soybean rotation. Thus, according to these results, farm

TABLE 15-3. Comparison of Alternative Cropping Systems, Urbana, Illinois, 1980–1982

	Conventional Tillage			Reduced Tillage		
Crop Rotation	High Pest Control	Medium Pest Control	Low Pest Control	High Pest Control	Medium Pest Control	Low Pest Control
Continuous Soybeans[a]						
Avg. yield (bu/acre)	44.4	36.2	29.1	42.0	36.6	33.1
Avg. net return ($/acre)	68.54	105.97	77.40	55.17	109.45	112.15
Lowest net return[b]	−7.22	12.81	−24.06	−21.67	−1.79	−48.33
Continuous Corn[c]						
Avg. yield (bu/acre)	139.5	130.4	114.2	126.3	129.1	107.9
Avg. net return ($/acre)	106.08	114.34	110.44	88.74	123.39	106.16
Lowest net return[b]	−47.92	−55.30	−100.36	−67.28	−24.87	−26.44
Corn-Soybean Rotation						
Avg. corn yield (bu/acre)	139.3	133.7	132.7	136.9	132.1	122.3
Avg. soybean yield (bu/acre)	46.40	43.90	40.80	46.10	46.20	40.20
Avg. net return ($/acre)	92.96	138.52	153.50	97.96	150.77	150.43
Lowest net return[b]	31.98	67.40	86.40	24.46	88.97	78.75

SOURCE: L. R. Zavaleta, B. Eleveld, et al., *Income and Risk Associated with Various Pest Management Levels, Tillage Systems, and Crop Rotations: An Analysis of Experimental Data*, University of Illinois Agricultural Economics Research Report 191, April 1984.

[a] The price for a bushel of soybeans is assumed to be $6.80.

[b] The lowest net return is estimated as the average net return minus two times the standard deviation. The chance of a lower outcome is less than 3 percent.

[c] The price for a bushel of corn is assumed to be $2.68.

managers who are averse to risk are better off if they control pests with low to medium levels of pesticide and grow corn and soybeans in rotation. From the results of this study, it is not possible to establish clearly whether conventional or reduced tillage provides the greatest advantage.

The "safety-first" rule (Chapter 8) can be used to compare the riskiness of the net returns for these 18 alternative cropping systems. Using this rule, the manager would choose the cropping system that has the highest average net return, provided that there is an acceptably low probability that the return will be below some minimum level. Suppose that the minimum net return necessary to meet cash requirements is $50 per acre. All but four of the cropping systems have lowest net returns below this level, indicating some probability that this criterion would not be satisfied. The four alternatives involving the corn-soybean rotation with conventional and reduced tillage and low and medium pest control measures all have lowest net returns above the $50 level. Three of these four have average net returns in the range of $150 to $154 per acre. Based on these profitability and risk considerations, these three alternatives are almost equivalent.

Summary

Designing the farm's cropping system includes several interrelated decisions regarding crop selection, soil conservation, crop rotations, fertilization, and irrigation, among others. The enterprise budget, which summarizes the crop's estimated revenue and expenses, is the primary tool for analyzing the economic implications of these decisions. Although the gross margin (revenue less operating expenses) indicates the profitability of the enterprise for annual planning, the return to management (revenue less total expenses) must be positive for long-run profitability.

Understanding land capabilities is fundamental to crop production management. These capabilities affect the suitability of crops, yield potential, and soil conservation requirements. Soil conservation is of concern not only to farmers but to society as a whole. Therefore, the government provides incentives for adopting conservation practices. Because of the significant implications of soil erosion, it is important that farmers seek all the information available in assessing the economic advantages of soil conservation practices.

Multiple-crop rotations can help to maintain the soil's productivity by improving organic matter, fertility, and soil structure. They are also sometimes recommended to control weeds, insects, diseases, and erosion. When considering these benefits from multiple-crop rotations, the alternative methods for achieving these same benefits should be analyzed. For example, it may be more profitable to use chemical weed control and continuous-crop rotation than to use a multiple-crop rotation to control weeds.

Applying fertilizer up to the point where its MVP is equal to its MIC will maximize profitability. However, information on the specific yield increase resulting from another unit of fertilizer often is not known with certainty. Experimentation and test plots will provide the needed information. The yield effects of the interaction of plant nutrients can also be analyzed using the marginal principle to determine the maximum profit combination of nutrients.

In certain farming areas where irrigation is required for successful crop production, water is an input to be managed in the same way as land and fertilizer. However, the response of yield to water applications is difficult to determine. Not only is the amount of water applied important, but the timing of the applications as well. Responses to irrigation also depend on other crop production practices, such as fertilization. Determining the cost of irrigation also presents problems. In some areas, water is available at a fixed cost up to a certain maximum amount of water per acre; if less is used, the cost is still the same. In other locations, the cost is directly related to water use. The type of irrigation system used also affects the costs of applying additional amounts of water.

The process for developing the cropping system involves consideration of (1) land and water resources, (2) crop alternatives, (3) crop rotations, (4) crop sequencing, (5) profitability, and (6) resource requirements. The evaluation of alternative cropping systems requires extensive data indicating the yields, expenses, and net returns resulting from various combinations of crop rotations, tillage methods, pest control levels, and other components of the cropping system. In addition to the average net returns resulting from the cropping system, the risk associated with the net returns should be considered. Farmers may forgo a cropping system with a higher average net return and choose one with a lower probability of loss if risk is a major concern.

Making decisions regarding the components of the cropping system is often complicated by incomplete data, risk, and the continual development of new crop production practices requiring evaluation. Such is the challenge of management.

Recommended Reading

RAE, A. N. *Crop Management Economics.* New York: St. Martin's Press, 1977.

CHAPTER 16

Machinery Management

Machinery constitutes one of the more important capital inputs in farm production. The farm manager must make many decisions regarding machinery: the number and size of tractors, trucks, combines, and other equipment to use; whether to own or hire custom services; when to replace machinery; whether to purchase new or used machinery; and whether to lease rather than purchase. Many of these decisions involve proportionality; that is, are the machinery, land, and labor being combined in the best proportions? Is there too much machinery relative to available land and labor? Alternatively, would it be more profitable to vary labor or land to utilize a given set of machinery more effectively? Arriving at the optimal proportion becomes a special problem with machinery because it is acquired in large single units; it is impossible to purchase one third of a truck or one half of a swather.

This chapter provides some examples of machinery decisions and illustrates the use of economic principles and farm management tools in making them. In the material that follows, typical farm machinery decisions are presented and procedures are suggested for dealing with these problems on an individual farm basis.

General recommendations for machinery use for a large number of farmers may be misleading when translated into specific individual applications. Timeliness of operations is an example. Although data are available that indicate the importance of performing operations on time, it is often difficult to apply these data to an individual farm situation. Similarly, it may be difficult to evaluate the potential of used machinery without knowing something about the mechanical ability of the farmer or farm workers.

Individual farms and farmers vary with respect to the amount of acreage farmed, the crops grown, the value of leisure time, risk preferences, and the opportunity cost of labor. Therefore, budgeting techniques applied to individual farm decisions provide better analyses and lead to better results.

Budgeting Machinery Costs

Modern farming technology requires the use of large, expensive machinery. Owning and operating this machinery is an important component of total farm expenses. To make informed decisions about the management of this expensive input, farmers must know the costs involved, including the costs of current machinery and of any alternatives being considered.

Machinery costs are typically classified into two categories: ownership costs and operating costs. Ownership costs include depreciation, interest on the investment, taxes, insurance, and housing, if appropriate. These costs are fixed, because they are incurred no matter how frequently the equipment is used. Operating costs, on the other hand, vary directly with equipment use. They include fuel, lubricants, repairs, and labor.

To assist in budgeting machinery costs, a machinery cost worksheet can be completed for each piece of machinery used. The worksheet in Table 16-1 can be used to estimate the annual ownership and operating costs associated with each machine. The first step is to describe the machine (model, horsepower, size) and indicate its cost when new, estimated life, salvage value, and annual use. Farmers who have no experience operating the machinery to be budgeted can get the data from machinery dealers, neighboring farmers, or the local extension agent.

Ownership Costs

Ownership costs include average annual depreciation, interest, taxes, insurance, and housing costs. A description of these costs and guidelines for calculating them follow.

TABLE 16-1. Machinery Cost Worksheet

Machine _____   New Cost $_____
Estimated Life _____ years   Salvage Value $_____
Annual Use _____ hours, acres

	Annual Cost	Cost Per Hour or Acre
Ownership (Fixed) Costs		
1. Depreciation	$_____	$_____
2. Interest	_____	_____
3. Taxes	_____	_____
4. Insurance	_____	_____
5. Housing	_____	_____
Total ownership cost	$_____	$_____
Operating (Variable) Costs		
1. Fuel: _____gallons multiplied by $_____per gallon	$_____	$_____
2. Engine oil, lubricants, filters	_____	_____
3. Repairs (parts, tires, labor)	_____	_____
4. Operating labor: _____hours multiplied by $_____per hour	_____	_____
5. Other costs (twine, wire, etc.)	_____	_____
Total operating cost	$_____	$_____
TOTAL MACHINE COST	$_____	$_____

Depreciation.   Depreciation is the loss of value due to age, use, and obsolescence. Although depreciation is partly influenced by use, it is considered an ownership cost. To calculate average annual depreciation costs, the salvage value is subtracted from the new cost and then this figure is divided by the number of years the machine is expected to be used (estimated life), which should be consistent with the salvage value. A piece of equipment that is expected to be used for only a few years and then replaced would have a higher salvage value than one used for a longer period of time.

Interest.   Interest costs are estimated by multiplying the average investment by the interest rate. *Average investment* is the sum of the new cost and the estimated salvage value divided by 2. If the equipment is financed, the interest rate is the loan rate; otherwise, it is the rate of return on the farmer's next best investment alternative, or the opportunity cost of capital.

Taxes.   The personal property tax is a cost based on the value of the machinery, but tax rates vary widely from one locality to another. The local tax rate multiplied by the average investment or the assessed value should approximate the average annual taxes on the machine.

Insurance.   Insurance is the cost of bearing the risk of loss from fire, wind, or other hazards. This risk may be borne by the farmer, or an insurance company can be paid to bear it. The average annual insurance cost typically ranges from 0.5 to 1.0 percent of the average investment.

Housing.   Depending on the climate, housing may extend the life of the machine and reduce repair and maintenance costs. Housing costs are estimated by first determining the square footage required to house the equipment and then multiplying this estimate by the average annual cost of providing a square foot of shelter. This average annual cost should include depreciation, interest, repair, tax, and insurance costs for the building. If machinery is not housed, depreciation and repair costs should be adjusted accordingly.

Total Ownership Cost.   The total annual ownership cost of the machine is the sum of depreciation, interest, taxes, insurance, and housing costs. This total is divided by the number of acres or hours of annual use to calculate the average ownership cost per acre or per hour. Because the ownership costs are fixed, the ownership cost per acre or per hour decreases as machine use increases.

Operating Costs

Operating costs include fuel, oil, lubricants, repairs, operating labor, and other costs. A description of these costs and guidelines for calculating them follow.

Fuel Costs.   Fuel costs vary with the size and type of engine as well as with the workload; estimates are based on records and personal experience. Annual fuel costs are determined by multiplying the gallons of fuel consumed per hour by the fuel price per gallon and then multiplying this figure by the hours of use.

Engine Oil, Lubricants, and Filters.   Estimated costs of engine oil, lubricants, and filters should be based on past experience and current prices. Annual oil costs equal the number of gallons of oil used per year multiplied by the current per-gallon price. Annual lubricant and filter costs are also based on current prices. Because these costs vary directly with fuel costs, they can be estimated as 15 percent of fuel costs when actual costs are not available.

Repair Costs.   Repair costs are affected by the type of machinery, the skill of the operator, the nature of the work, the hours of use, and so forth. They should be based on past records and personal experience and should include the value of the operator's labor used in making repairs as well as the expenditures for parts and service. With normal use of machinery, annual repair costs will average between 2 and 4 percent of its new purchase price.

Operating Labor.   The annual labor costs of operating a machine are the estimated hours of labor per year multiplied by the hourly labor charge, which should reflect the opportunity cost of the operator's labor. Because of the labor required for travel and equipment preparation, the estimated hours of labor may exceed the hours the equipment is actually used.

Other Costs.   Other costs, such as those for baler twine, wire, and so forth, that are required to operate the machine should be estimated on an annual basis. Past experience can serve as a guide in making these estimates.

Total Operating Cost.   The total operating cost of the machine for the year is the sum of fuel, engine oil, lubricant and filter, repair, operating labor, and other costs. This total is divided by the annual use (number of acres or hours) to determine the average annual operating cost either per acre or per hour of use. The total operating cost per acre or per hour stays relatively constant despite annual increases in machine use.

Total Machine Cost

The total average annual cost of owning and operating a machine is the sum of the annual ownership cost and the annual operating cost. It is important to remember that this is the *average* annual total cost over the estimated life of the machine. The actual annual costs will vary from year to year. Included in this average annual cost is a return to machinery investment and an operator labor charge. The total cost either per acre or per hour of use can also be calculated; it decreases as total use increases. The rate of decrease is dependent on the relationship between the ownership and operating costs.

Per-Acre Machine Cost

To find the cost of a field operation such as plowing, where both a tractor and an implement are involved, a cost budget like the one in Table 16-1 is prepared for each machine item. Then the costs for the two can be summed to get the total cost. If per-hour costs are calculated, they can be converted to per-acre costs by dividing the hourly costs by the number of acres that can be plowed in one hour. The following formula is used to calculate the number of acres covered by a machine in one hour:

$$\text{Acres per hour} = \frac{\text{width (ft)} \times \text{speed (mph)} \times \text{field efficiency (\%)}}{825}$$

The field efficiency factor allows for such things as adjusting the machines, overlap, turning, and anything else that might slow or stop the operation. For example, the field efficiency of a cultivator might be 80 percent, whereas that of a seed drill could be 65 to 70 percent. Such factors as field size and shape, driving patterns, moisture, and crop yield will also affect field efficiency.

The acres per hour for a 10-foot chisel plow operating at 4.5 miles per hour with an efficiency of 75 percent would be calculated as follows:

$$\text{Acres per hour} = \frac{10 \times 4.5 \times 75}{825} = 4.1$$

Dividing the cost per hour for the tractor, plow, and labor by 4.1 would give the cost per acre for this chisel plowing operation.

To summarize, finding the cost per acre of a field operation such as disking or harvesting involves first preparing annual cost budgets for each machine used. The total of the annual costs for the machines is then divided by the number of acres for which the machines are used. Alternatively, if a machine, such as a tractor, is used for more than one operation, the cost per hour (from the cost budget) is divided by the number of acres covered per hour.

If managers find that their machinery costs per acre are higher than the averages for their area, they should study the operation and look for ways to make improvements. They should keep in mind that high machinery costs may be due to large, expensive machinery, but the cost per unit (per acre) is the important figure to examine. When machinery is being substituted for labor, the combined labor and machine cost per unit is the factor to consider.

Substitution of Machinery for Labor

The use and development of agricultural machinery during the past 50 years has had a significant impact upon the agriculture industry. Machinery is used primarily to replace labor, but some machinery is also a substitute for land and management: Investment in a self-propelled sprinkler irrigation system increases land productivity and eliminates the labor necessary to move pipes; the manager who uses a computer as a management aid is substituting machinery for management time.

The substitution principle explained in Chapter 2 applies to this type of situation. In some of the earlier illustrations of the principle, output was held constant. However, the principle has application even if output is varied. The use of machinery frequently results in greater output, and this factor must be taken into account. Consideration should also be given to whether sufficient land and working capital are available for optimal utilization of the additional machinery.

The opportunity cost of any labor saved by substituting machinery should also be evaluated. If the labor saved is put to productive use in expanding the farm business, the opportunity cost of labor may be relatively high. On the other hand, if the labor is not put to productive use, a different situation results. Farmers, of course, may devote the labor savings to leisure activities. This may be a rational decision, and they may be willing to pay the cost.

With economic principles providing the logical framework to use in considering all the relevant variables, a partial budget can be constructed for making machinery management decisions. The economic principles most frequently used for making farm machinery decisions are the substitution principle, the opportunity cost concept, and the distinction between fixed and variable costs.

For example, the decision facing Jim Johnson is whether to trade his current tractor for a larger one or for another of the same size. He plans no other changes in the farm operation. The larger tractor would reduce the hours of labor required to produce his crops; it would also give a small increase in average yields as a result of more timely operations.

Jim Johnson's choice of tractors will affect his machinery ownership and operating costs. Because of increased depreciation and interest, ownership costs would be higher for the more expensive, larger tractor; operating costs would depend on the number of hours the larger unit would be operated relative to the smaller one. Mr. Johnson expects the larger unit to be used 50 fewer hours per year, for a total of 500 hours, compared to 550 hours for a tractor of the same size as the current one.

The decision is analyzed in the partial budget shown in Table 16-2. Based on the estimated costs of the two tractors, the expected yield increase, and the cost

TABLE 16-2. Partial Budget Example: Whether to Buy a Larger Tractor

1. *Added Revenue*			
Increased crop yields		$ 400	
2. *Reduced Expenses*			
Same size tractor:	Ownership	7100	
	Operating	5300	
	Labor ($5/hr)	2750	
3. Added revenue plus reduced expense			$15,550
4. *Added Expenses*			
Larger tractor:	Ownership	7800	
	Operating	5400	
	Labor ($5/hr)	2500	
5. *Reduced Revenue*			
None		0	
6. Added expenses plus reduced revenue			15,700
7. Difference (change in return to management)			$ − 150

charged for Mr. Johnson's labor, purchasing a larger tractor would not be profitable. It would decrease his return to management by $150. However, this difference is not especially significant. That is, if Mr. Johnson valued his labor at $8 per hour rather than at $5, this difference would be zero. Likewise, if the revenue from increased crop yields were $550 rather than $400, purchasing a larger tractor rather than one of the same size would not affect the return to management.

Two additional factors for Mr. Johnson to consider are more difficult to reduce to dollars and cents. One is the extra convenience and satisfaction he would get from having the larger tractor. The second is that the larger tractor would permit him to farm additional land that he might lease or purchase in the future.

The partial budget, constructed in accordance with economic principles and buttressed by data from farm records, is an appropriate tool for analyzing such decisions. Care must be taken, however, to consider all parts of the farm business, as well as the revenue and expense items affected by such a change.

Machinery Selection

Farm machinery is manufactured in a wide range of sizes. Choosing the tractor size and the complement of implements must be related to the size of the farm, the crops to be grown, and the cost of labor. The problem of selecting the appropriate size of machinery for a particular farming operation is complicated by the need to consider cost relationships and timeliness.

Cost Relationships

The relationships between ownership costs and operating costs and the annual use of machinery are important considerations in determining the size of machinery to use. Operating costs per acre are relatively constant as the number of acres farmed increases, and they do not vary greatly with the size of machinery. Although larger tractors consume more gallons of fuel per hour, they also do more work per hour. As a result, these two factors largely offset each other.

The per-acre ownership costs decrease as the number of acres farmed increase. The ownership costs also vary across different sizes of machinery; the ownership costs per acre of larger machines are higher than those of smaller ones. However, the difference between the per-acre ownership costs of a larger machine and a smaller one decreases as more acres are farmed. These relationships are particularly important because ownership costs account for the greatest share of total machinery costs.

Selecting the most profitable machinery size involves an analysis of tradeoffs: Will the larger machinery reduce labor and timeliness costs enough to offset the higher ownership costs? The most profitable machinery size balances the low ownership cost advantage of a small machine against the timely operation advantages of a larger one.

Timeliness

Timeliness refers to the ability to complete an operation during the optimal time. The timeliness of a field operation can have economic value; for example, late planting or harvesting reduces yields. Performing the operation before or after this best time can cause an economic loss as a result of reduced crop yield, quality, or both. Therefore, smaller machinery covering fewer acres per hour can be very costly. The inability of a machine to complete a job within the optimal time period (untimeliness) can be considered as a charge (cost) against the machine.

The value of timeliness is most easily recognized during harvest, but other operations also have timeliness aspects. Timeliness is critical to tillage operations, because soil must be worked when it is in a certain moisture range; cultivation and the application of pesticides and fertilizer must be performed at the right time. But timeliness in harvesting is probably most critical. The ideal time for harvesting horticultural crops may be only a few hours, when the tenderness, flavor, and color are optimal and before significant quality deterioration occurs. Forage and grain crops also have an ideal harvest time, but it may extend over several days before serious losses occur.

Machine size must be balanced against timeliness. Generally, larger machinery costs more per acre to own and operate. However, if the costs of untimely operations are considered, larger machinery may actually cost less because crops can be planted and harvested closer to the optimal times, thereby increasing yields over those that could be achieved with smaller machinery.

In summary, by considering yield and revenue losses due to untimely field operations as costs, the problem of selecting the best size of machinery can be analyzed by comparing the total machinery costs per acre.

An Example

An Iowa study compared machinery complements of varying sizes used on farms ranging from 100 to 1000 crop acres.[1] The results for three of the machinery sets are presented. The components of these three sets are described in Table 16-3.

Table 16-4 compares the per-acre costs of the smallest machinery set for farms of 100 and 1000 acres. Increasing the farm size increases the timeliness costs for this set of small machinery. For example, the timeliness cost is only $1.83 per acre on the 100-acre farm but increases to $147.22 per acre on the 1000-acre farm.

The relationships among the machinery costs per acre for the three sizes of machinery are illustrated in Figure 16-1. The curves indicate how the per-acre costs change for each machinery set as it is used over a larger crop acreage. The cost per acre of the smallest machinery alternative (four-row small) decreases from 100 to 600 acres and then increases up to 1000 acres. The reason for the increasing costs is the cost associated with timeliness, or not completing the field operations on time.

[1] W. Edwards and M. Boehlje, *Farm Machinery Selection in Iowa Under Variable Weather Conditions,* Iowa State University Cooperative Extension Service and Agricultural Experiment Station Special Report 85, March 1980.

TABLE 16-3. Size and Composition of Three Alternative Machinery
Sets for Iowa Farms

Selected Components	Machinery Sets		
	4-Row Small	6-Row Large	8-Row Extra Large
First tractor	85 hp	125 hp	145 hp
Second tractor	55 hp	75 hp	85 hp
Moldboard plow	4 × 16 in.	6 × 16 in.	7 × 16 in.
Planter	4 × 30 in.	6 × 30 in.	8 × 30 in.
Combine	75 hp	100 hp	125 hp
Corn head	2 × 30 in.	3 × 30 in.	4 × 30 in.
Soybean head	10 ft	13 ft	15 ft

SOURCE: W. Edwards and M. Boehlje, *Farm Machinery Selection in Iowa Under Variable Weather Conditions*, Iowa State University Cooperative Extension Service and Agricultural Experiment Station Special Report 85, March 1980.

The costs per acre of the larger machinery sets continually decrease over the 100 to 1000 acre range, but at decreasing rates with increasing acreage. The larger machinery sets permit the operations to be performed on time or in season. This capability becomes particularly important in certain years when weather conditions reduce the number of normal working days. In other words, the larger machine may permit flexibility, an important concept in uncertain circumstances, as discussed in Chapter 8. The additional cost of the larger machinery is offset by the insurance it provides against the years when bad weather disrupts field operations. The larger machinery also affords flexibility in the event that the manager wishes to acquire additional land.

TABLE 16-4. Average Machinery Costs per Acre for the
Four-Row Small Machinery Set for Iowa Farms

	Farm Size	
	100 Acres	1000 Acres
Fixed costs	$233.79	$ 24.31
Fuel and lubrication	12.33	6.70
Repairs	1.84	9.31
Labor	9.25	9.25
Timeliness	1.83	147.22
Tax savings (−)	45.53	93.21
Total	$213.52	$103.58

SOURCE: W. Edwards and M. Boehlje, *Farm Machinery Selection in Iowa Under Variable Weather Conditions*, Iowa State University Cooperative Extension Service and Agricultural Experiment Station Special Report 85, March 1980.

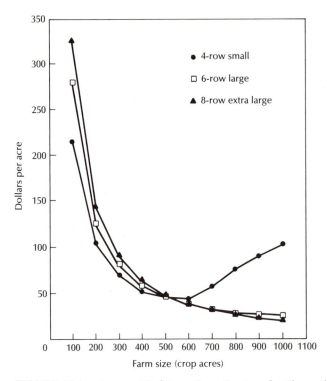

FIGURE 16-1. Average Machinery Costs Per Acre for Three Alternative Machinery Sets on Iowa Farms.

SOURCE: W. Edwards and M. Boehlje, *Farm Machinery Selection in Iowa Under Variable Weather Conditions*, Iowa State University Cooperative Extension Service and Agricultural Experiment Station Special Report 85, March 1980.

From 100 to 500 acres, the smallest machinery set (four-row small) has the lowest average cost of the three machinery sets. The cost curves of the four-row small and the six-row large machinery sets intersect between 500 and 600 acres. The six-row large machinery set is the least-cost alternative for 600 to 700 acres, and then the eight-row extra large set has the lowest cost for 800 to 1000 acres.

Machinery sizes larger than those with the lowest cost per acre can be justified for other reasons, such as limited availability of labor, extreme weather variability, or rapid crop deterioration. Larger machines require less labor per acre and cut timeliness losses during the squeeze of harvesting and planting periods in years when weather is particularly bad. Thus, risk-averse farmers may prefer larger machines.

The Iowa study suggests that total machinery costs, including timeliness losses, do not vary from year to year as much as might be expected. The variations were relatively small except when the smaller sets of machinery were used on the largest acreage levels. So, a smaller than optimal set of machinery may not expose a manager to a significantly higher level of risk.

To help choose the optimal size of a farm's set of machinery, it is useful to construct a calendar of operations. This calendar will indicate the tasks to be performed in a particular month and the amount of time required. Reference to weather records will establish the number of days that are normally suitable for performing these operations. The adequacy of different machine sizes in completing these operations can then be assessed by comparing the number of days required and the number of days normally suitable. Consideration should also be given to those years when unfavorable weather will limit available time more than is normally the case. The calendar of operations can also be used to show the advantage of night work during those periods when timeliness of operation is important.

Economizing on Machinery Costs

One of the major problems associated with machinery ownership is keeping machinery costs per acre at competitive levels. Farmers realize that it is necessary to spread the fixed costs of owning machinery over a considerable number of acres, because the variable costs per acre of operating machinery remain relatively constant regardless of the number of acres farmed. As a result, average machinery costs generally decline rapidly until a considerable volume is produced.

Many farmers tend to invest heavily in modern farm machinery without being able to expand production sufficiently to justify the added investment. Research shows that overinvestment tends to decrease as the acreage farmed increases. The implication is that if small operations are to compete with larger ones, they must either turn to custom operation or increase the acreage farmed. It may be impractical for some farmers to increase their volume of production greatly, yet they would like to enjoy the advantages of using modern machinery. There are a number of ways in which they can realize these advantages.

Hiring Custom Services

One way to economize on machinery costs is to have certain work done on a custom basis. On those farms where the jobs require a small amount of time and where the machinery is expensive, hiring custom work may be justified. Again, timeliness of operation is important. The local situation must be considered in this light. One common complaint about custom work is that the custom operator may not be available when the work should be done. Although this problem may occur, consideration should also be given to whether the work would be performed in a more timely manner if the machine were owned. Owning a machine does not ensure timeliness of operation, because other enterprises may compete for the same machine. Again, the relative costs and returns can be estimated by the use of the partial budget.

An example of the type of analysis that can be used is illustrated in Table 16-5.[2] The relative costs of owning a hay baler are compared to those of custom

[2] J. Plath and W. Ford, *Owning Versus Custom-Hiring Hay Harvesting Machinery, Columbia Basin, Washington*, Washington State University Extension Bulletin 1265, March 1984.

TABLE 16-5. The Average Annual Costs of Owning
and Operating a Hay Baler

Fixed costs, annual	
Depreciation	$1917
Interest	1613
Taxes	194
Insurance	65
Housing	204
Total annual fixed costs	$3993
Variable costs per ton	
Repairs	$0.40
Twine	1.60
Labor	0.55
Variable tractor costs	1.25
Total variable costs (per ton)	$3.80

SOURCE: J. Plath and W. Ford, *Owning Versus Custom-Hiring Hay Harvesting Machinery, Columbia Basin,
Washington*, Washington State University Extension Bulletin 1265, March 1984.

hiring. In order to determine the number of tons necessary to justify owning a
baler, it is necessary to know the cost of custom work. In this example, the cost is
$12.50 per ton. To arrive at the number of tons necessary to break even:

$$\$12.50 - \$3.80 = \$8.70/ton$$

$$\frac{\$3993}{\$8.70/ton} = \text{approximately 460 tons}$$

The essentials of the preceding problem can be understood more readily by
referring to Figure 16-2. The average fixed cost curve is obtained by dividing the
total annual fixed costs of $3993 by the number of tons of hay. Obviously, fixed
costs per ton of hay decline as the tonnage increases. The average fixed cost plus
the average variable cost per ton gives the average total cost per ton. The average
total cost curve intersects the cost of custom work at 460 tons. At this point, the
costs of the two are equal. Of the $12.50, $8.70 is fixed cost and $3.80 is variable
cost. When hay production is greater than this amount, ownership is less costly
than custom work. With less production, custom work is less costly.

Before deciding to hire custom service, farm managers should consider some of
the advantages and disadvantages, not all of which can be measured and budgeted.
One major advantage of custom service is that ownership costs are eliminated and
the investment capital thus saved can be used elsewhere. Another advantage for
certain jobs is that the custom operator may furnish skilled labor and specialized
machines that may not be readily available to the farmer.

For large jobs, it is usually cheaper for farmers to own the machine, especially
if they cannot make good use of the labor they release by hiring custom service.

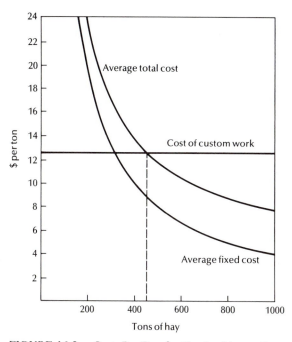

FIGURE 16-2.   Costs Per Ton for Owning Versus Custom-Hiring a Hay Baler.

SOURCE: J. Plath and W. Ford, *Owning Versus Custom-Hiring Hay Harvesting Machinery, Columbia Basin, Washington,* Washington State University Extension Bulletin 1265, March 1984.

Although custom operators generally do good work, occasionally the quality may be poor, resulting in yield losses or reduced product quality. Also, the use of custom operators may entail some risk of carrying noxious weeds or diseases from farm to farm.

To prevent any misunderstanding of the terms and conditions of an agreement to do custom work, there should be a written contract. The contract should specify the rate to be charged for the work done (such as dollars per hour, per acre, or per bushel); how the work is to be done; where and how much work is to be done; when the work is to be performed or completed; and when payment will be made. If different rates are used, based upon yield for example, this fact should be specified in the contract. Minimum standards and incentives based on quality should also be explained in the contract. The written contract need not be a long, complicated document, but it should contain sufficient detail so that both those doing custom work and those hiring it completely understand the terms and conditions of the agreement.

Other Economizing Measures

Joint ownership is another way to reduce machinery investment costs. Two or more farmers may buy an expensive piece of equipment if the volume on any one farm is insufficient to justify sole ownership. For this arrangement to be successful, certain rather obvious conditions must exist. First, there must be an attitude of mutual trust between the participants. Agreement must be established at the outset as to priority in the event that a conflict arises over who will use the equipment when. There should also be a prearranged procedure for dissolving the arrangement in an orderly manner if future circumstances so dictate. Properly conceived, joint ownership of certain pieces of equipment can result in an efficient use of capital resources.

Some machinery dealers offer short-term rentals. This arrangement may also be desirable for farmers if the machine is expensive and they do not have a significant need for it. For example, some grain producers rent trucks to haul harvested grain from the field to storage. The trucks may be needed for only three or four weeks. Although the rental cost may be high, it may be less than cost of year-round ownership. The analysis to compare the rental rate with the cost of ownership is similar to the analysis of the preceding custom hiring problem.

Machinery Replacement Decisions

When to replace machinery is probably one of the most difficult machinery decisions. Often replacement investments are made according to established schedules, or they are postponed until they become "must" expenditures. A study of Washington farmers found that the most frequently mentioned rule for replacing old machinery was whenever the old machine could no longer be depended upon to perform the required job.[3] Another important factor affecting replacement is the level of net farm income. To predict sales, farm machinery manufacturers keep a close watch on farmers' net incomes.

Following are some reasons for replacing farm machinery:[4]

1. Ownership and operating costs are greater than alternatives; repair costs are increasing. The total cost of owning and operating the machine exceeds that of a newer machine.
2. The current machine is not dependable. Reliability involves economic and risk considerations, especially when timeliness is important. The costs of breakdowns include not only the parts, service, and labor for repairs but also the reductions in quality and quantity of crop yield.
3. The machine is obsolete. Due to new developments or improvements in machine or farming practices, the current machine is no longer the best one to do the job.
4. Machinery capacity is too small; the acreage has increased, the crops have changed, or timeliness is a more important factor. The capacity of the current machine is not adequate to meet the increased requirements, and operations cannot be completed on time.

---

[3] D. E. Umberger, N. K. Whittlesey, and M. E. Wirth, *Machinery Investment Practices of Washington Farmers*, Washington Agricultural Experiment Station Bulletin 737, August 1971.

[4] Adapted from W. K. Waters and D. R. Daum, *Farm Machinery Management Guide*, Pennsylvania State University Extension Service Special Circular 192, no date.

5. Tax implications. There may be tax advantages associated with replacing the old machine with a new one, particularly in a high-income year.
6. The farmer wants a new machine and can afford it. Noneconomic considerations, such as the prestige of having new machinery, also affect replacement decisions.

The annual total costs of owning and operating a machine vary over its life. Generally, the total annual costs decrease during the first years of the machine's life and then increase as it gets older. Early in the life of a machine, the annual costs decrease with use because of decreasing ownership costs, mainly depreciation. However, as the machine accumulates hours of use, the repair rate increases; at some point, the increasing repair and down-time costs will more than offset the decreasing ownership costs. At this point, the annual costs of owning and operating the machine start to increase, and consideration should be given to replacing it.

The replacement decision is one of finding the economically optimal time in the life of the old machine to make the investment to replace it. The problem is more easily understood in terms of like-for-like replacements, that is, when the investment is for a new machine that will do the same job as the discarded machine and only the costs of owning and operating the two machines are considered. However, in the analysis of the replacement problem, it is not necessary for the replacement to be a duplicate machine. Added and reduced revenues can be considered, as well as added and reduced expenses. Considering the typical cost relationship, the like-for-like replacement decision becomes one of whether it would be less costly to replace the machine now or to defer replacement until some future date.

The first step in deciding whether to replace is to estimate the costs of owning and operating the current machine for the next year. This process involves projecting how much the machine will depreciate in value next year, the interest on its current value, and repairs that will be required next year, as well as the other costs (see the worksheet in Table 16-1). Comparing this total amount to the projected average annual cost of a new machine will indicate which is less costly. If the replacement will do the same work as the current machine, the current machine should not be replaced as long as its estimated cost for the next year is less than the average annual cost projected over the life of the new machine.

In analyzing this comparison, the decision against replacement should not imply that the old machine is to be kept for several years. It is reasonable to expect that the old machine could again be considered for replacement one year from now at the latest. The improper comparison gives erroneous results, usually in favor of replacing now.

Careful analysis requires budgeting the costs of replacing a machine now with the alternatives of waiting for one or more years. For example, if the analysis shows that the average annual cost of the new machine would be less than next year's cost of the current machine, the best decision may not be to replace now. The analysis is not complete until other alternatives have been considered. The alternative of waiting for more than one year should also be evaluated. Because of fluctuating repair costs, it is possible that the alternative of waiting for two years would be less costly than replacing now. For example, if a major overhaul is expected next year, repair costs should be considerably lower the following year.

Uncertain future costs complicate the comparison of alternatives, but farmers who are familiar with their machinery can anticipate future repairs more accurately. If farmers knew the actual cost of repairs ahead of time, it would be easier to make machinery replacement decisions. Records of past repairs sometimes help in predicting future repairs. Keeping a record book for each machine and recording data on usage, repairs (parts and labor), fuel consumption, and hourly cost can be useful in deciding when to trade and for budgeting other decisions.

When should replacement be considered? Because of fluctuating repair costs, no general rule can be established. As a guideline, however, replacement should be considered after a machine has reached half of its normal life—except, of course, for "lemons," which should be considered for replacement immediately. After the machine has reached half of its normal life, replacement should be considered yearly or when major repairs are anticipated.

There are other important factors to consider in making machinery replacement decisions. One example is reliability. As a machine ages, its reliability decreases. So, in critical operations such as planting and harvesting, when timeliness losses are potentially high, trading sooner may be justified. A new machine will probably be more dependable than the current one.

Obsolescence may also indicate that it is time to replace the current machine. Farm machinery becomes obsolete as technology advances. A new machine may boost production by improving weed control, reducing fuel consumption, or increasing field efficiency. Such benefits should be evaluated by the manager in comparing the costs of the current machine with those of the new one.

The Iowa data presented earlier illustrate how machinery costs are affected when farm size increases. Even though the machinery is still relatively new and dependable, it may be cost effective to replace it with larger machinery as a result of an increase in the number of acres farmed.

Income taxes may also influence replacement. Farmers in higher income tax brackets should consider replacing machinery more frequently. However, taxes should not be a major factor. The objective should be to increase after-tax income rather than to reduce taxes. Replacing machinery too frequently can be a very expensive way to reduce income taxes.

Other factors, such as credit availability, interest rates, inflation, and personal preference, may affect machinery replacement decisions. Although these decisions are complex, they are important, involving large expenditures and having significant implications for timeliness, labor efficiency, and crop yields.

New Versus Used Machinery

Farmers are frequently faced with the decision of whether to buy new or used machinery. Again, the correct decision depends on an individual farmer's circumstances. There are, however, certain considerations that must be taken into account on any farm. Purchasing used machinery is one method of reducing the investment in machinery. Some farmers do not have the problem of limited capital, but many, especially beginning farmers, do. Buying used machinery may be a more efficient use of their limited capital.

In general, buying used rather than new equipment results in higher variable costs relative to fixed costs. Depreciation, interest on the investment, and insurance are higher with new equipment. Repairs, fuel and lubrication costs, and labor probably are higher with used equipment. Used equipment generally is less dependable than new, and breakdowns may reduce yields either in quality or in quantity.

Because breakdowns are more likely with used equipment, the farmer's ability as a mechanic is a factor to be considered. Skillful mechanics may be able to operate used equipment without many delays. Repairing and servicing machines in the off season is also important. If trouble can be spotted in advance, costly delays during the busy season can often be avoided.

New machines are often more convenient. They may have features that may not add greatly to output but that make them more enjoyable to operate. Also, satisfaction is gained from using a piece of new equipment; put another way, the prestige value of new equipment is greater.

A simple example may highlight the principal components of making a cost comparison. Assume that a farmer is considering two combines, one used and one new. The used machine is expected to do the same job as the new one but will not last as long. The cost comparison can be illustrated as follows:

	New	Used
Original cost	$100,000	$60,000
Life	Ten years	Six years
Annual depreciation	$ 9,800	$ 9,600
Interest (12%)	6,120	3,744
Repairs	2,000	3,600
Insurance	200	140
Taxes	740	460
Total annual cost	$ 18,860	$17,544

In this example, the used machine seems to have the advantage. If interest is ignored, the new machine costs less than the used one, primarily because of the lower repair costs. However, if the opportunity cost of the additional capital invested in the new machine exceeds 5.4 percent, the used machine is less costly.

There are some risks involved in purchasing used machinery: There is no way to be certain of its condition or dependability. One way to assess this risk is to determine why the unit is for sale. Used machinery is usually put on the market because (1) it no longer has the dependability required by the owner or (2) it no longer meets the needs of the owner's operation (the farm has been expanded or crops have been changed). The first reason should cause the buyer to be cautious; the second may indicate that the machine is still reasonably dependable and carries a low risk. Reputable sources, such as the local farm machinery dealer or a neighbor, are usually low-risk sources of used farm machinery. Auctions and other sales where there is little opportunity to inspect the machine should be avoided.

Machinery Leasing

Farm machinery and equipment may be acquired by purchasing (often with borrowed funds) or by financial leasing, which provides a way for farmers to acquire the use of a capital asset without owning it. In recent years, machinery leasing has generated more interest, although it is still not widely used.

There are basically two types of leases for depreciable farm assets: operating leases and financial leases. Under an operating lease or rental agreement, the farmer obtains the use of the equipment for a relatively short period of time, for example, a tractor for three months. The rental cost is based on an hourly, daily, weekly, or monthly rate. Maintenance and repair costs are typically borne by the lessor, and purchase options are not available to the lessee.

A financial lease should not be confused with short-term renting, or an operating lease. *Leasing*, as used in this section, refers to a relatively long-term arrangement (one year or more) that gives the lessee exclusive use of the leased property over the period of the lease. Thus, the lease is a method for acquiring long-term control of property. It is a contract that essentially provides financing to the lessee. The contract (lease) is usually written for the useful (depreciable) life of the asset being leased, and the lessee is normally, but not always, responsible for repairs, maintenance, and insurance. The lessor receives the income tax benefits associated with owning the asset, but the lessee can deduct the lease payments.

Some financial leases permit the lessee to purchase the asset at the end of the lease period, but only at a fair market value determined when the lease ends. The disadvantage of this type of lease is that the lessee does not share in any gain in the value of the asset resulting from inflation. Other lease arrangements have a fixed purchase price option at the end of the lease—for example, 10 percent of the original cost of the asset.

Farmers expanding their operations might lease rather than purchase machinery because of credit limitations and high interest rates. Also, farmers with heavy debt loads might sell their machinery, use the proceeds to pay off some of their debts, and then lease the machinery from the purchaser. Favorable tax treatment of investments increases the interest in, and availability of, financial leases for machinery. Investors see farm machinery as providing an opportunity for a highly leveraged investment that generates tax benefits.

For farmers, leasing may have the advantage of reducing cash flow requirements. The first lease payment is normally less than what the down payment would be if the asset were purchased. Also, lease payments may be lower than the principal and interest payments associated with purchasing. This situation is possible when the lessor passes the tax benefits on to the lessee in the form of lower lease payments. Through financial leasing, farmers can access capital that would otherwise be unavailable to them.

A large number of companies offer leasing services to farmers. Some of these companies lease only products manufactured by their parent companies. Others lease products produced by a variety of manufacturers. Some leasing companies are locally owned and managed, and others are part of national or regional businesses. Examples of businesses offering leases on agricultural machinery and equipment include equipment manufacturers and dealers, commercial banks, production credit associations, and various firms established specifically to offer leasing services.

Although virtually all types of depreciable agricultural assets might be financed through leasing, machinery and equipment are the items most often leased at the present time.

An Example

The appropriate method for comparing the costs of leasing with those of purchasing machinery is net present value analysis (Chapter 7). For this example, the farmer has the alternatives of leasing a $50,000 machine or purchasing it. The after-tax discount rate is assumed to be 10 percent.

The net present value of cash outflows for purchasing the machine are shown in Table 16-6. For this example, the farmer is assumed to have a 32 percent marginal tax rate. The tax savings are based on the investment tax credit and the deductions for depreciation. The after-tax cash outflows are multiplied by the 10-percent discount factor to arrive at the net present value of the cash outflows. For this example, this amount is $34,059.

The net present value of the cash outflows under the lease arrangement are shown in Table 16-7. The lease runs for five years, with annual prepaid lease payments of $10,000. At the end of the lease, the farmer can buy the machine for $5000. The tax savings result from the tax deduction of the lease payments. When the after-tax cash flows are discounted back to the present, the net present value amounts to $32,673, or $1386 less than the purchase alternative. Under these conditions, it would be less expensive to lease.

The cash flow implications of leasing rather than purchasing can be examined by comparing the after-tax cash flows of the two alternatives. The cash outflows for purchasing include the down payment, principal and interest payments, and tax savings associated with financing the purchase. At the beginning of the first year, the lease alternative requires only $10,000 as a prepaid lease payment, com-

TABLE 16-6. Net Present Value of Cash Outflows for Purchasing a $50,000 Machine

Year	Purchase Cost	Income Taxes[a]	After-Tax Cash Outflow	Discount Factor[b]	Present Value
0	$50,000	$ 0	$50,000	$1.0000	$50,000
1	0	−7,280	−7,280	0.9091	−6,618
2	0	−3,344	−3,344	0.8264	−2,763
3	0	−3,192	−3,192	0.7513	−2,398
4	0	−3,192	−3,192	0.6830	−2,180
5	0	−3,192	−3,192	0.6209	−1,982
				Net present value	$34,059

[a] Income tax savings include an investment tax credit at 10 percent of the machinery cost and depreciation deductions at a 32 percent marginal tax rate.
[b] Factors for 10 percent interest are derived from Appendix Table 2 (Chapter 7).

TABLE 16-7. Net Present Value of Cash Outflows for Leasing a $50,000 Machine

Year	Lease Payments	Income Taxes	After-Tax Cash Outflow	Discount Factor[a]	Present Value
0	$10,000	$ 0	$10,000	$1.0000	$10,000
1	10,000	− 3,200	6,800	0.9091	6,182
2	10,000	− 3,200	6,800	0.8264	5,620
3	10,000	− 3,200	6,800	0.7513	5,109
4	10,000	− 3,200	6,800	0.6830	4,644
5	5,000[b]	− 3,200	1,800	0.6209	1,118
				Net present value	$32,673

[a] Factors for 10 percent interest are derived from Appendix Table 2 (Chapter 7).
[b] Purchase price at 10 percent of the original purchase price.

pared to the $12,500 down payment required for purchasing. However, the after-tax cash outflows for the next three years are higher for the lease option.

Conclusion

Leasing, in this example, is a more costly way to acquire the use of a machine. However, leasing may not always be more expensive, especially if manufacturers are promoting leasing to market more machinery, if interest rates are high, or if the farmer has a low marginal tax rate and cannot use the tax benefits of ownership. Also, the lease payments may be more easily accommodated by the cash flow available than payments to purchase the machinery.

Machinery leasing is getting more attention. In some cases, it can be advantageous to both the lessor and the farmer. Thus, leasing is an option that farmers should consider when evaluating the feasibility of adding or replacing machinery and equipment. A well-structured leasing plan can improve cash flow situations and reduce the costs of acquiring machinery and equipment. Leasing is likely to be more attractive to farmers in low income tax brackets and with higher opportunity costs for capital. In any event, before entering into a leasing agreement, farmers should consider the profit, cash flow, and tax implications of the lease agreement.

Summary

Machinery costs account for a major portion of the expenses on most U.S. farms. Thus, machinery management decisions related to machinery acquisition, size, and replacement have important implications for the profitability and success of the farm business. Machinery costs are usually categorized as either ownership costs (depreciation, interest on the investment, taxes, insurance, and housing) or operating costs (fuel, lubricants, repairs, and labor). Annual ownership costs tend to be fixed regardless of the amount of use the machine receives; annual operating costs, on the other hand, are variable and increase in direct proportion to use.

The problem of selecting machinery of the appropriate size for a particular farm involves analyzing the tradeoffs between the lower labor and timeliness costs and the higher ownership costs. Although larger machines cost more to own than smaller ones, they cover more acres per hour, thus reducing labor costs and improving timeliness. Timeliness is the ability to complete operations at the optimal time in order to minimize yield and revenue losses. When selecting machinery, these losses due to untimely operations should be considered, along with the ownership, operating, and labor costs of the alternative machinery sets.

Because of the magnitude and lumpiness of machinery costs, it is important for farmers, particularly those with smaller operations, to consider hiring custom services to economize on their use of machinery. In addition to the costs of owning the machine relative to custom hiring, factors such as timeliness, quality of work, the opportunity cost of the farmer's labor, and requirements for special skills and machinery should be included in the analysis of this decision. Joint ownership of machinery by two or more farmers and short-term rentals are other possibilities for economizing.

Machinery replacement decisions are complicated by the many considerations involved: increasing costs for the current machine, reliability, obsolescence, a change in the farm operation, tax implications, and noneconomic factors. The current machine should be kept if its projected cost for the next year is less than the average annual cost over the life of a similar replacement. To justify replacement, the average annual cost of the new machine must be less than the alternatives of keeping the current machine for one, two, or three years. The additional comparisons must be made because of the year-to-year variability of repair costs.

Purchasing used machinery is a viable alternative for farmers with mechanical skills, but there are risks involved. The risk of buying a "lemon" can be reduced by knowing the machine's history and buying from a reputable source. Again, the way to analyze this decision is to construct a budget consistent with economic principles, using data from farm records.

Leasing machinery is an alternative to purchasing it. Long-term leases can be viewed as another way to acquire control of machinery and equipment. The lease-or-purchase decision should be evaluated using net present value analysis. Farmers facing high interest rates, low marginal tax rates, tight cash flows, and favorable lease rates are more likely to find leasing advantageous.

Recommended Reading

STEVENS, D. M., and D. E. AGEE. *Using Farm Machinery Efficiently.* University of Wyoming Agricultural Experiment Station Bulletin 482R, September 1977.

CHAPTER 17

Animal Production Management

Managers of animal production enterprises must make decisions about the feeding, breeding, and health maintenance programs to be established; the level of investment to be made in facilities; marketing; and so forth. Managing animal enterprises begins with enterprise budgeting, the first topic of this chapter. The considerations involved in selecting animal enterprises are presented next, followed by a discussion of the decisions involved in the management of animal enterprises.

Animal Enterprise Budgets

Similar to crop enterprise budgets, animal enterprise budgets summarize the revenue and expenses for the year and help managers to plan for the future, to assess current profitability, and to secure credit. These budgets also help them select enterprises and production systems from among various alternatives. For example, a farmer might develop two enterprise budgets to compare two systems for producing the same animal species.

It is important to specify the unit of analysis in the budget. Animal enterprise budgets are usually prepared on a per-head basis. For example, a swine enterprise budget might be prepared on a per-sow or per-litter basis; a beef cow-calf enterprise budget might provide estimates for one beef cow; or the budget might be for one ewe or one hen. In some cases, the unit of analysis might be consistent with a logical enterprise size, for example, 100 beef cows, 48 sows, or 10,000 laying hens.

The animal enterprise budget, typically prepared for one year, indicates the resources required, the products produced, and the prices of the products. Microcomputers can facilitate budget preparation, and the data for constructing the budget can be gathered from past records, from typical budgets provided by university extension services, and from other animal producers and advisers with knowledge of feed requirements, expected rates of gain, labor requirements, and so forth.

Preparing Animal Enterprise Budgets

The economic principles discussed in Chapter 2 should be used to guide the budgeting process, and a good understanding of the production process should be applied as well. For example, an understanding of how the nutrient requirements of animals change during their lives is basic to the application of economic principles in determining feed requirements.

Budgets for animal enterprises follow the same general format as crop budgets, but there are some differences. With animal enterprises, there are often multiple products, such as calves and cull cows for a beef cow enterprise. For a dairy enterprise, the multiple products might be milk, calves, and cull cows. Another difference is the inclusion of the costs of raised or purchased replacement animals to maintain the breeding herd over time. Also, farm-produced inputs used in the enterprise must be valued. Valuing farm-produced feed, for example, is often complicated, because pasture and crop residues may have little or no value if not used by the animals.

The title of the budget should describe the enterprise. An example of a format that might be used is presented in Table 17-1.[1] The six sections of the budget—revenue, operating expenses, gross margin, ownership expenses, total expenses, and return to management—are explained in the following sections.

Revenue.   To complete the revenue section of the budget, the commodities produced by the enterprise are listed first. These commodities include the animals produced for sale, as well as animal products such as milk and wool. The output-per-unit estimates of milk and wool are typically based on historical data and trends.

Depending on the enterprise, estimating animal production may be a little more complicated. The dairy budget in Table 17-1 is an example. The number of cull cows sold depends on the replacement rate and the death loss. If 30 percent of the dairy cows are replaced each year and the death loss is 4 percent, then there is 0.26 cull cow available for sale each year for each cow in the herd (0.30–0.04 = 0.26). To convert this amount to pounds of cull cows sold, the average weight per animal sold is multiplied by 0.26 to give the number of pounds sold per cow in the herd.

The expected number of bull calves (B) born annually per cow is calculated using this formula:

$$B = 0.5 \left[\frac{12P}{C} + R \right]$$

where P is the proportion of cows that produce a calf each year,
C is the average calving interval (months), and
R is the proportion of dairy cows replaced each year.

---

[1] This example is adapted from B. M. Buxton et al., *Milk Production: A Four-State Earnings Comparison.* U.S. Department of Agriculture Agricultural Economics Report No. 528, February 1985.

TABLE 17-1. Sample Enterprise Budget for Dairy Production, Pacific Northwest, 1986

Item	Units	Price per Unit	Quantity	Value or Cost per Cow
1. *Revenue*				
Milk sales	cwt.	$ 14.65	165.0	$2417
Cull cow sales	pounds	0.44	336.0	148
Bull calves	head	75.00	0.48	36
Replacement heifers	head	1250.00	0.16	200
Manure production	head	38.00	1.0	38
Total Revenue				$2839
2. *Operating Expenses*				
Hay	tons	109.00	6.6	$ 719
Concentrates	tons	168.00	3.4	571
Other feed	head	16.00	1.0	16
Pasture costs	head	56.00	1.0	56
Repairs	head	96.00	1.0	96
Vet., med., breeding	head	61.00	1.0	61
Dairy supplies	head	34.00	1.0	34
DHIA testing	head	13.00	1.0	13
Milk hauling	cwt.	0.48	165.0	79
Hired labor	hours	6.00	31.0	186
Miscellaneous	dollars	81.00	1.0	81
Oper. capital interest	dollars	0.12	166.7	20
Total Operating Expenses				$1932
3. *Gross Margin (Return Above Operating Expenses)*				$ 907
4. *Ownership Expenses*				
Deprec. on facilities	head	202.00	1.0	$ 202
Int. on facilities	dollars	0.12	1350.0	162
Interest on cattle	dollars	0.12	2000.0	240
Taxes and insurance	head	44.00	1.0	44
Operator labor	hours	10.00	18.0	180
Land charge	head	70.00	1.0	70
Total Ownership Expenses				$ 898
5. *Total Expenses*				$2830
6. *Return to Management*				$ 9

For example:

$$B = 0.5 \left[\frac{12\,(0.8)}{(13)} + (0.3) \right] = 0.52$$

Subtracting a death loss of 7.5 percent gives the number of bull calves for sale, or 0.48 per cow.

The expected number of heifer calves born annually per cow is calculated in the same way. The production system budgeted in Table 17-1 assumes that heifers are not sold until they are two years of age, and then only those not needed for herd replacement are sold. To find the number of two-year-old heifers for sale, the death loss must first be subtracted from the 0.52 heifer calves born per cow. From birth until two years of age, the death loss is estimated to be 11.5 percent, which leaves 0.46 head. Next, the 0.30 heifers selected for replacement must be subtracted, making the replacement heifers available for sale equal to 0.16 head.

Although the data and the calculations would differ, the approach for estimating the number of any other animal species for sale would be the same. For example, a farmer with a hog production enterprise must consider the number of litters farrowed per sow per year, the average number of pigs weaned per litter, the mortality rate between weaning and sale, the rate of replacement for sows, and the death loss of sows.

The average weights and prices for all animals and animal products sold must be estimated to calculate the revenue. These prices should be realistic estimates based on past trends and current expectations.

A by-product of animal enterprises is manure. If this manure can be substituted for the fertilizer used in crop production, it may have some value that should be included in revenue.

Operating Expenses.   The procedure for listing operating expenses for animal budgets is essentially the same as for crop budgets. These are the expenses that vary with the size of the enterprise. To estimate the cost of feed, a major expense for most animal production enterprises, the quantities required for the various types and weights of animals must be determined. All purchased feed should be priced at its estimated cost, and farm-produced feed should be priced at its opportunity cost. For example, the price placed on farm-produced corn for a feeder calf is the net market price that would be received if the corn were sold. Pasture cost estimates can be based on the opportunity cost of renting the pasture to another farmer or rancher. Alternatively, as is done in the dairy budget (Table 17-1), the cost of maintaining the pasture can be included in the operating expenses and a land charge included in the ownership expenses.

The operating expenses also include veterinarian and medical costs, breeding fees, marketing costs, utilities, and other expenses the operator would avoid if the enterprise were discontinued. There is usually some machinery operating expense associated with some aspects of animal production, such as feed hauling and manure handling. Building repairs that result from using the buildings to house animals should also be listed in the operating expense section.

The costs of feeder pigs and feeder calves purchased for resale are included in the operating expenses. The expenses of purchased breeding stock can be handled in one of two ways:

1. The purchase of breeding stock may be included as an operating expense. Then, when cull animals are sold, the proceeds from selling purchased breeding stock are included in revenue, just like those of raised breeding stock.

2. Depreciation can be charged for purchased breeding animals and included as an ownership expense. If this is done, the revenue from cull sales is not included unless there is a gain (or loss) over the book value (cost less depreciation). In other words, the expenses for breeding stock are handled just as they are for machinery.

The best method to use depends on whether purchased breeding animals are a major expense and on the degree of difficulty of each method. If breeding stock is mostly purchased, it is more appropriate to charge depreciation. If only a few bulls, for example, are routinely purchased each year, then it is easier to treat their purchase cost as an expense.

The cost of hired labor, usually considered an operating expense, should be based on the wages and fringe benefits paid per hour. During the year, operating capital is needed to purchase feed and to pay other operating expenses. Even if these inputs (such as feed) are raised, a cost is incurred for the operating capital tied up in inventories. The interest on operating capital is the opportunity cost of that capital multiplied by the average amount of operating capital required over the year.

Gross Margin. The gross margin for the enterprise is the difference between the revenue and the total operating expenses. For the dairy enterprise budget shown in Table 17-1, it is $907 per cow. The gross margin represents the contribution to the payment of ownership expenses and to profitability, which is measured as the return to management.

Ownership Expenses. Ownership expenses for animal production include the ownership costs (depreciation, interest, taxes, and insurance) for facilities such as buildings and equipment. These expenses are calculated in the same way as they are for machinery used in crop production (Chapter 15); that is, they are based on the new or replacement cost, salvage value, and useful life. If the facilities are shared by more than one enterprise, then the expenses should be prorated among the enterprises, based on their relative use of the facilities.

The operator's labor and unpaid family labor used in the enterprise should be estimated and valued at its opportunity cost, or what could be earned by using this labor elsewhere.

Budgets for animal enterprises often include a section for listing the average number of each type of animal in the breeding herd; that is, there are separate lines for mature females, replacement females, and males. This procedure further describes the enterprise and provides estimates of the investment in breeding stock, so that the appropriate ownership costs can be estimated.

The ownership expenses associated with a breeding herd normally include interest on the investment in the animals, property taxes, and insurance. For dairy, beef, sheep, and swine breeding herds in which replacement females are raised and the age distribution is relatively constant, the costs of raising replacements (feed, veterinary expense, and so forth) are normally included in the operating expense section. Thus, no depreciation is charged for the raised breeding stock. Interest, insurance, and personal property taxes are calculated as they are for machinery.

If replacement breeding animals are purchased rather than raised, a charge for depreciation is included in the ownership expenses unless the purchase cost of the replacement has been included as an operating expense. Estimating depreciation for purchased breeding stock is analogous to estimating depreciation for machinery; that is, it is found by dividing the difference between the cost of replacements and the sale price of cull animals by the number of years the animals are kept in the herd.

The charge for any owned land used by the animal enterprise can be calculated as it is for crop enterprises (Chapter 15).

Total Expenses. Total expenses, the sum of the operating and ownership costs, can be used to calculate the average total cost of the commodity produced (Chapter 2). For animal enterprises, this calculation is often complicated by the multiple products produced, which for milk production are cull cows, bull calves, heifers, and manure. To find the average cost per hundredweight of milk produced, the total net expenses for producing milk must be calculated first. The revenue from the other products is subtracted from the total expenses. For the dairy example (Table 17-1), this difference is $2408 per cow ($2830–$422). The estimated average cost of milk produced is the total net expenses divided by the estimated milk output per cow. This amount would be $2408 divided by 165 hundredweight, or $14.59 per hundredweight.

When using this enterprise budget to make decisions about whether to expand or reduce the size of the enterprise, this estimate of total expenses should be used cautiously. This total includes both fixed and variable expenses. That is, some of the expenses will change with the change in the size of the enterprise, whereas others will remain constant. For example, the interest on cows will be affected by a change in herd size. On the other hand, interest and depreciation on facilities would not be affected if the herd size were reduced or if it were expanded and excess capacity were available. The operating and ownership categories are not necessarily consistent with the distinction between fixed and variable expenses for a specific decision. When using the enterprise budget for decision-making purposes, each expense item should be examined to ascertain whether it is fixed or variable in relation to the decision at hand. An example is provided at the end of Chapter 5.

Return to Management. The return to management is the net return after all expenses are subtracted from revenue. Expenses include a charge for the capital invested in the enterprise and the cost per hour of labor used. A positive return to management indicates a profitable enterprise, provided the assumptions and estimates made in preparing the budget are accurate.

Factors Affecting Animal Enterprise Profits

Several factors affect the profitability of an animal enterprise, and they vary in importance from one enterprise to another. Before considering some examples of enterprise-specific factors, a relatively general factor should be discussed.

As described in Chapter 5, one of the primary factors in the analysis of a farm business is size, or volume; another is enterprise efficiency, which affects the profit (return to management) per enterprise unit. The enterprise should be large enough to (1) take advantage of modern technology and economies of size and (2) provide the operator with an adequate income. If the enterprise is not large enough to accomplish these two objectives, the manager should emphasize expansion—that is, of course, if the profit margin is positive.

On the other hand, if expenses are high and the profit margin is narrow or negative, expansion may only exacerbate the problem. In this case, the manager should focus on increasing efficiency. This is not the same as reducing total expenses; rather, it involves reducing the cost per unit of output. Reducing the cost per hundredweight of milk, beef calf, or market hog may involve an increase in the total expenses per cow or sow if the increase results in greater milk output per cow, a higher percentage calf crop per cow, or more pigs per litter.

Following are some enterprise-specific examples of factors affecting the profitability of beef and hog enterprises.

Beef Enterprises.   Three important factors affecting the profitability of a beef cow-calf enterprise are (1) the calf crop weaned as a percentage of cows, (2) the average weaning weight of calves produced, and (3) the average cost of maintaining the cow for the year. The following formula expresses the relationship of the cost per pound of calf produced to these three factors:

$$\frac{\text{Cost per cow}}{\text{Calf crop} \times \text{weaning weight}} = \text{cost per pound}$$

For example:

$$\frac{\$230}{0.85 \times 450 \text{ pounds}} = \$0.60 \text{ per pound}$$

The calf crop percentage must be consistent with the cost calculations; that is:

$$\frac{\text{Number of calves weaned}}{\text{Number of cows exposed}} = \text{calf crop percentage}$$

$$\frac{140 \text{ calves weaned}}{170 \text{ cows kept}} = 82 \text{ percent}$$

Two common mistakes are often made in calculating calf crop percentages. Figuring on the basis of cows calving rather than cows exposed (available for breeding) fails to take into account the death loss of cows prior to calving, cows losing calves during pregnancy, and cows that were not bred. Figuring on the basis of live calves born or those tallied at branding time misses the calves lost before weaning.

This method for analyzing production costs points out what a manager may need to do to reduce production costs in view of any given cattle price situation. If production per cow is too low in a particular ranch environment and operating costs are not too high, then an additional investment in management, nutrition, and selective breeding might result in a lower cost of production. If, however, the pounds of calf produced per cow are already high and the cost of production is still above the expected market price for calves, then the emphasis should be on reducing the cost per cow.

Good year-round management, proper nutrition practices, and disease prevention are all essential to getting a high calf crop percentage. Results from such efforts can usually be seen within a single year. Although the nutrition of the cow herd has a definite effect on weaning weights, breeding and selection are also important and may be less expensive.

Hog Enterprises. The data in Table 17-2, from records kept by 187 hog producers, demonstrate the wide variation in enterprise profitability and indicate the relative importance of various factors affecting profitability.

Feed is the major expense in hog production. Records show that feed cost is 55 to 70 percent of the total cost of producing a market hog. An increase of $0.50 in the cost per hundredweight of feed can add $2.00 to the cost of producing a hundredweight of pork. Labor, health, buildings, and equipment are next in importance after feed costs.

Production costs reflect management and efficiency: To make comparisons and locate areas where costs can be reduced, managers must keep records; identifying the problem areas, knowing where to reduce expenses, and emphasizing efficiency improvements are important aspects of good animal production management.

Selection of Animal Enterprises

Several factors are important in the selection of animal enterprises. The land, labor, capital, and management resources available will not be equally well adapted to all animal enterprises. Risk and size considerations will also narrow the list of possible alternatives.

Resource Availability

The availability of certain resources has a significant effect upon which animal enterprises are possibilities, just as it does upon crop enterprises. Farm acreage may be limited, with little possibility for acquiring additional land. Under such conditions, an enterprise that provides a high gross income per unit of land input, such as dairy or cattle feeding, should be considered for testing by the budgeting method. A cow-calf beef operation probably would not be appropriate unless there is a long pasture season.

Feed availability is, of course, related to land availability if the feed is farm produced. An abundant supply of low-cost feed, farm produced or purchased, gives an enterprise a competitive advantage. Enterprises that have a high grain requirement, such as cattle feeding, have an advantage when they are located near grain production areas. Conversely, cow herds are most profitable where low-cost forage is available.

The availability of labor resources is also relevant to the selection of an animal enterprise. A dairy operation, for example, has a rather high labor requirement

TABLE 17-2. Comparison of Farrow-to-Finish Hog Enterprises by Profit Margin,
187 Iowa Farms, 1983

Factor	High Margin, 62 Farms	Average for 187 Farms	Low Margin, 62 Farms
Profitability			
Profit margin per cwt. pork produced ($)	2.77	−5.21	−13.28
Hourly return to labor and management ($)	11.82	2.70	−4.26
Annual percent return on capital (%)	20.65	2.57	−12.58
Size, Production, and Prices			
Average sow herd size (number of head)	114.00	107.00	94.00
Total number of market hogs sold (head)	1373.00	1279.00	1058.00
Average weight of market hogs sold (lbs.)	229.00	229.00	229.00
Average price per cwt. of market hogs sold ($)	47.79	47.08	46.30
Efficiency			
Number of pigs weaned per litter	7.93	7.92	7.78
Number of litters weaned per sow per year	1.82	1.79	1.75
Number of pigs weaned per sow per year	14.42	14.16	13.69
Total pounds of feed per cwt. of pork produced (lbs.)	375.00	387.00	400.00
Cost of ration per cwt. ($)	7.30	7.70	8.00
Feed cost per cwt. of pork produced ($)	27.01	29.49	31.95
Hours of labor per sow in the herd (hours/sow)	20.60	23.66	26.06
Total cost per cwt. of pork produced ($)	42.97	48.33	54.11

SOURCE: W. R. Wood, K. S. Hendrix, et al., "Can You Reduce Your Livestock Production Costs?"
Better Farming—Better Living, Purdue University Cooperative Extension Service, 1985.

throughout the year. The same is true of poultry and, to a lesser extent, hogs. The labor requirements of certain feeding operations may be more seasonal and have greater possibilities for integration with cropping operations.

Because various enterprises require different labor skills, the availability and acquisition of the needed skills must be considered. For example, because the number of pigs per litter is an important factor in determining the profits of a hog enterprise, dependable and moderately well-skilled labor is necessary at farrowing time. Dairy production, on the other hand, requires highly skilled labor year round.

Different animal enterprises have different capital requirements. Certain enterprises might suit a particular farm quite well in many respects yet be out of the question simply because sufficient capital is not available. Obviously, such enterprises would not be included in a detailed planning analysis. For example, feeder cattle have higher capital requirements per $100 of gross income than swine. With enterprises such as broilers, it may be possible to obtain nearly complete financing by contract farming.

Finally, the demands of various animal enterprises on management must be considered. Clearly, all enterprises are not equal in this respect. Progress in nutrition, disease control, and breeding is making animal production more of a science and less of an art: The conditions under which poultry is raised indicate that considerable progress has been made in this direction for this enterprise. Milk production can now be more specified and detailed than was possible a few years ago; advances in nutrition and mechanization have also further routinized cattle feeding and hog production.

Risk Considerations

Animal and animal product enterprises subject to considerable price variability may be avoided by farmers who are heavily in debt or who cannot afford a loss in a particular year. All animal enterprises are not equal in the stability of production either, because biological and physical forces such as disease and weather do not affect all animal enterprises in the same way. Therefore, the risk and uncertainty associated with animal production should be considered.

One of the most important factors influencing the riskiness of an enterprise is the type of cost structure associated with it. This point can best be illustrated by the cost structure of a cattle feeding enterprise where the feeder is purchased compared to the cost structure of a cow operation where the calf is fed to slaughter weight. The feeding operation illustrated in Table 17-3 is contrasted to an operation where the feeder is raised on the farm (Table 17-4). In this case, the feeder cannot be raised as efficiently as it can be purchased from another source.

With these price relationships, it would be more profitable to buy the feeder than to raise it. But consider the situation in which the finished animal can be sold for only $50, rather than $60, per 100 pounds. In both cases, the farmer would be losing money on the enterprise. However, if the animal were purchased, there would be insufficient income, by $5 per animal, to cover the operating expenses that must be met on a year-to-year basis. If the animal were raised, $200 per head above operating expenses would remain. Because only the operating expenses must be met in a particular year, the amount allowed for ownership expenses, such as depreciation on equipment and interest on the cow investment, could be used for family living without affecting the operation in the short run.

This example illustrates how a difference in cost structure can affect the riskiness of an animal operation. Ownership expenses have to be met eventually if the business is to be maintained, but an enterprise that allows the manager to choose

TABLE 17-3. Enterprise Budget for the Cattle Feeding Example

Revenue	
1000-pound steer at $60 per hundredweight	$600
Operating Expenses	
Cost of feeder (700 pounds at $55 per hundredweight)	385
Feed	100
Miscellaneous	20
Total Operating Expenses	$505
Ownership Expenses	
Operator's labor	$ 40
Facilities and equipment	20
Total Ownership Expenses	$ 60
Total Expenses	$565
Return to Management	$ 35

when they can be met has greater flexibility. The operator who is in a position to survive one or two lean years probably should choose the enterprise that will offer the greatest long-run profit. The operator who cannot survive poor years should consider animal enterprises that minimize losses while maintaining the prospect of a more certain, though less substantial, profit. Considerable variation exists among enterprises in this respect. When fixed costs are a large proportion of total costs, it is possible to remain in business longer than when most of the costs are of a cash or variable nature, provided payments can be met on loans and mortgages.

TABLE 17-4. Enterprise Budget for the Calf Production and
 Feeding Example

Revenue	
1000-pound steer at $60 per hundredweight	$600
Operating Expenses	
Raising feeder to 700 pounds	$180
Feed	100
Miscellaneous	20
Total Operating Expenses	$300
Ownership Expenses	
Raising feeder to 700 pounds	$175
Operator's labor	80
Facilities and equipment	20
Total Ownership Expenses	$275
Total Expenses	$575
Return to Management	$ 25

Size of the Enterprise

Chapter 10 showed that unit costs vary considerably as farm size is increased. The same set of principles affect animal enterprises. A 60-sow swine herd may be an uneconomical operation for a farm. Yet if the number of sows were doubled to 120, a swine enterprise might represent the best use of resources on the farm. Therefore, it is important that farm managers understand the relationship of the size of herd or flock to the cost of production.

In chapter 10 an example was presented that was based on a study of dairy farms in California: The estimated average costs indicate significant economies of size up to a cow herd size of 750. It should be realized that this example is for illustrative purposes only. The economic size of an animal enterprise is not the same in all parts of the country. Dependent on technological and economic conditions, it changes rapidly. Therefore, it is important that the size–cost relationships of the animal enterprises being considered are based on the individual farm's situation.

There is still a place for the animal enterprise that is supplementary or complementary to other farm enterprises. Such enterprises use resources that would otherwise go to waste. For example, a winter-feeding cattle operation may use labor that would not otherwise be employed. A sheep enterprise may yield manure that will permit a more economical production of some crop. This situation has led many farm management experts to advise that farmers have animal enterprises that are either small enough to take advantage of supplementary or complementary relationships or large enough to realize cost efficiencies. Rather than follow a general rule, however, managers should carefully evaluate the farm's resources to determine the size and type of animal enterprise best adapted to that farm. This analysis can be done by using budgets to estimate the effect of different sizes of an animal enterprise on net income.

Feed Ration Decisions

Once the type and size of animal production enterprise have been determined, a number of operating decisions must be made. Decisions regarding feed rations are particularly significant.

Feed costs represent a substantial portion of the production costs for all animal enterprises. The principles of nutrition interact with economic forces to determine the most profitable rate of feeding and the composition of the ration. Nutritionists have determined the various nutrients that must be present in animal feeds for the growth, reproduction, maintenance, and production of animals and animal products. Different feeds contain different quantities of these nutrients. Therefore, the nutritional requirements of the animals must be supplied from a variety of feeds that contain varying proportions of the essential nutrients.

Feeds, of course, can be classified into nutrient groups. Some are largely protein feeds, for example, soybean, cottonseed, and linseed meals. Others are energy sources, such as the various grain concentrates, although the grains contain varying proportions of carbohydrates and protein.

To the extent that one feed can be substituted for another at a constant rate within a nutrient classification, the classification has economic meaning. That is, if 1.18 pounds of soybean meal will substitute for 1 pound of linseed meal, regardless of the proportion of the two that is used, then they substitute at a constant rate. Two feeds in different classifications, such as corn and soybean meal, will usually substitute at a diminishing rather than a constant rate. The economics of nutrition depend largely upon these relationships between feeds, which were described in Chapter 2.

Table 2-4 illustrates these principles. Columns 1 and 2 give the combinations of alfalfa hay and grain concentrate that will produce, on the average, a 300-pound gain on steer calves. The marginal rate of substitution (MRS) is given in column 3. As indicated earlier, this is determined as follows:

$$\frac{\text{Pounds of replaced resource}}{\text{Pounds of added resource}}$$

When hay is increased from 1000 to 1100 pounds, the amount of grain is reduced from 1316 to 1259 pounds. Therefore, the MRS is as follows:

$$\frac{1316-1259}{1100-1000} = 0.57$$

Each added pound of hay substitutes for 0.57 pound of grain.

In order to obtain the most economic combination, the prices of the two feeds must be known. If alfalfa is $50 per ton and grain is $0.05 per pound, then the inverse ratio is as follows:

$$\frac{0.025 \text{ (alfalfa price per pound)}}{0.05 \text{ (grain price per pound)}} = 0.5$$

Based on these prices and the relationship between the two feeds in Table 2-4, approximately 1200 pounds of alfalfa hay should be fed to 1208 pounds of concentrate to equate the MRS and the inverse price ratio.

There are two major considerations that are not included in the analysis but that should be recognized: the quality of the product produced (important if the animals are to be slaughtered) and time (if a high alfalfa ration relative to concentrate is fed, it will take longer to put on a 300-pound gain than if the amount of concentrate is higher).

A big challenge in animal feeding research is to identify the extent to which the principal feeds will substitute for each other. Once this relationship has been determined, relative prices may be used in determining the most economical ratio. Obviously, considerable analysis is required because the composition of feeds is affected by the environmental conditions under which they are produced. The various state experiment stations are continuously conducting experiments from which such determinations can be made.

Farmers buy considerable quantities of feeds premixed. In such cases, the feed companies must evaluate different combinations of feedstuffs and take into consideration relative prices when making their mixes. Farmers, of course, must decide whether it is more economical to mix their own feed. Again, reference to the results obtained by an independent research agency such as an experiment station appears to be the best procedure.

It should not be concluded that farmers change, or should change, their ration on a daily or weekly basis in response to price changes. However, relative feed prices do change significantly from one season to the next, and the relative proportions of different feeds in the ration become an important farm management decision. Alert managers should be aware of the feed substitution possibilities for their animal enterprises.[2]

Evaluating a Production Practice

This section illustrates the application of partial budgeting to a specific decision regarding an animal production practice, namely, whether to provide supplemental feed to unweaned beef calves. After a beef calf is 90 days of age, mother's milk usually will supply about half of the nutrients it needs for maximum growth. A common practice is to allow beef calves to remain with their mothers on pasture throughout the grazing season without supplemental feed. Some producers, however, creep-feed grain to the calves during the last 30 to 90 days before weaning. A creep feeder allows the calves to reach the supplemental feed while excluding adult animals. Supplementing the forage and milk available to unweaned beef calves with additional feed may be profitable particularly in drought years and in months when forage is limited. Feeding trials indicate that the response to supplemental feeding is greatest when natural forage is lacking.

Each rancher must determine whether creep feeding pays, considering the costs and returns specific to each operation. To analyze whether creep feeding will be profitable in a given year, the rancher can use a partial budget. Looking only at those costs and revenue items that will be affected by the decision to offer supplemental feed to calves, the rancher needs to estimate the added revenue, reduced expenses, added expenses, and reduced revenue. The rancher must answer a number of questions to complete a partial budget in order to decide whether to creep-feed calves. An example is shown in Table 17-5.

How Much Additional Gain?   The usual range in extra weaning weight gained as a result of creep rations is 30 to 60 pounds. In this example, the additional weight gain is assumed to be 40 pounds. As indicated under "Added Revenue," the calves will weigh 420 pounds, in comparison to 380 pounds if they are not creep fed ("Reduced Revenue").

---

[2] A more complete discussion of the formulation of feed rations for farm animals is beyond the scope of this book. Linear programming is commonly used to find the least-cost combination of feed ingredients that satisfies the nutritional requirements of particular animals. For an example of a least-cost dairy ration, see S. T. Sonka, *Computers in Farming: Selection and Use*, New York: McGraw-Hill Book Company, 1983, pp. 168–171. The use of electronic spreadsheets and microcomputers for ration formulation is presented in R. J. Lane and T. L. Cross, *Spreadsheet Applications for Animal Nutrition and Feeding*, Reston, Va: Reston Publishing Company, 1985.

TABLE 17-5. Example of a Partial Budget to Evaluate the Economics of Supplemental (Creep) Feeding of Beef Calves

Added Revenue		
Sale of creep-fed calves (420 pounds at $0.66)	$277.20	
Reduced Expenses		
Forage for calves (240 pounds at $0.03)	7.20	
		$284.40
Added Expenses		
Creep ration (320 pounds at $.09)	$ 28.80	
Use of equipment	0	
Labor for feeding	0	
Reduced Revenue		
Sale of calves not creep fed (380 pounds at $0.68)	258.40	
		$287.20
Change in Return to Management		$ – 2.80

How Much Feed?   The quantity of supplemental feed to use can be determined by estimating the cost of the additional feed per additional pound of gain as reported in the various research findings. However, the estimates of feed conversion range from 6.9 to 13.4 pounds of additional feed per pound of additional gain. This large variation is due to several factors, one of which is the tradeoff that the calf makes between the normal consumption of forage and the supplemental feed ration. For the purposes of the budget example, the conversion is assumed to be eight pounds of supplement feed per pound of additional gain. Assuming a cost of $0.09 per pound of feed, this puts the total cost for the extra ration at $28.80 per calf.

What Is the Value of Forage Saved?   In a supplemental feeding program, the calf is substituting the added ration for some of the normal forage requirement. Again, there are no clear answers about the quantity of forage that will be saved, but a reasonable estimate is that each pound of supplemental grain consumed may result in a savings of 0.5 to 1.0 pound of forage (dry matter basis). Using the midpoint of that range and assuming a reduction in forage consumption of 0.75 pound per pound of supplemental ration fed, the calves would consume 240 pounds less forage or equivalent grazing per head. If the value of this forage is $0.03 per pound, then there would be a savings, or reduced expense, for forage amounting to $7.20 per head. Depending on the quality and value of the forage saved by supplemental feeding, this estimate will vary from ranch to ranch.

What Is the Value of the Gain?   First, the price the rancher would receive for the calves if they were not given supplemental feed must be estimated. At $0.68 per pound and 380 pounds, the "Reduced Revenue" would amount to $258.40. What price would the rancher receive from the sale of creep-fed calves weighing 40 pounds more, or 420 pounds per head? Normally, these heavier calves would bring a lower price. The price difference for the heavier calves depends on the year and even varies among lots of cattle. Observing the market, the rancher must estimate the price for the heavier, creep-fed calves compared to lighter calves without creep feeding. For this example, the price is assumed to be $0.02 per pound less ($0.66 per pound) for the creep-fed calves, giving a total "Added Revenue" of $277.20 per calf.

What Are the Other Expenses?   Other expenses associated with supplemental feeding include labor, the creep feeder, and the use of equipment such as feed bunks, gates, and vehicles for transporting the feed. In the budget example, these inputs are assumed to be fixed. The appropriate charge for labor would depend on the opportunities for employing the labor elsewhere. For large operations, there may be additional costs for repairs and fuel to transport the feed. If the equipment were not available, the rancher would have to budget expenses for depreciation and interest on the investment.

Does It Pay?   The profitability of creep feeding, then, is the difference between the sum of "Added Revenue" and "Reduced Expenses" minus the sum of "Added Expenses" and "Reduced Revenue." In this example, the "Change in Return to Management" amounts to − $2.80 per head. Therefore, in this case, the practice of supplemental feeding would not be profitable. Ranchers should consider these assumptions and make projections regarding the profitability of supplemental feeding, given their specific situations.

A common mistake made in evaluating management practices such as supplemental feeding, implanting, backgrounding, or simply holding calves longer is to overvalue the extra weight gain. Giving the added weight the same value as the starting weight can be misleading. In the example in Table 17-5, if the creep-fed calves had been valued the same as those not creep fed, the return to management per head would be overestimated by $8.40 and the practice would appear to be profitable when it is not.

To find the value of additional weight gain, the following procedure can be followed:

420-pound calf at $0.66 per pound	$277.20
380-pound calf at $0.68 per pound	− 258.40
Difference	$ 18.80
$18.80/40 pounds = $0.47 per pound	

If the rancher could produce the extra weight at a cost of less than $0.47 per pound, the practice would be profitable. When calf prices are high, it is easy to forget that the extra weight sells at a much lower price. Of course, other factors such as degree of fatness, feedlot cost of gain, and slaughter cattle price outlook will affect the value of this added weight on calves.

Evaluating Alternative Production Systems

One of the more important decisions made in the management of animal production enterprises is the selection of the production system—the combination of buildings, equipment, and practices used to house, feed, and care for the animals. A given production system can have a significant long-run impact on the farm's capital investment, operating expenses, and labor requirements.

Because of the high capital and labor requirements associated with some animal enterprises, farm managers are interested in the capital and labor tradeoffs for different production systems. However, selecting the optimal production system is complicated by the number of systems available and their differences in labor efficiency and capital requirements. In deciding on the best system, the manager will want to substitute capital (increased mechanization) for labor as long as the labor savings are sufficient to pay for the use of the additional capital.

There are two ways to use labor more efficiently: Combine it with the proper amount of capital in the form of milking machines, automatic feeders, hay making equipment, and so forth; or make it more productive without increasing capital by simplifying the chores to be performed as much as possible.

Determining the proper amount of capital to combine with labor is frequently a complex problem. If the question is whether to substitute capital for labor when output is unchanged, the problem is relatively simple. A rather straightforward partial budget can be constructed to analyze it. However, when capital in the form of labor-saving facilities is added, the size of the animal enterprise can frequently be increased. Or if the enterprise is not enlarged, labor may be available to expand some other part of the farming operation.

To illustrate how such a decision can be evaluated, the results of a study of three confinement systems of hog production are presented.[3] These three systems represent different levels of labor input and capital investment: (1) open-pen, solid floor finishing with individual sow farrowing units; (2) modified open-front, partially slotted-floor finishing with enclosed slotted floor farrowing; and (3) enclosed, totally slotted-floor finishing with the same farrowing facility as that used in the second system.

Each system was designed for farrowing 32 sows six times per year for a total of 192 litters. With 7.5 pigs weaned per litter, each system would produce 1440 hogs annually. The three systems are as follows:

A. OPEN-PEN, SOLID-FLOOR FINISHING SYSTEM—(OP).
1. *Individual sow farrowing units.* Thirty-two individual sow units (7.5 by 8 feet) are spaced about 8 feet apart on a concrete pad enclosed by a metal fence. Each unit is equipped with an infrared gas heater and an

---

[3] A. G. Mueller and R. P. Kesler, "Investment Analysis of New Confinement Hog Systems," University of Illinois Department of Agricultural Economics, January 1979.

individual feeder and waterer. The pigs are weaned at five to seven weeks of age and then moved to the nursery.

2. *Low-profile, open-front shed nursery with an outside run.*   The nursery accommodates 240 pigs at a time. The shed (2.5 to 3 square feet per pig) and the outside area (10 to 12 square feet per pig) are paved with concrete. The pigs are moved to a finishing unit at a weight of 50 to 70 pounds.

3. *Low-profile, open-front shed finishing with an outside run.*   This unit accommodates 480 pigs to market weight and is similar to the nursery facility, with 4 to 4.5 square feet of shed area and 14 square feet of outside area per pig.

B. MODIFIED, OPEN-FRONT, PARTIALLY SLOTTED-FLOOR FINISHING SYSTEM—(OF).

1. *Slotted-floor farrowing house.*   The 32-sow farrowing house has a slotted floor and farrowing crates. All feeding and watering are done in the crates. A space heater is used to maintain an even temperature. The pigs are weaned at four to six weeks of age.

2. *Totally slotted-floor nursery building.*   The controlled-environment, 240-pig nursery building with a totally slotted floor has eight 10 by 10.5-foot pens. The pigs are self-fed in the building until they weigh 50 to 70 pounds.

3. *Open-front, partially slotted-floor finishing building.*   This 112 by 32-foot building, with a 10-foot, slotted-floor feeding area inside, has a capacity for 480 pigs. The building is insulated, and radiant heat is used in the sleeping area.

C. ENCLOSED, TOTALLY SLOTTED-FLOOR FINISHING SYSTEM—(TS).

1. *Slotted-floor farrowing house.*   Same as OF.

2. *Totally slotted-floor nursery building.*   Same as OF.

3. *Enclosed, slotted-floor finishing building.*   This controlled-environment unit for 480 pigs measures 120 by 30 feet.

4. *Enclosed, slotted-floor, sow-confinement building.*   The building provides 13 square feet of sleeping area per sow, with individual sow-feeding stalls. The sows are moved into the confinement building about two weeks after breeding.

The OF system is the most commonly used system for hog production in the Midwest. The TS system, with the breeding herd in enclosed confinement, reduces the labor input and allows better management control over this important phase of the production cycle. The sows are kept on pasture for the other two systems. The OP system, with individual sow farrowing units, reduces the large capital investment required for modern confinement hog production. One of the advantages of this system is that farm workers without extensive construction experience can build the facilities.

Investment costs were estimated from construction costs for erected and equipped, ready-to-use buildings. Table 17-6 shows the initial investments for the three production systems and the annual ownership costs per litter and per hundredweight. The annual ownership costs are based on a useful life of 10 years.

TABLE 17-6. Investments and Annual Ownership Costs for Buildings and Equipment for Three Building Systems—Farrowing Six Times per year, 192 Litters

Investment and Cost	Building System[a]		
	OP	OF	TS
Farrowing, 32 sow space	$ 32,000	$ 57,600	$ 57,600
Nursery	9,600	23,000	23,000
Finishing	26,400	61,200	79,200
Sow facilities	7,400	7,400	41,500
Feed processing	18,000	18,000	18,000
Manure handling	14,000	15,400	15,400
Total Investment	$107,400	$182,600	$234,700
Per Litter	$ 559	951	1,222
Annual Ownership Costs[b]			
Per litter	$ 101	$ 171	$ 220
Per hundredweight of Pork	$ 5.92	$ 10.07	$ 12.94

SOURCE: A. G. Mueller and R. P. Kesler, "Investment Analysis of New Confinement Hog Systems," University of Illinois Department of Agricultural Economics, January 1979.

[a] The building systems are
  OP = Open-pen, solid-floor finishing system
  OF = Modified, open-front, partially slotted-floor finishing system
  TS = Enclosed, totally slotted-floor finishing system
[b] Based on depreciation at 10 percent of the initial investment; 8 percent interest on the average investment; repairs at 3 percent of the initial investment; and taxes and insurance at 1 percent of the initial investment.

Table 17-7 shows the estimated costs to produce one hundredweight of pork for the three building systems. Production costs range from $39.23 for the OP system to $44.72 for the TS system. In this analysis, feed conversion, feed prices, cash operating expenses, interest on feed and livestock, and farm overhead expenses were held constant for the three systems. One reason for this assumption is that experimental data and field observation provide little evidence of any substantial improvement in the annual performance of hogs in various production systems on a year-round basis. Another reason is that it allows attention to be focused on the labor substitution effect of the three systems.

If labor is assumed to be a variable input and costs $6.50 per hour, the labor savings are not sufficient to offset the high capital investment costs for the OF and TS systems. Another way to analyze the profitability of the three systems is to assume that labor is fixed and to calculate the return to labor and management for one person (2000 hours of labor) at various selling prices for the hogs produced. The results are presented in Table 17-8. The greater number of hogs produced per person in the OF and TS systems multiply the returns to labor and management at hog prices above $50 per hundredweight. On the other hand, if prices drop below $50, those farmers who invest in the OF and TS systems will be subject to greater risk.

TABLE 17-7. Estimated Cost per Hundredweight of Pork for Three
Building Systems

| | Building System | | |
	OP	OF	TS
Feed Costs[a]	$23.64	$23.64	$23.64
Nonfeed Costs			
Labor at $6.50 per hour	$ 4.97	$ 4.21	$ 3.44
Buildings and equipment			
(Table 17-6)	5.92	10.07	12.94
Other expenses	4.70	4.70	4.70
Total Nonfeed Costs	$15.59	$18.98	$21.08
Total Costs	$39.23	$42.62	$44.72
Hours of Labor			
Per hundredweight	0.76	0.65	0.53
Per litter	13.00	11.00	9.00

SOURCE: A. G. Mueller and R. P. Kesler, "Investment Analysis of New Confinement
Hog Systems," University of Illinois Department of Agricultural Economics,
January 1979.
[a] Based on a corn price of $2.20 per bushel and a protein supplement price of $12.50
per hundredweight with a feed conversion of 6.2 bushels of corn and 80 pounds of
protein supplement per hundredweight of pork produced.

This budgeting analysis may be used as a guide in planning investments in hog
production systems. Uncertainty about future hog market price levels, possible
changes in construction costs, and feed costs must be carefully weighed, along
with the financial position of the farm business. Not all of these sources of risk
can be specified precisely in the decision-making process. However, the range of

TABLE 17-8. Returns to Annual Labor for One Person at Varying Hog
Price Levels with Three Building Systems

| Selling Price of Hogs (cwt.) | Building System[a] | | |
	OP	OF	TS
$40	$15,027	$ 8,920	$ −4,831
45	28,117	24,390	14,039
50	41,207	39,860	32,909
55	54,297	55,330	51,779
60	67,387	70,800	70,649
65	80,477	86,270	89,519

SOURCE: A. G. Mueller and R. P. Kesler, "Investment Analysis of New Confinement
Hog Systems," University of Illinois Department of Agricultural Economics,
January 1979.
[a] Assumes that 2000 hours of labor are available per year. A total of 154 litters are
produced in the OP system, 182 in the OF system, and 222 in the TS system.

outcomes associated with the variability inherent in these factors can be analyzed, as is done for price risk in Table 17-8.

Cyclical and Seasonal Prices

One of the characteristics of animal production is the cycles of total production, which are closely associated with cyclical price movements. These price and production movements have a direct bearing on farm management decisions.

Price and production cycles are the result of the decisions of many animal producers, who decide to increase or decrease production on the basis of the current year's prices. For example, when cattle prices start to move up, heifers are kept for breeding and cows that normally would be sold for slaughter are kept for continued production of calves. The higher cattle prices stimulate production, but the production increase is realized only after a few years have elapsed. In the case of hog production, where the feed–hog price ratio is quite important to decision making, highs and lows occur more frequently because of the shorter response time. For hog production, the time lapse between the decision to expand and the result is about one year. For cattle, it is about two years.

Because these cycles can be observed over time, the information can be put to use in making farm management decisions. Farm managers should not try to outguess these cycles or try to produce for market only when the cycles are at an extremely high point. Even price forecasters who spend their time studying market forces and trends are unable to predict the high and low points with accuracy. However, cattle producers may wish to curtail their operations somewhat if they learn, for instance, that the number of cattle in the nation is extremely high. Conversely, if they believe prices will rise during succeeding years, they may handle their breeding stock differently. Such decisions should be based on a careful study of market forces and the farm's cost structure. Current prices are a poor basis for decisions of this kind.

Like information on crop yields, technological change, and other planning data, information on future price movements is subject to error. Thus, price information should be treated in the same way as other sources of uncertainty. Errors in cyclical price projections are probably no greater than those in some of the other variables.

A valuable use of such information is in timing the addition or elimination of an animal enterprise. A farmer may decide to delay the addition of an animal enterprise if the price of breeding stock is unusually high. Conversely, an enterprise may be added sooner than planned if it appears that the initial investment can be significantly reduced by purchasing animals when the price is unusually low. Or a farmer may decide, on the basis of long-run profits, to eliminate an animal enterprise; however, if price prospects are bright, the farmer may decide that it is profitable to postpone the decision.

The preceding situations apply to animal enterprises that involve breeding stock. Managers of feeding enterprises are even more justified in adjusting their operation from one year to the next on the basis of current conditions. It makes little sense to maintain the same feeding program every year when different prices prevail for feeder stock, as well as for market livestock, to say nothing of the price and availability of feed. However, managers who vary their feeding programs should not do so only on the basis of current prices or what other managers seem to be doing. Market outlook information is available from various university extension services and the U.S. Department of Agriculture. No system of price forecasting has been developed that is highly reliable, yet people who specialize in the study of such problems are more likely than others to predict the correct direction of prices.

Animal and animal product prices have also been noted for their seasonal highs and lows. If demand remained constant, the only thing that could cause a change in price is a change in supply, but the quantity of animals and animal products that go on the market is not constant throughout the year. Because of lush spring pastures, maximum milk production usually occurs in May and June, resulting in lower summer prices and higher winter prices. Because farrowing hogs is more difficult during the winter and early spring, fewer hogs go to market in July and August, resulting in more favorable prices then. Therefore, the lows and highs of seasonal prices are usually the result of production conditions. However, seasonal demands for certain items, such as turkeys at Christmas and Thanksgiving, may also influence seasonal prices.

This situation leads to one of the more complex problems of animal production management: When should animals and animal products be marketed? First, seasonal highs cannot always be predicted. Although the seasonal high for hogs has historically been in the summer months, this pattern may not hold for a particular year; in addition, there is some evidence that seasonal highs and lows are gradually being leveled out. But assuming that such seasonal differentials do exist and can be predicted, the cost of producing at various times of the year must also be considered. Therefore, the analysis of these marketing decisions may become very complicated.

If adequate information is available regarding the present production pattern, the problem is simplified considerably. It should be possible to estimate cost changes that will be associated with a change in the production pattern. Not all costs will change, of course, so that many parts of the farm business need not be taken into consideration.

Marketing agencies have long been aware of the seasonal aspects of agricultural production. They like to have an even flow of agricultural products into the marketing channels throughout the year to permit their labor and equipment to be used more efficiently. Achieving this consistent supply of input is an important motive for marketing firms to become involved in contract farming and vertical integration.

The production of broilers has become a year-round operation. Many hog producers are moving away from twice-a-year farrowing and are marketing hogs throughout the year. Various pricing and quota devices have been used to even out the quantity of milk products on the market during the year. Some packers have their own feedlots, which permit them to integrate their feeding operations with farmer marketing. It may be that seasonal animal production is changing. However, as long as the costs of production and the prices received are not constant throughout the year, deciding when to produce will remain a farm management decision.

Summary

Selecting and managing animal production enterprises involve several interrelated decisions regarding feeding, management practices, production systems, and marketing. The enterprise budget is the primary tool for analyzing the factors that affect the profitability of animal production. The appropriate efficiency measures depend on the enterprise. For example, the calf crop percentage, the average weaning weight, and the cost of cow maintenance are important in cow-calf operations.

The animal enterprise selected depends on resource availability, risk considerations, and size. The availability of land, feed, labor, capital, and management determines how well an animal enterprise will fit the farm business. An important risk consideration is the cost structure of the enterprise, that is, the relationship between operating (variable) and ownership (fixed) expenses. Economies of size are important for some animal enterprises, but supplementary or complementary relationships are also relevant.

An understanding of economic principles, as well as of the production relationships involved, is needed to make decisions about feeding and other management practices. An evaluation of alternative production systems often entails an analysis of the substitution of capital for labor. For the hog production example presented earlier, when labor is considered a variable cost, the system with the higher labor requirement and lower capital investment is most profitable. However, when labor is fixed and the enterprise is increased in size to use the labor available, the system requiring less labor per litter and a higher capital investment is most profitable, assuming that hog prices exceed $50 per hundredweight.

Farm managers should understand the implications of cyclical and seasonal patterns of animal prices. Although these patterns cannot be predicted with absolute accuracy, the tendencies may influence the timing of enterprise production, expansion, contraction, or elimination. Again, a budget should be used to compare the profit prospects for these animal production management decisions.

PART V

Future Opportunities in Agriculture

Agricultural production and other sectors of the economy are becoming increasingly interdependent. Structural economic changes and new technology must be reckoned with in farm management. Part V is devoted to these considerations.

CHAPTER 18

The Changing Business of Farming

Young people desiring to enter farming as an occupation, as well as established farm managers planning the long-run future of their businesses, should recognize that farming is a changing business. This proclivity for change is dramatically illustrated by contrasting the first half of the 1970s to that of the 1980s.

During the early 1970s, export markets were expanding, farmers were being encouraged to expand production, land values were increasing, and interest rates were relatively low. During the early 1980s, export markets were contracting, policymakers were considering measures to curtail production, land values were decreasing, and interest rates were relatively high. The mood of optimism that prevailed in the 1970s was replaced by one of pessimism in the 1980s. Successful farm managers must be able to take these reversals in stride, and they do so by making decisions and planning with a long-run perspective. They must understand the long-run trends and developments affecting farm management in order to build a viable and profitable business.

This chapter begins with a discussion of the changing structure of agriculture and the future of the family farm. The factors contributing to these changes are explored. New technology has had and will continue to have a major impact on agriculture. A relatively new consideration is the increasing international interdependence of U.S. agriculture. Agricultural policies as well as monetary, fiscal, and tax policies also continue to shape the future of agriculture. But the effects of these forces are difficult to predict, and uncertainty will continue to characterize farm management. Whether agriculture will continue to be a growth industry or enter a period of decline has important implications for young people planning to enter agricultural careers. There will be opportunities for those meeting the qualifications presented in the final section.

The Changing Structure of Agriculture

Farms are becoming fewer and larger, and production has become more concentrated among larger operations. Will this trend continue? How will the family farm be affected?

Farm Size Trends

Since 1950, U.S. agriculture has changed significantly, and more changes are expected. The most obvious changes have been in the size and number of farms. The U.S. Department of Agriculture estimated that there were 4.0 million farms in 1960 and 2.4 million in 1980 (Table 18-1), and that there will be only 1.8 million by the year 2000.[1] As the number of farms decreases, the size of the average farm increases.

In 1984, commercial farms with annual gross sales of more than $100,000, representing only 15 percent of all farms, generated 73 percent of all gross farm income (see Table 18-2). These farms are likely to be large enough to provide full-time employment for their operators and adequate returns to investment and resources. Most of these farms were managed by the farm family and provided their principal source of income. About one-third of the farms in this group had sales of more than $250,000.

About 48 percent of all farms in 1984 generated less than $10,000 in farm product sales, the total of which represented less than 3 percent of all gross farm income. The composition of this group is diverse. Few of these farmers are exclusively dependent on their farm income; most supplement their income with off-farm earnings; several work almost exclusively off the farm. Some of these farms are retirement operations, others merely rural residences or hobby farms that sell enough agricultural products to be called farms.

Those farms selling between $10,000 and $100,000 worth of farm products account for the remaining 37 percent of all farms and 24 percent of all gross farm

[1] William Lin, George Coffman, and J. B. Penn., *U.S. Farm Numbers, Sizes, and Related Structural Dimensions: Projections to Year 2000*, U.S. Department of Agriculture Economics, Statistics, and Cooperatives Service Technical Bulletin 1625, July 1980.

TABLE 18-1. Number of Farms and Average Acres per Farm, 1950 to 1984

Year	Number of Farms (1000s)	Average Size (Acres)
1950	5648	213
1960	3963	294
1970	2949	372
1975[a]	2521	418
1980	2433	425
1981	2434	422
1982	2401	427
1983	2370	430
1984	2328	438

SOURCE: U.S. Department of Agriculture, *Economic Indicators of the Farm Sector: National Financial Summary, 1984*, Economic Research Service ECIFS 4–3, January 1986, pp. 40, 68.

[a] For 1975 and the following years, a farm is defined as any place that sells or normally would sell $1000 worth of agricultural produce.

TABLE 18-2. Distribution of Farms and Gross Cash Farm Income by Value
of Agricultural Products Sold, 1984

Value of Agricultural Products Sold ($)	Number of Farms (1000s)	Percent of All Farms	Percent of Gross Cash Farm Income
Less than 10,000	1122	48.1	2.9
10,000–39,999	516	22.2	7.8
40,000–99,999	353	15.2	16.0
100,000–249,999	229	9.9	24.4
250,000–499,999	77	3.3	17.7
500,000 and over	31	1.3	31.2
Total	2328	100.0	100.0

SOURCE: U.S. Department of Agriculture, *Economic Indicators of the Farm Sector: National Financial Summary, 1984,* Economic Research Service ECIFS 4–3, January 1986, pp. 40, 43.

income. Some of their managers are part-time farmers; some are nearing retirement; others in this group, the younger people who rely on farming for their primary income, will either expand and move into the larger commercial class, quit farming, or join the ranks of part-time farmers working off the farm to supplement their income.

These data demonstrate that there is great diversity in size, ownership, and production on U.S. farms. Small part-time operations represent a large percentage of the total number of farms, but owner-operated family farms still predominate. These changes in the structure of agriculture are the result of the interaction of many factors such as technology, demand, and government policy, which are discussed later in this chapter.

The Family Farm

Because of the trend toward larger farms and fewer farmers, some people contend that family farming is a thing of the past. They argue that much of the control of farming is passing to people off the farm and that the farm as a place where the family can work together as a unit (employing their own labor, management, and capital) is fast disappearing.

This issue is charged with emotion, half-truths, and misinformation. Everyone is for family farms; political speeches often include a call to save the family farm. This attitude is not hard to understand. Americans have progressed rapidly from a primarily agricultural nation to an industrial one, yet they are still nostalgic for the rural way of life. The issues should be clearly understood.

At one time, it was appropriate to define the family farm in terms of land ownership. If the farm was owned, operated, and managed by the family, it was a family farm. But this definition is too exclusive because many prosperous and efficient farms are operated by tenant families.

Another definition may be framed in terms of labor: A farm is a family farm when the family is the major source of labor. The use of hired labor depends on the type of farm. For instance, certain fruit and vegetable farms that require seasonal labor for harvesting and other operations would be omitted from this classification.

Perhaps a more useful concept is that a farm is a family farm when it provides the family's principal source of income and is managed by the family. This definition, although arbitrary, does include tenant farmers, farm families that supply all labor, and farmers who must occasionally hire labor. Vertical integration, or contract farming, also fits within this definition, because the family still has the final word in deciding whether to contract. In this sense, the family controls the management of the farm. The family farm, then, is one that is managed by and provides the principal support for the farm family. This is a broad definition. It includes the Wisconsin dairy, the Illinois corn-hog farm, the Kansas wheat farm, and the Wyoming cattle ranch.

Today's family farm is very different from what it was three decades ago, and tomorrow's family farm will be larger, more mechanized, and more productive than today's. Regardless of our definition, the family farm will account for the vast majority of farms and the largest share of U.S. agricultural production at least for the remainder of the twentieth century.

Although the long-term trend toward fewer and larger farms is slowing, it will not change direction. This trend is being set not by nonfarm corporations or wealthy investors but by family farmers themselves. In recent years, about three-fourths of the farmland sold was purchased by established farm operators.[2] The family farm, rather than resisting change, has been the cause of it.

Technology and the Structure of Agriculture

A technological revolution has been underway in U.S. agriculture for many years. It has had an enormous impact on the structure of agriculture, and apparently the trend has not yet run its course. The *structure of agriculture* refers to the number and size of farms making up the agricultural sector, as well as to the ownership and control of farm resources and the business arrangements used.

The creation, development, and adoption of technology is a complex subject that is probably not completely understood by anyone. The United States has consciously fostered technical change in agriculture through public policies supporting agricultural research and education. Corporations have grown and flourished by participating in this new technology. They have supplied inputs to those working on the land and have marketed the agricultural products that have poured from the farms in increasing volume over the years. Thus, the increased output

[2] U.S. Department of Agriculture, *Agricultural Land Values and Markets: Outlook and Situation Report*, Economic Research Service CD-90, August 1985.

resulting from technological change on the farm and in agribusiness can be attributed to both public policy and the private sector.

Much of the technology used in agriculture is produced off the farm. Inputs purchased from off the farm have increased almost threefold since the early 1900s, and agribusiness has been a growth industry. These off-farm interests include the vast input industries of machinery and chemicals, as well as the marketing firms that handle agricultural products. Because the relationship is so mutually dependent, there is an incentive for the off-farm firm to acquire control of the farm operation. This trend has led to the development of vertical integration, that is, the coordination of decision making at two or more stages of production, by contracting, by outright ownership, or by a combination of the two. Poultry production and marketing is an example of a farm enterprise that has become vertically integrated.

Technological improvements have also permitted the individual farm manager to control more land, livestock, and inputs and to produce in greater volume. During the last three decades this technology has resulted in the horizontal growth of farm businesses—in contrast to vertical integration. Farmers have tended to reduce the number of commodities produced and to produce more of the remaining commodities. Successful expansion of the farm business depends largely on the ability of the farm manager to routinize production and delegate authority. This increase in managerial capacity has been facilitated by the availability for purchase of inputs produced off the farm.

As noted, the adoption of technological improvements has resulted in larger and larger farms, regardless of how size is measured. Why has this result occurred? Is it not possible that technology can make life easier and more profitable for the small farmer rather than always working to the advantage of the larger one?

To understand why new technology usually results in increased farm size, the principles discussed in Chapter 10 must be utilized. In that chapter it was shown that average cost curves for many types of farming decline rapidly at first and then often become fairly flat. In other words, after a certain size has been reached, further increases in size usually will not give additional cost savings. Under such circumstances, farms of varying size can exist side by side, and this is just what occurs in practice. However, the net income of smaller farms can be increased by increasing the volume if the per-unit costs do not rise. Thus, there is an incentive to increase size even though the small farms are able to produce for the same average total cost as the larger ones. And those farmers who have the ability to produce in larger volume efficiently are the ones who survive.

Most of the new technology permits a larger unit to be managed by the single farmer and by the farm family. This fact is best illustrated by farm machinery, which comes in many different sizes. But it is also true of many inputs that are less lumpy and that cost much less per unit, such as fertilizer and premixed feeds. Besides the possible cost savings from volume purchasing, standardized inputs also make it easier to routinize production and delegate authority.

It is possible that the small farms can be competitive on a cost basis but may not be able to generate enough income to meet the goals of the family. This situation explains why many part-time farm managers, who supplement their income by working off the farm, seem to be flourishing, whereas many mid-size farm operators experience financial difficulties. The small-farm manager is not entirely dependent on the farm for income, but the mid-size farm manager may be. Certainly it is true that, relative to mid-size farms, the number of very large and very small farms has increased in recent years.

The farmer as an entrepreneur and manager has changed with improvements in technology. The amount of schooling a farmer receives has increased every decade, and no doubt the amount of information mastered by the farm manager has grown as well. This increasing education, of course, adds to the farmer's capacity to manage an increasingly larger business.

The most recent technological revolution in agriculture—indeed, in business generally—is the efficiency with which information is collected, analyzed, and transmitted. Computer technology is such that its benefits can be realized by smaller as well as larger farm operators. The computer increases the managerial capacity of anyone who can use it.

Thus, new technology probably heightens the competition among farmers. They tend to bid against one another for land to buy or rent, as well as for other factors of production. The farmer with the greatest capacity for managing larger units wins the competition, and farms inevitably get larger.

It does not follow that the sole proprietor or the family farm will always win the competition. However, for most types of agriculture to date, the family farm as either a sole proprietor or a small corporation has been most competitive. Nevertheless, in some types of agriculture, the large corporation has become predominant.

International Interdependence

U.S. agriculture has become an international business highly dependent on trade. Foreign demand for U.S. crops rose rapidly during the 1970s, greatly increasing the role of agricultural exports in the national economy. Nearly one-fifth of total U.S. crop output goes into the export market.[3] Exports represent 40 percent or more of the total market for U.S.-produced wheat, soybeans, corn, and cotton. About 20 percent of the fruit, vegetable, and nut production is exported. For livestock production, only about 4 percent is exported.

The United States now accounts for nearly half of the world grain trade, compared to about 30 percent in the early 1950s. Thus, U.S. and world commodity markets have become increasingly integrated. This integration has both advantages and disadvantages. The growth in agricultural markets has increased farm income and employment. But U.S. agricultural commodity prices are now subject to political and economic changes throughout the world, over which U.S. producers have no control.

Major changes in the international economy have increased the interdependencies of U.S. agriculture and the international markets. International trade has

---

[3] A. C. Manchester, *Agriculture's Links with U.S. and World Economies*, Economic Research Service Agricultural Information Bulletin No. 496, September 1985.

grown during the past three decades. Also, the development of international financial markets that facilitate foreign investment in the U.S. economy, as well as the investment of U.S. dollars abroad, has increased the interdependence among economies.

The floating exchange rate system, established in the early 1970s, allows fiscal and monetary policies to affect the agricultural sector through the effects of currency exchange rates on the export demand for agricultural commodities. As a result, agriculture is no longer isolated. Changes in U.S. economic policies, as well as the economic policies of other nations, have important implications for the well-being of U.S. agriculture.

The greater dependence on trade, development of international financial markets, and floating exchange rates have resulted in a great deal of instablility in international markets for both commodities and capital. The economic policies of the United States, as well as the policies of other nations, have also contributed to this instability.

As a result of this internationalization of U.S. agriculture, farm managers should keep themselves informed about the international markets and institutions that affect their businesses in order to make informed decisions. Events in other parts of the world, such as drought in the Soviet Union, and the economic policies of other nations, such as the monetary policies of Japan and the agricultural policies of France, have significant implications for the U.S. agricultural sector. The world economy in which farmers manage their businesses has become interdependent and dynamic.

The Public Policy Environment

More than ever, farmers must be aware of the trends that are external to agriculture but that may influence its future more than the decisions made within agriculture. The family farm remains dominant, but if it is to remain so in the future, it must constantly adapt to agriculture's changing environment.

Agriculture is subject to a wide spectrum of public policies. Agricultural policies are designed to alleviate chronic problems encountered in the production and marketing of agricultural products. Fiscal, monetary, and tax policies also affect agriculture and its structure, particularly in recent years. Public policies affect the structure of agriculture because they have different effects on different kinds of farm businesses.

Agricultural Policies

For many decades, the federal government has intervened regularly in agriculture. The reasons have been many and varied, but the fact is that there are many government programs designed to provide assistance to agriculture, including research, education, credit, and price supports.

The question frequently raised is whether these various government interventions are biased in favor of larger farms. Whether the bias is deliberate or not, in practice many government programs, especially price supports and commodity programs, tend to benefit large producers and land owners more than smaller farmers and tenants. But even in the absence of government programs, market forces reward efficiency and size for the reasons previously outlined. This does not excuse government programs that may be biased to benefit one group more than another, but it does suggest that the long-term trend with respect to farm size probably cannot be reversed by reforming farm programs.

Fiscal and Monetary Policies

Fiscal and monetary policies influence the financial positions of farmers.[4] For example, in the early 1980s, the U.S. economic policy shifted to a more restrictive monetary policy and a more expansive fiscal policy. The change in the policy of the Federal Reserve System exerted greater control on the supply of money. The expansive fiscal policy was initiated by the Economic Recovery Tax Act of 1981, resulting in an unprecedented tax decrease. The combination of these policy shifts slowed inflation, increased interest rates, and generated record-level deficits in the federal budget.

The higher interest rates also attracted large sums of money from foreign investors who found safe and profitable investments in the United States. As a result of this inflow of currency, the U.S. dollar increased in value relative to other currencies. The stronger dollar not only made U.S. commodities more expensive to importers, it also attracted foreign producers into international markets because their cheaper currencies allowed them to compete with U.S. farm exports. As a result, demand for U.S. agricultural exports declined and commodity prices fell.

Because the U.S. agricultural sector has become increasingly capital intensive, the profitability of farm businesses is more susceptible to these higher interest rates, which increase interest costs to farmers. These interest cost increases combined with the decline in export demand threaten the financial health of many farmers, particularly those who are heavily in debt.

An important characteristic of agriculture is the great differences in the effects on various farmers. A U.S. Department of Agriculture study estimated that 11.6 percent of all farms had debt to asset ratios of 40 to 70 percent as of January 1, 1985[5]. Based on 1985 interest rates and net returns, operations with debt to asset ratios of more than 40 percent are likely to have inadequate cash flow to meet all of their obligations. These managers are experiencing serious financial problems that will require refinancing, loan repayment extensions, or liquidation of some assets.

Another 7.3 percent of all farms were estimated by the U.S. Department of Agriculture to have debt to asset ratios above 70 percent. These operations are almost certain to have serious cash shortfalls, and some in this category are technically insolvent. Others have financial problems requiring major adjustments to survive.

[4] W. D. Dobson, "How Macroeconomic Policies Affect U.S. Farmers," *Purdue Agricultural Economics Report*, Purdue University Agricultural Economics Department, May 1985.

[5] U.S. Department of Agriculture, *Financial Characteristics of U. S. Farms, January 1985*, Economic Research Service Agricultural Information Bulletin No. 495, July 1985.

Other effects of these fiscal and monetary policies, such as reductions in farmland prices, lower land lease rates, and decreases in prices of other farm assets, plus gains in production efficiency, will permit U.S. farmers to produce at a lower cost in the years ahead. Lower costs will eventually make U.S. agricultural exports more competitive in international markets.

Fiscal and monetary policies affect the well-being of U.S. farm producers directly through inflation and interest rates and indirectly through changes in international markets and export demand. Because of the increased use of capital in agriculture, its greater dependency on purchased inputs, and the importance of exports in the marketing of agricultural production, U.S. farmers have an important stake in the development of fiscal and monetary policies.

Tax Policies

Tax rules play a significant role in farm management decisions such as purchasing machinery, scheduling the marketing of commodities, and choosing business arrangements. Because of the impact of tax rules on decision making, federal tax laws have influenced structural changes in agriculture. For example, tax rules that allow larger deductions for larger farmers may encourage farm growth. Tax policies may also promote the adoption of new technology, the substitution of capital for labor, and the exploitation of economies of size. If tax laws have different effects depending on the size, tenure, and enterprises of the farms, the structure of agriculture will be influenced. By drawing in outside sources of capital, tax policies may increase the separation of the ownership and operation of farm businesses.

Substantial nonfarm, tax-motivated investments were attracted to agriculture during the 1960s.[6] Citrus and almond orchards were popular investments; however, changes in the tax rules eliminated tax benefits for these two crops, so investment interest shifted to wine grapes and walnut and pistachio orchards. Cattle feeding was the most popular investment in the 1970's. Tax shelters in breeding cattle also received considerable publicity, although the impact was not nearly as significant.

A recent U.S. Department of Agriculture study indicates that tax policies from 1956 to 1978 stimulated net investment in agricultural equipment and facilities.[7] The study also found that certain tax provisions, particularly the investment tax credit and interest deduction, stimulated net investment in agriculture beyond what it would have been without these tax provisions. Nearly 20 percent of net investment in agricultural equipment during the 1956–1978 period is attributed to these tax policies.

To the extent that tax-induced investment in agriculture increases production and lowers prices, tax policies contradict agricultural policies seeking to control production and support prices. However, because changes in tax laws can be frequent and unpredictable, their direct effects on agricultural structure are often difficult to predict.

[6] H. F. Carman, "Taxation as a Factor in Economies of Size," *Farm-Size Relationships, with an Emphasis on California*, University of California Giannini Foundation of Agricultural Economics Project Report, December 1980, pp. 55–75.

[7] J. Hrubovcak and M. LeBlanc, *Tax Policy and Agricultural Investment*, U.S. Department of Agriculture Economic Research Service Technical Bulletin No. 1699, June 1985.

Farming in an Uncertain World

The history of agriculture demonstrates that the welfare of farmers is often greatly affected by forces over which they and the industry have little or no control. World events, monetary and fiscal policies, and the health of the economy are all powerful forces affecting agriculture but are largely outside the influence of those in agriculture. Farmers need to recognize that these outside shocks will occur frequently. Some will be temporary, but others will influence agriculture for a long time. An important aspect of effective farm management is the development of a strategy that permits adaptation to these outside forces and influences.

In recent years, two developments outside agriculture have had a major impact on the structure of the industry. One was the rapid growth in world demand for agricultural products, primarily grain, during the 1970s. This trend was accompanied by rapid domestic inflation and a monetary policy that kept the rate of interest so low that it did not reflect the inflation in prices. The growth in demand increased the price of many farm products. Inflation and low interest rates made the purchase of land attractive. Farms grew rapidly because of these forces, and many farmers incurred debt that could not be serviced when the rate of growth in exports declined and when U.S. monetary policy changed and interest rates were freed to reflect inflation as well as the supply and demand for money.

Few people, regardless of how expert or how well informed they may be, can foresee such developments and plan accordingly. But individuals can judge the risks associated with making certain plans on the assumption that existing trends will continue indefinitely. Some very marginal land was brought into production apparently on the assumption that exports would continue to grow at a rapid rate. A great deal of debt was incurred with the expectation that inflation would continue. An important part of farm management in the future will involve contingency planning for such trends. How confident are managers that current trends will continue? Which ones are most likely to change? What will be the consequences if present trends do not continue?

The increasing size of farm businesses magnifies the risk that individual producers must bear. When most of the production inputs are purchased either from other farms or from nonfarm firms, fluctuating input prices become a serious matter. Similarly, the more specialized the farm, the more dependent the farmer is on product prices; a small percentage change in broiler prices, for example, may determine whether production is profitable or highly unprofitable. As a consequence, marketing management is emphasized to reduce the risks involved in purchasing production inputs and selling output. In common with other industries, U.S. agriculture has developed markets that are better suited to minimizing severe price risks than the small open market that was typical of traditional agriculture. These more highly developed markets permit the use of such marketing management techniques as forward contracting, hedging, and options.

Whether sole proprietorships will prevail as the dominant producing unit in American agriculture will depend on the relative strength of the forces previously identified. If the production process can be routinized and authority and responsibility can be delegated, farm size will continue to increase. If economies continue to be realized from horizontal integration and growth, there will be rewards to the sole proprietor in U.S. agriculture. However, if the rewards for horizontal expansion decline relative to those for vertical integration, there will be an incentive for the nonfarm firm to invest and participate in the decision making of the farm business.

Farming: Growth Opportunity or Declining Industry?

In the early 1970s, almost overnight it seemed, the world was confronted with an oil crisis and shortages of food supplies. Many experts argued that the industrial nations had entered an era of shortages. Agricultural exports from the United States increased sharply. A soybean embargo was even imposed to protect domestic consumers. U.S. farmers increased their output in response to the greater demand. But they did such a good job of responding that the supply increased more rapidly than the demand, and farm prices and incomes were depressed. What of the future? Can farmers anticipate a growing market for their output over the long haul, or was the situation that existed in the early 1970s unique?

In approaching this question, it is important to try to understand why demand grew so rapidly in the 1970s. Among the more important factors is that the centrally planned economies of Eastern Europe, especially the Soviet Union, made a political decision to import grain in order to improve the diets of their citizens. Their own agricultural enterprises had been less successful than they had hoped. Large quantities of grain imports became necessary if their people were to increase the amount of animal products in their diets. Other countries, such as Taiwan and Korea, experienced rapid economic growth during this period. Higher incomes increased the demand for food in general and the composition of diets shifted toward more animal products, thus strengthening the world demand for agricultural products, especially feed grains. In some countries, such as Poland, increased imports during this period were financed by debt rather than by increased exports.

Most scholars who have studied the international trade situation for agricultural products have concluded that the combination of circumstances that existed during the 1970s is not likely to recur in this century. Although most experts expect agricultural exports to continue to grow, they do not foresee a resumption of the rapid growth of exports that occurred during the early 1970s. Agricultural production is expanding significantly in many parts of the world.

Domestic demand projections are also conservative. Even though the U.S. population will continue to grow and enjoy improved incomes, it is doubtful that the domestic demand for food will increase dramatically. When domestic and international considerations are combined, a consistent but slow rate of growth in the demand for agricultural output seems probable. Historically, improvements in production have occurred more rapidly than increased demand, with the result that prices of farm products have declined in comparison to those of other goods and services. There is no consensus at present among the experts regarding the probable rate of technical change in U.S. agriculture for the next several decades. Thus, persons entering farming today cannot be assured of rising real prices for the products they will sell. If their income from farming is to grow over time, it will be heavily dependent on superior farm management.

A Career in Farming

Young people should first realistically appraise their opportunities both on and off the farm before pursuing farming as an occupation. This analysis requires knowledge of the availability of off-farm positions, the competition for these positions, and the qualifications required for future farm managers.

Opportunities for New Farmers

The opportunities to enter farming are not as numerous as they were 20 years ago. However, because of the number of farmers reaching retirement age, the openings represent a higher percentage of the total number of farms, and will continue to do so until the 1990s. Nevertheless, although the percentage of openings will be higher, the actual number will tend to decrease. As farms become fewer and larger, fewer new farmers will be needed to replace current operators. According to a recent U.S. Department of Agriculture study, 452,000 new farmers under 35 years of age entered farming between 1974 and 1984.[8] Between 1984 and 1994, the number of new operators under 35 years of age is estimated to be 405,000. This number is expected to decrease to 284,000 by 1994 to 2004.

The competition for these new farming opportunities will change over the next two decades. High birth rates during the 1950s and early 1960s provided three to four times as many young farm adults as there are farming opportunities.[9] Fortunately, not all farm youth want to become farmers. The key point here is the trend: Although the absolute number of farming opportunities will decline with the decreasing number of farms, declining family size will reduce the number of potential entrants more rapidly. Therefore, competition will not be as keen as before.

Another source of competition for these entrants will be managers who are trying to expand their operations. Because of their farming experience and stronger financial position, these managers will have an economic advantage. Among young farmers trying to get started, the advantage will be with those whose families have a good farm business, which serves as a springboard to their own farming careers. About 82 percent of farmers today were raised on a farm; thus, in starting, they were likely to have had family assistance or an inheritance.[10]

The percentage of women in farming is increasing, even though the number is still small. There were 122,000 farms operated by women in 1982, an increase of 8 percent from 1978. As a percentage of total farm operators, female farmers increased from 5 percent to 11 percent from 1970 to 1980. These statistics, however, understate the increasingly critical role of women as part of the family farm

[8] Lin, Coffman, and Penn, *U.S. Farm Numbers, Sizes, and Related Structural Dimensions.*

[9] E. L. LaDue, "Financing the Entry of Young Farmers," *Agricultural Finance Review* 39 (1979): 101–122.

[10] George Coffman, U.S. Department of Agriculture economist, quoted in "Who Will Be Able to Start Farming," *Prairie Farmer*, February 2, 1980, p. 90b.

management team. The wives of farmers are devoting more time to the farm business in all capacities. This changing role of the woman on the farm has occurred in response to increasing managerial demands on the business.

Qualifications for Farming

There are no legal or educational requirements for becoming a farmer. Anyone who wants to farm can try. There are, however, substantial barriers to overcome:[11]

1. *Capital requirements.*   Related to the larger farm is the sharp increase in the use of capital; investments per farm have more than doubled each decade.[12] The prospective farmer without adequate capital finds it more difficult to begin a farm operation. This is particularly true of persons who must start with little family assistance.

2. *Managerial requirements.*   As farms have become larger and more complex, so have managerial requirements. Modern farming methods require a higher degree of technical knowledge and skill. Furthermore, skills in financial management, marketing, and personnel management are becoming more important. Acquiring such skills involves a significant commitment to formal training as well as practical experience.

3. *Risk-bearing ability.*   One of the most common characteristics of beginning farmers is a limited ability to assume risk. The net worth of beginning farmers is typically low, and they do not have the experience to cope with wide fluctuations in income. Thus, the ability and capacity to cope with change and uncertainty will continue to be important. Risk management is often made even more challenging by the desire to provide a reasonable standard of living for a young and growing family.

4. *Resource control.*   An additional barrier is limited access to enough resources to build a viable business. The beginning farmer is often looking for land to lease. But with land already leased to established farmers with stronger financial positions, it is difficult for the beginning farmer to bid competitively. Because of the risk involved, cash lease arrangements and short-term leases present special problems for beginning farmers, who therefore run the risk of losing control of their land bases after one or two years. Innovations in financing to permit the nonfarmer to invest in agriculture can be expected.

Another way to determine the qualifications required of beginning farmers is to examine the characteristics of those who have been successful. Interviews with successful farmers in Indiana and Minnesota indicated that they had the following characteristics:[13]

[11] Michael Boehlje and Kenneth Thomas, "Entry Into Agriculture: Barriers and Policy Alternatives," *Journal of the American Society of Farm Managers and Rural Appraisers*, 43(October 1979): 20–25.

[12] E. G. Stoneberg and William Edwards, *Iowa Farm Costs and Returns—1977*, Iowa State University Cooperative Extension Service FM-1754, June 1978.

[13] John A. Watzek "Factors Related to Success or Failure on Getting Started in Farming," unpublished M.S. thesis, Purdue University, 1970; and Kenneth H. Thomas and Harald R. Jensen, *Starting Farming in South Central Minnesota: Guidelines, Financial Rewards, Requirements*, University of Minnesota Agricultural Experiment Station Bulletin 499, 1969.

1. They had a good education before launching a career in farming.
2. They tended to invest extra income rather than to spend it.
3. They had small families to support.
4. They selected intensive enterprises, such as hog production.
5. They had operations large enough to obtain economies of size.
6. They had low debt relative to equity, with the borrowing capacity to enable them to control more resources.

The person intending to enter agriculture must have access to adequate capital, control of a sizable resource base, the willingness and ability to assume risk, and well-developed managerial skills. Opportunities are available to beginning farmers with excellent managerial abilities. There are few opportunities for those with only average management skills.

Summary

Agriculture is a diverse and changing business. There are many different types of farming systems and a considerable range in the size of operations. The structure of agriculture will continue to change. The number of farms is expected to decrease with the increase in the average size of farms. A small percentage of all farms will produce most of the nation's agricultural output.

Although no one can predict the economic future, two conservative conclusions appear to be warranted. First, the family farm will continue to be the primary producing unit of most agricultural commodities through the end of this century, and it will become larger and more complex, requiring higher capital investment and highly skilled management. Second, different commodities will experience different types of development. In the production and distribution of some products, the sole proprietorship may well disappear. The West Coast has experienced developments of this type, and other regions can expect more such occurrences.

The U.S. agricultural sector is very resilient. It will survive and adjust to the changes in technology and in its environment. However, the more relevant question for farm business managers is, who will the individual survivors be? The answer to this question will depend largely on the decisions made by individual managers, which in turn depend on the managers' knowledge and skills.

There will be opportunities for young people to enter farming; however, those who control adequate amounts of capital will have an advantage. One way to achieve this goal may be to participate in the family farm business. Another is to participate in joint ventures.

Young people who want to enter agriculture must have realistic expectations. Enthusiasm is needed, but management skills are also important. Beginning farmers should carefully consider their goals, keeping them realistic and consistent with their management abilities.

Glossary

Accounting The principles and procedures for systematically recording business transactions and for summarizing, reporting, and interpreting the results.

Accounts Payable Money owed by an individual or business to others for goods received or services rendered.

Accounts Receivable Money others owe to an individual or business for products sold or services provided.

Accrued An amount that has accumulated, but has yet been paid; for example, interest on a savings account accrues or accumulates day-by-day but is not paid until the end of the period. Also, the unpaid income tax owed for the year just ended is an accrued tax liability.

Accrual Method of Accounting An accounting method by which revenue and expenses are recorded when they are earned or incurred regardless of when the cash is received or paid. The accrual method matches expenses to the revenue earned during the year (see Cash Method of Accounting).

Amortization A method for repaying loans whereby the borrower makes periodic payments so that the principal will be completely repaid when the loan matures. There are two types of amortization plans: the decreasing-payment plan and the level-payment plan.

Animal Unit A common denominator for measuring animal feed requirements where one animal unit is equivalent to the feed requirement for a 1000-pound mature beef cow.

Annuity A series of uniform, periodic dollar payments. Annuities can be specified for a certain number of periods, or for the lifetime of an individual.

Appraisal The valuation, especially of land, buildings, machinery, and equipment, made by an expert in estimating such values.

Appreciation The increase in the value of a capital asset (for example, land) due to external influences such as inflation.

Assets Items of value owned by a business or person. Assets are classified as current, intermediate, or fixed.

Average Cost (AC) The total cost of production divided by the total number of units produced.

Average Physical Product (APP)   The total physical product divided by the units of variable input used.

Balance Sheet   A financial statement that summarizes the assets, liabilities, and net worth of an individual or business as of a specific date.

Book Value   The undepreciated value of an asset shown in the business's records. It is found by subtracting the accumulated depreciation from the original cost of the asset.

Borrowing Capacity   The maximum amount of credit available to an individual or business.

Break-even   The volume of output required for revenue to equal the total of fixed and variable expenses; the price or yield at which the gross margin of one enterprise equals the gross margin of another enterprise.

Budget   A logically consistent device for examining alternative plans for the business and estimating the profitability of each alternative.

Business Analysis   A system of procedures and ratios a manager uses to evaluate the performance and financial strength of the business.

Capital   Nonhuman resources used in production that last more than one year; the total economic value of the resources available for use in the business.

Capital Expenditures   Payments made to purchase assets that will be used for more than one year.

Capital Gain   The amount by which the sale value of an asset exceeds its book value. Capital gains income is sometimes treated differently than ordinary income for tax purposes.

Capitalization   The process for determining the present value of a series of payments to be received indefinitely.

Capital Loss   The amount by which the book value of an asset exceeds its sale value.

Capital Receipts   Payments received from the sale of capital assets.

Cash   Currency on hand or the balance in a checking account.

Cash Costs   Out-of-pocket expenditures involving the exchange of currency or a check for goods and services.

Cash Flows   Cash receipts (inflows) or cash expenses (outflows). Examples of cash inflows are proceeds from sales, borrowed money, and income from custom work. Examples of cash outflows are payments for fertilizer, principal and interest, and family living expenses.

Cash Flow Budget   A detailed estimate of the projected cash inflows and outflows over a future period of time used to evaluate the financial feasibility of a plan.

Cash Flow Statement   A financial statement that summarizes the actual sources of cash and uses of cash for a given period of time, a month or a year.

Cash Lease   A type of lease agreement where the tenant pays a cash amount for the use of land (see Lease).

Cash Method of Accounting   An accounting method by which revenue and expenses are recorded when the cash is received or paid. Farmers are allowed to use the cash method of accounting for reporting taxable income (see Accrual Method of Accounting).

Coefficient   A number showing a constant relationship between two variables.

Competitive   The production relationship between two products where an increase in the production of one decreases the output of the other.

Complementary   The production relationship between two products where an increase in the production of one also increases the output of the other.

Compounding   The mathematical process used to find the value of an amount of money at some future date when its present value is known. The process involves calculating interest on the accumulated interest as well as on the original balance.

Contingent Income Taxes   The estimated amount of income taxes that would be due if assets were liquidated at the market value shown on the balance sheet.

Corporation   A legal business entity separate and distinct from its owners and managers that can own property and conduct business. The owners receive shares of stock representing their contributions of resources and select officers to manage the business.

Cost   A charge that should be made for an item used in the production of goods or services. There are cash costs, noncash costs, variable costs, fixed costs, and opportunity costs.

Credit   The ability to obtain money, goods, or services in exchange for a promise to repay it by a specific future date.

Credit Reserve   Unused borrowing capacity available as a source of liquidity for unforeseen obligations and investment opportunities (see Borrowing Capacity).

Crop Share Lease   (see Share Lease).

Current Assets   Cash and other assets that will be used up within the next year or that can be converted to cash without disrupting the business.

Current Liabilities   Liabilities that are due within the next 12 months.

Custom Work   Farm work performed for a charge involving the use of machinery and labor provided by someone other than the farm operator.

Debt   An obligation to pay in the future (see Liabilities).

Debt Service   The cash required to meet principal and interest payments.

Deflation   A general decrease in prices that increases the purchasing power of a dollar.

Depreciation   The decrease in the value of an asset due to age, use, and obsolescence; the pro-rated expense of owning the asset.

Diminshing Returns   The characteristic that if one input is increased incrementally with all the other inputs held constant, then after some point the resulting increases in output become smaller and smaller.

Discounting   The mathematical process used to find the present value of an amount of money today when its value at some future date is known; discounting is the inverse of compounding (see Compounding).

Diversification   A management strategy of producing a number of different commodities in order to use resources more efficiently or to reduce risk.

Double-Entry Accounting   The system of bookkeeping in which every transaction is recorded as a debit in one or more accounts and as a credit in one or more accounts such that the total of the debit entries equals the total of the credit entries.

Down Payment   The part of the purchase price paid at the time the contract is made.

Economics   The study of how people, individually and in groups, allocate scarce resources among competing uses to maximize satisfaction over time.

Economies of Size   The situation where the total cost per unit of output decreases as a result of increases in the size of the business.

Efficiency   A ratio of output to input. Economic efficiency refers to the ratio of the value of output to the cost of input. Production efficiency refers to the ratio of physical output to physical input.

Enterprise   A segment of the farm business that can be isolated by accounting procedures so that its revenue and expenses can be identified.

Enterprise Accounting   A record keeping procedure that determines the detailed revenue and expenses for each enterprise.

Enterprise Budget   A detailed list of all estimated revenue and expenses associated with a specific enterprise.

Equimarginal Returns Principle   If a scarce resource is to be allocated among two or more uses, the highest total return is obtained when the returns to the last unit of the resource used in each alternative are equal.

Equity   The owner's claim or right to the assets of a business (see Net Worth).

Expenditure   Cash paid for the purpose of obtaining goods or services.

Expense   A cost of operating the business during a specific time period or associated with revenues earned during that period.

Factor of Production   An input or resource used in production, such as seed, fertilizer, labor, and land.

Family Farm   A farm business that provides the family's principal source of income and is managed by the family.

Farm Management   The field of study concerned with the decisions that affect the profitability of the farm business.

Financing   Acquiring control of assets by borrowing money.

Fixed Assets   Real estate and permanent improvements that make an on-going contribution to the business for at least ten years. Selling fixed assets would seriously disrupt the business.

Fixed Costs   Costs that do not vary, or cannot be avoided, by changing the amount produced.

Flexibility   A management strategy of maintaining the ability to make adjustments in the operation of the business in response to changing conditions in order to reduce risk.

Goal   A target or desired condition that motivates the decision maker.

Gross Margin   The difference between the revenue and the variable production cost for one unit (one acre or one animal) of an enterprise.

Income   The difference between revenue and expenses, which is referred to as net income; gross income refers to total revenue.

Income Satement   A financial statement that summarizes all revenues and expenses to determine the net income or net loss for a given period of time, usually a year.

Incorporation   The legal process of bringing a corporation into existence.

Inflation   A general increase in prices that decreases the purchasing power of a dollar.

Input   A factor or resource used in production, such as seed, fetilizer, labor, and land.

Interest   The amount paid for the use of another person's money.

Intermediate Assets   Assets that support production for one to ten years and are not easily converted to cash without affecting the operation of the business.

Intermediate Liabilities   Liabilities due after one year but within ten years of the date of the balance sheet.

Inventory   An itemized listing of commodities held for sale, commodities in the process of production, and inputs to be used in production; for example, feed, seed, farm supplies, animals to be sold within the year, crop production in storage, and growing crops.

Investment   The process of adding assets to a business.

Joint Venture   Any business arrangement whereby two or more parties contribute resources to, and engage in, a specific business undertaking.

Land   The portion of the earth's surface over which ownership rights may be exercised.

Land Contract   An arrangement for financing the purchase of real estate where the buyer gets the use of the property and the seller agrees to deliver the title when the agreed principal and interest payments have been made.

Lease   The granting of the right to use an asset by one person (the lessor) to another (the lessee) for a specified period of time in exchange for a cash payment or a share of production.

Leverage   The relationship between the total liabilities and the net worth of a business. The higher the ratio of debt to net worth, the greater is the leverage.

Leveraging   A strategy of acquiring assets with a large proportion of borrowed funds.

Liabilities   Obligations or debts owed by a business or person to others. Liabilities are classified as current, intermediate, and long-term.

Lien   The claim given to a lender to have the debt satisfied out of property belonging to the debtor if repayment is not made as agreed.

Limited Partnership   A partnership consisting of at least one general partner, who is responsible for the management and liabilities of the business, and at least one limited partner, whose liability is limited to his or her investment.

Linear Programming   A mathematical technique used to find profit maximizing combinations of production activities or cost minimizing combinations of ingredients subject to a number of linear relationships that constrain the activities or ingredients.

Line of Credit   A financing arrangement whereby the lender agrees to extend credit to the borrower up to a pre-set maximum amount.

Liquidate   To convert to cash; to sell.

Liquidity   The ability of a business to generate sufficient cash to meet its financial obligations when they come due; the ease with which assets can be converted to cash.

Long-Term Liabilities   Mortages, land contracts, and other liabilities that will mature (be completely repaid) more than ten years from the date of the balance sheet.

Loss   The excess of expenses over revenues for a period.

Marginal Cost (MC)   The change in total cost resulting from the production of one more unit of output.

Marginal Input Cost (MIC)   The change in total cost resulting from the use of one more unit of input in production.

Marginal Physical Product (MPP)   The change in the total physical product resulting from a one unit change in the amount of input used.

Marginal Revenue (MR)   The change in total revenue resulting from a one unit increase in output. In a perfectly competitive market, it is the price of the product.

Marginal Tax Rate   The change in the total income tax resulting from a one dollar increase in taxable income.

Marginal Value of Product (MVP)   The change in total value of production resulting from the use of one more unit of input. In a perfectly competitive market, it is the marginal physical product multiplied times the price of the product.

Market Value   The value of an asset based on the estimated amount that would be received from selling the asset, after deducting all expenses of the sale.

Mortgage   An arrangement for financing the purchase of real estate where the buyer gets legal title and use of the property and the lender has the property as security to assure the loan is repaid as agreed; a lender's claim on property owned by the borrower.

Net Farm Income   The total revenue generated by farm operations minus total expenses incurred for a given period of time; the return to the farmer's unpaid labor, management, and net worth invested in the farm business.

Net Income   The total of farm and nonfarm revenue earned, minus farm and nonfarm expenses incurred, for a given period of time.

Net Worth   The difference between the value of total assets and total liabilities. Net worth represents the owner's claim on the assets of the business.

Noncash Costs   Costs that do not involve a cash payment; for example, wages for the operator's own labor is a noncash cost.

Note   (see Promissory Note).

Obsolescence   The condition of becoming out-of-date or useless as a result of new discoveries, improvements, or changes in consumer demand.

Operating Expenses   Expenses incurred in the usual production cycle, such as seed, fuel, feed, and hired labor.

Opportunity cost   The cost of using a resource based on what it could have earned using it in the next best alternative use.

Output   The quantity of goods and services produced.

Overhead   Expenses incurred in the operation of the business that cannot conveniently be attributed to the production of specific commodities or services.

Partial Budget   A budget that includes only those revenue and expense items that would change as a result of the proposed change in the business. The components of the partial budget are added revenue, reduced expense, added expense, and reduced revenue.

Partnership   A business with two or more owners who contribute their resources and share management responsibilities, profits, and losses.

Prepaid Expense   Services or products for which payment has been made but have not yet been used up or received.

Principal   The original amount of a loan; unpaid balance of the purchase price.

Product   A bundle of physical, service, and other attributes that satisfies consumer wants and needs.

Production   A process that transforms one or more inputs into one or more products.

Production Function   The graphical or mathematical representation of the physical relationship between the amount(s) a production input(s) used and the amount of output produced when other inputs are held constant.

Profitability   The ability of a business to generate earnings adequate to reimburse the opportunity cost of the resources used (see Opportunity Cost).

Promissory Note   A formal written promise to pay on demand or at a future time an amount of money with interest.

Proprietorship   A business owned and managed by one individual person or family.

Real   The modifier used with monetary measures such as net income and prices to indicate that the effect of inflation has been removed and that the purchasing power of the money is constant.

Real estate   Land and improvements, including buildings, standing timber, orchard trees, and so forth.

Real Rate of Interest   The rate of interest earned after subtracting the average rate of inflation.

Receipts   Money received from sales of commodities produced, services rendered, or capital items sold.

Replacement Cost   A method of valuing an asset based on the cost of replacing the asset with another that will render similar services.

Resource   An input or factor used in production, such as seed, fertilizer, labor, and land.

Revenue   Gross proceeds from the sale of products and services, government program payments, and other business operations. The proceeds may be in the form of cash, accounts receivable, and increases in inventories.

Risk   The decision-making situation where there is a chance of an unfavorable outcome for one or more of the alternative actions being considered.

Sales Contract   An arrangement for financing the purchase of assets such as machinery or equipment where the buyer gets use of the asset and the seller keeps title to the asset until the agreed payments on principal and interest have been made.

Salvage Value   The remaining value of an asset at the end of its useful life.

Share Lease   A type of lease agreement where the tenant gives a share of the production for the use of the land (see Lease).

Simple Interest   The method of calculating interest as a percentage of the outstanding balance times the number of years (or fraction of a year) that the amount was outstanding.

Solvency   The ability of a business to pay all its liabilities if all the assets were sold; the relationship between total assets and total liabilities.

Sunk Cost   A cost associated with an irreversible past decision.

Supplementary   The production relationship between two products where an increase in the production of one does not affect the other.

Technology   In agriculture, the scientific methods, techniques and systems used in modern farm production and marketing of farm products.

Tenant   A business or person who leases real estate.

Tenure   The act or right of holding real estate; for example, land on a farm may be owned, leased, or partly owned and partly leased.

Term Debt   (see Term Loan).

Term Loan   A loan repaid with periodic payments over a stipulated time period of more than one year.

Total Cost (TC)   The sum of variable and fixed costs incurred in production.

Total Physical Product (TPP)   The total output produced as the result of the use of a specific combination of inputs.

Total Value of Product (TVP)   The value of the total physical product (see Total Physical Product).

Uncertainty   The situation where the manager does not know the future outcomes of decisions. Because of uncertainty, managers must consider risk (see Risk).

Value of Farm Production   Gross revenue minus livestock and feed purchases, because the latter are produced by other farms.

Variable Costs   Costs that change with the amount produced. If the manager decides to cease production, these costs are avoidable.

Working Capital   Current assets minus the current liabilities. It is the capital available to finance the routine operations of the business.

Index